Residential
STRUCTURE & FRAMING

Practical Engineering and Advanced
Framing Techniques for Builders

From the Editors of
The Journal of Light Construction

With contributions from Robert Randall, P.E.,
Paul Fisette, Frank Woeste, P.E., Carl Hagstrom,
Harris Hyman, P.E., and others.

Copyright © 1999 by Hanley Wood, LLC.
All rights reserved.
First printing: July 1999
Second printing: June 2001
Third printing: February 2004

Cover Photograph by Rick Arnold and Mike Guertin
Cover Illustrations: Tim Healey, Lena Savage

Editor: Steven Bliss
Production Editor: Josie Masterson-Glen
Article Editors: Sal Alfano, Ted Cushman, Clayton DeKorne, Dave Dobbs, Don Jackson, Carl Hagstrom

Production Manager: Theresa Emerson
Graphic Designers: Annie Clark, Lyn Hoffelt, Judy McVicker
Illustrator: Tim Healey

International Standard Book Number 1-928580-17-3
Library of Congress Catalog Card Number 99-095244
Printed in the United States of America

A Journal of Light Construction Book

The Journal of Light Construction is a tradename of Hanley Wood, LLC.

The Journal of Light Construction
186 Allen Brook Lane
Williston, VT 05495

INTRODUCTION

Every good carpenter has a strong sense of structure — what framing details will create a sturdy building and which will feel flimsy or sag over time. Building on that instinctual knowledge with a fuller understanding of the engineering principles involved will make any tradesperson a better, more confident builder. That's the goal of this book, which is adapted from articles originally published in *The Journal of Light Construction*. To that end, we've made every effort to translate the work of the engineer into practical principles and guidelines that builders can understand and put to use on the job site.

In addition, we've included the best framing articles from *The Journal* focusing on the tools, materials, and techniques that can bring greater efficiency to high-quality custom work.

To get the best information available, we sought out leading production framers — the real framing experts — to learn about the lightning fast techniques they've developed and honed to perfection and how to apply them to custom work.

Like all material in *JLC*, the information in this book comes directly from the field through practicing building and remodeling professionals with many years of job-site experience. So whether you're a seasoned professional or a tradesperson just starting out, we think you'll find the structural insights and framing pointers in these pages of real practical value.

Steven Bliss
Editorial Director

Table of Contents

Section One: Structural Design

Chapter 1. Sizing Joists, Beams, and Columns

How To Read Span Tables
Paul Fisette 2

Beyond Code: Preventing Floor Vibration
Frank Woeste and Dan Dolan 6

Calculating Loads on Beams and Headers
Paul Fisette 8

Sizing Simple Beams
Harris Hyman 10

Simple Approach to Sizing Built-Up Headers
Robert Randall 14

Sizing Stiff Floor Girders
Frank Woeste and Dan Dolan 17

When Columns Buckle
John Siegenthaler 19

Misplaced Load Paths
Robert Randall 22

Chapter 2. Materials and Connections

Structural Sheathing: Plywood vs. OSB
Paul Fisette 24

Strength of Nails
Harris Hyman 26

Box vs. Common Nails
Frank Woeste 28

Using Metal Connectors
Robert Randall 29

Frequently Asked Framing Questions
Christopher DeBlois, Scott McVicker, Robert Randall,
and Frank Woeste 32

Chapter 3. Roof Structure

Holding the Roof Up
Robert Randall 42

Framing With a Raised Rafter Plate
Robert Randall 43

Resisting Wind Uplift
Robert Randall 45

Straight Talk About Hips and Valleys
Robert Randall 47

Chapter 4. Seismic Framing and Wind Bracing

Plywood vs. Let-In Bracing
Philip Westover 54

Stiffening Garage Door Openings
Harris Hyman 56

Earthquake Design
Richard Mayo 58

Shear Wall Construction Basics
Jim Hart 60

Installing Seismic Framing Connectors
Jim Hart 64

Hold Down Problems and Solutions
Scott McVicker 70

Fixing Shear Wall Nailing Errors
Scott McVicker 73

Chapter 5. Exterior Decks

Deck Support: Making the Crucial Connections
Christopher DeBlois 78

Overhanging Decks: How Far Can You Go?
Harris Hyman 79

Section Two: Field Techniques

Chapter 6. Floor Framing

Leveling the Deck
Ron and Roger Whitaker 82

Floor Framing With Wood I-Joists
Ned Murphy 85

Floor Framing With Open-Web Trusses
Charles Wardell 87

A Fix for Bouncy Floors
Robert Randall 93

Chapter 7. Wall Framing

Layout Tricks for Rough Openings
Carl Hagstrom 98

Framing Rake Walls
Eric Dickerson 102

Site-Built Panelized Walls
Bill Nebeker 105

Building Stiff Two-Story Window Walls
Christopher DeBlois 109

Bracing Foam-Sheathed Walls
Paul Fisette 113

Framing for Corner Windows
Robert Randall 116

Chapter 8. Roof Framing

Layout Basics for Common, Hip, and Jack Rafters
Rob Dale Gilbert 122

Laying Out Unequally-Sloped Gables
Carl Hagstrom 127

Joining Unequally Pitched Hips and Valleys
Carl Hagstrom 131

Stacking Supported Valleys
Will Holladay 136

Building Doghouse Dormers
Carl Hagstrom 142

Roof Framing With Wood I-Joists
Curtis Eck 145

Flat Roof Framing Options
Don Gordon 151

Chapter 9. Remodelers' Specialties

Roughing In for Kitchens and Baths
Jim Hart 158

Tying Into Existing Framing
David Schwartz 162

A Second Story in Five Days
Chuck Green 166

Case Study: A Retrofit Ridge Beam
Silas Towler and Rick Schneider 169

Jacking Old Houses
Mike Shannahan 172

Chapter 10. Pickup Work

Fast Fascia Techniques
Mike Stary 178

Framing Recessed Ceilings
Don Dunkley 181

Framing a Simple Radius Stair
Robert Thompson 183

Section Three: Engineered Lumber

Chapter 11. Trusses

Installing Gable Roof Trusses
Rick Arnold and Mike Guertin 188

Installing Hip Roof Trusses
Paul Bartholomew 193

Truss Bracing Tips
Paul DeBaggis 197

Critical Bracing for Piggyback Trusses
Frank Woeste 199

Making Room at the Top
Gary Rowland 202

Site-Built King Truss for Ridge Beam Support
Robert Randall 205

Chapter 12. Engineered Lumber

Wood I-Joist Fundamentals
John Siegenthaler 210

Sizing Engineered Beams
Paul Fisette 211

Working With Laminated Veneer Lumber
Ned Murphy 214

On Site With Parallam
Patricia Hamilton 217

Building With Glulams
Eliot Goldstein 220

Engineered Studs for Tall Walls
JLC Staff 224

Chapter 13. Steel Framing

Steel Beam Options
Harris Hyman 230

Strengthening Beams With Flitchplates
Robert Randall 232

Attaching Steel Beams and Columns
Stuart Jacobson 234

Installing a Steel Moment Frame
Patricia Hamilton 236

Hybrid Framing With Wood and Steel
Tim Duff 238

Index 245

Section One
STRUCTURAL DESIGN

Chapter One
SIZING JOISTS, BEAMS, AND COLUMNS

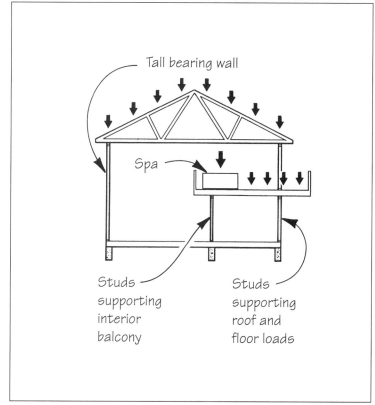

How To Read Span Tables 2

Beyond Code: Preventing Floor Vibration . . .6

Calculating Loads on
Beams and Headers 8

Sizing Simple Beams 10

Simple Approach to Sizing
Built-Up Headers 14

Sizing Stiff Floor Girders 17

When Columns Buckle 19

Misplaced Load Paths 22

How To Read Span Tables

Wood is naturally engineered to serve as a structural material. The tree is fastened to the earth with its roots, the foundation; the trunk supports the weight of the branches, as a column; and it bends when loaded by the wind, as a cantilever beam. Wood's mechanical properties are complex, but if you understand a few basics of lumber strength you can easily size uniformly loaded joists and rafters with span tables.

Stiffness and Strength

A good set of tables includes a number of variables, the most basic of which are *stiffness* and *strength*. A house frame has to resist dead loads (the weight of materials), live loads (the weights imposed by use and occupancy), snow loads, and wind loads. Beams, studs, joists, and rafters must be strong enough *and* stiff enough to resist these loads.

Stiffness. A set of second-story floor joists can be strong enough to support all dead and live loads yet still be too bouncy. The joists won't break, but the first-story ceiling plaster may crack as the occupants walk across the second floor.

Stiffness requirements for joists or rafters are limited by their maximum allowable *deflection*, which is set by code. Deflection limits vary for different parts of the house and are based on the live loads experienced in each room. They're expressed as a fraction: the clear span in inches (*L*) over a specified number.

Typical code-prescribed deflection limits are L/360 for all floors and any rafters with plaster on their underside, L/240 for rafters with drywall attached, and L/180 for rafters with no plaster or drywall. A floor joist that's appropriately selected to span 10 feet with an L/360 limit will deflect no more than 1/3 inch (120 inches ÷ 360) under its maximum design load.

The measure of a material's stiffness is "modulus of elasticity," or *E*. It's expressed in pounds per square inch, or psi. A material with a higher E value is stiffer. For example, No. 2 eastern white pine has an E value of 1,100,000 psi, while No. 2 hem-fir, which is stiffer, has an E value of 1,300,000 psi.

Strength is obviously important, too: Joists and rafters must be strong enough not to break when loaded. Strength is expressed as "extreme fiber stress in bending," or *Fb* (Figure 1).

Loads cause structural members like beams, joists, and rafters to bend. As a

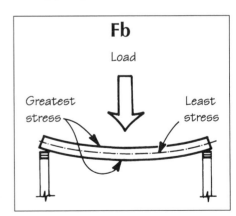

Figure 1. *As a structural member bends, the fibers along its upper and lower edges are highly stressed, while those near the center are not stressed at all. Fb ("extreme fiber stress in bending") is a measure of the stress that can be placed on these outermost (or "extreme") fibers; the higher the Fb, the stronger the wood.*

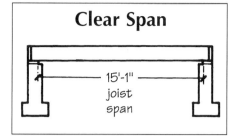

Figure 2. *When sizing joists, use the clear span — the length from support to support — not the full length of the joist.*

structural member bends, the wood fibers on its top and bottom edges are stressed more than the fibers along its centerline. The fibers along the top edge are squeezed in compression, while those along the bottom edge are stretched in tension. Fb is the design strength of those "extreme," or outermost, fibers; the higher the Fb, the stronger the wood.

How strong a structural member must be depends on the load it will carry. You can calculate the minimum design values required of a structural member by adding the live loads and dead loads carried by that member. The individual weights of drywall, strapping, floor joists, plywood, and carpet are listed in *Architectural Graphic Standards* and other reference books. But adding the weights of materials is rarely necessary except in unusual designs. The tables list a variety of average live and dead load combinations for floors, ceilings, and rafters. These combinations are more than adequate for most residential designs.

Other Considerations

Of course, stiffness and strength aren't the only factors that determine how a structural member responds to loading. That's why the tables also include several other variables. The ability to balance these lets you fine-tune a structure's cost and performance.

Depth of structural members. The deeper the joist or rafter, the more weight it can support. For example, 2x10 joists spaced 24 inches on-center often provide a stronger and stiffer floor assembly than 2x8s of the same grade and species spaced 16 inches on-center.

Lumber grade. A higher grade of a

CABO Table No. R–201.4
Minimum Uniformly Distributed Live Loads

Use	Live Load
Balconies (exterior)	60
Decks	40
Fire escapes	40
Garages (passenger cars only)	50
Attics (no storage with roof slope not steeper than 3 in 12)	10
Attics (limited attic storage)	20
Dwelling units (except sleeping rooms)	40
Sleeping rooms	30
Stairs	40

CABO Table No. R–201.6
Allowable Deflection of Structural Members

Structural Member	Allowable Deflection
Rafters having slopes > 3/12 with no ceiling load	L/180
Interior Walls and Partitions	L**/180
Floors and Plastered Ceilings	L/360
All Other Structural Members	L/240

Notes:
L = span length   L** = vertical span

These tables are reprinted from the One and Two Family Dwelling Code, *Council of American Building Officials, Falls Church, Va.*

Figure 3. *Live loads and deflection limits are set by code. These tables are from the CABO One and Two Family Dwelling Code.*

AFPA Table F–2
Floor Joists with L/360 Deflection Limits

DESIGN CRITERIA:
Deflection – For 40 psf live load.
Limited to span in inches divided by 360.
Strength – Live load of 40 psf plus dead load of 10 psf determines the required bending design value.

Joist Size (in.)	Spacing (in.)	Modulus of Elasticity, E, in 1,000,000 psi							
		0.8	0.9	1.0	1.1	1.2	1.3	1.4	1.5	1.6
2x6	12.0	8–6	8–10	9–2	9–6	9–9	10–0	10–3	10–6	10–9
	16.0	7–9	8–0	8–4	8–7	8–10	9–1	9–4	9–6	9–9
	19.2	7–3	7–7	7–10	8–1	8–4	8–7	8–9	9–0	9–2
	24.0	6–9	7–0	7–3	7–6	7–9	7–11	8–2	8–4	8–6
2x8	12.0	11–3	11–8	12–1	12–6	12–10	13–2	13–6	13–10	14–2
	16.0	10–2	10–7	11–0	11–4	11–8	12–0	12–3	12–7	12–10
	19.2	9–7	10–0	10–4	10–8	11–0	11–3	11–7	11–10	12–1
	24.0	8–11	9–3	9–7	9–11	10–2	10–6	10–9	11–0	11–3
2x10	12.0	14–4	14–11	15–5	15–11	16–5	16–10	17–3	17–8	18–0
	16.0	13–0	13–6	14–0	14–6	14–11	15–3	15–8	16–0	16–5
	19.2	12–3	12–9	13–2	13–7	14–0	14–5	14–9	15–1	15–5
	24.0	11–4	11–10	12–3	12–8	13–0	13–4	13–8	14–0	14–4
2x12	12.0	17–5	18–1	18–9	19–4	19–11	20–6	21–0	21–6	21–11
	16.0	15–10	16–5	17–0	17–7	18–1	18–7	19–1	19–6	19–11
	19.2	14–11	15–6	16–0	16–7	17–0	17–6	17–11	18–4	18–9
	24.0	13–10	14–4	14–11	15–4	15–10	16–3	16–8	17–0	17–5
F_b	12.0	718	777	833	888	941	993	1043	1092	1140
F_b	16.0	790	855	917	977	1036	1093	1148	1202	1255
F_b	19.2	840	909	975	1039	1101	1161	1220	1277	1333
F_b	24.0	905	979	1050	1119	1186	1251	1314	1376	1436

Note: The required bending design value, F_b, in pounds per square inch is shown at the bottom of each table and is applicable to all lumber sizes shown. Spans are shown in feet–inches and are limited to 26' and less. Check sources of supply for availability of lumber in lengths greater than 20'.

EXCERPTED FROM SPAN TABLES FOR JOISTS AND RAFTERS, AMERICAN FOREST & PAPER ASSN., WASHINGTON, D.C. ACTUAL TABLE GIVES E VALUES UP TO 2,400,000 PSI.

Figure 4. *Given a design span of 15 feet 1 inch and a 16-inch joist spacing, first determine which size lumber will work. Then find the required Fb value at the bottom of the column.*

How To Read Span Tables

AFPA Table W–1
Design Values for Joists and Rafters
Visually Graded Lumber

These "F_b" values are for use where repetitive members are spaced not more than 24 inches. For wider spacing, the "F_b" values shall be reduced 13%.
Values for surfaced dry or surfaced green lumber apply at 19% maximum moisture content in use.

Species and Grade	Size	Design Value In Bending ("F_b")			Modulus of Elasticity ("E")
		Normal Duration	Snow Loading	7 Day Loading	
HEM-FIR					
Select Structural	2x10	1770	2035	2215	1,600,000
No.1 & Btr		1330	1525	1660	1,500,000
No.1		1200	1380	1500	1,500,000
No.2		1075	1235	1345	1,300,000
No.3		635	725	790	1,200,000
Select Structural	2x12	1610	1850	2015	1,600,000
No.1 & Btr		1210	1390	1510	1,500,000
No.1		1095	1255	1365	1,500,000
No.2		980	1125	1220	1,300,000
No.3		575	660	720	1,200,000
DOUGLAS FIR-LARCH					
Select Structural	2x10	1835	2110	2295	1,900,000
No.1 & Btr		1455	1675	1820	1,800,000
No.1		1265	1455	1580	1,700,000
No.2		1105	1275	1385	1,600,000
No.3		635	725	790	1,400,000
Select Structural	2x12	1670	1920	2085	1,900,000
No.1 & Btr		1325	1520	1655	1,800,000
No.1		1150	1325	1440	1,700,000
No.2		1005	1155	1260	1,600,000
No.3		575	660	720	1,400,000
SPRUCE-PINE-FIR					
Select Structural	2x10	1580	1820	1975	1,500,000
No.1/No.2		1105	1275	1385	1,400,000
No.3		635	725	790	1,200,000
Select Structural	2x12	1440	1655	1795	1,500,000
No.1/No.2		1005	1155	1260	1,400,000
No.3		575	660	720	1,200,000

EXCERPTED FROM DESIGN VALUES FOR JOISTS AND RAFTERS, AMERICAN FOREST & PAPER ASSN., WASHINGTON, D.C. ACTUAL TABLES LIST DESIGN VALUES FOR 2x4s, 2x6s, AND 2x8s.

Figure 5. *After determining what size lumber to use, turn to the tables in* Design Values For Joists and Rafters *to select a species and grade that meets the required Fb and E values. The tables shown here are excerpts from the hem-fir, Douglas fir-larch, and spruce-pine-fir tables.*

given species usually has a higher strength rating (Fb) and often a higher stiffness value (E), too.

Wood species. All species are not created equal. Southern pine, for example, is generally stronger and stiffer (higher Fb and E values) than spruce.

Duration of load. How long will the members be loaded? Full-time live loading (as with floor joists) serves as the benchmark value, so-called normal duration. Normal duration values are multiplied by 1.15 to yield snow-load values and by 1.25 for seven-day loading (explained below).

Over time, the load on a joist or rafter can cause it to bend permanently. This happens whether or not the load is continuous; the effect is cumulative. The *normal duration* Fb value assumes that, during its lifetime, a joist will be subjected to its *full design load* for a cumulative total of ten years. Using the normal duration Fb value for a given wood species ensures that the joist will not fail. In reality, actual loads on the joist are much less than the design loads. The cumulative effect of lighter loads drops off sharply as the load decreases, meaning that rarely are joists in danger of failure.

Likewise, the *snow loading* Fb value assumes that a roof will have to support the *design* snow load for a total of only two months during its lifetime. Snow load Fb values are increased 15% over normal duration values because shorter loading periods have less effect than loads of longer duration. This means, for example, that a 2x10 of a given species may have a higher assigned Fb value when used as a rafter than when used as a joist.

Seven-day loading assumes an even shorter loading period, and is applied in some code districts where there are no wind or snow loads on roofs. The "seven days" assumes that over the lifetime of the roof, construction workers may place full design loads on the roof for a cumulative total of a week — roofers storing shingles on the roof, for instance.

Calculations for normal duration, snow loading, and seven-day loading are automatically factored into the tables. You can apply them according to your local code.

What You Need

To use this information, you'll need three publications. The first is a building code book, which includes information about required grades, spans, bearing, lateral support, notching, etc. The *One and Two Family Dwelling Code* from the International Code Council (5203 Leesburg Pike, Suite 600, Falls Church, VA 22041; 703/931-4533; www.iccsafe.org) is a good choice. It has one appendix with span tables for joists and rafters and another appendix with design values for joists and rafters. Check with your local building department, as local codes may vary.

The other two publications are available from the American Forest & Paper Association (AF&PA, 202/463-2700; www.afandpa.org). They are *Design Values for Joists and Rafters*, which lists Fb and

AFPA Table 9.1
Span Tables for Joists and Rafters

Required compression perpendicular to grain design values ($F_{c\perp}$) in pounds per square inch for simple span joists and rafters with uniform load.

Span, ft	Bearing Length, in				
	1.5	2.0	2.5	3.0	3.5
8	119	89	71	59	51
10	148	111	89	74	63
12	178	133	107	89	76
14	207	156	124	104	89
16	237	178	142	119	102
18	267	200	160	133	114
20	296	222	178	148	127
22	326	244	196	163	140
24	356	267	213	178	152

Notes: 1) Bearing width is assumed to be 1.5".
2) Total uniform load is assumed to be 66.67 plf.
3) Alternate $F_{c\perp}$ values are possible by adjusting the tabulated values in direct proportion to the desired load. Adjustment factors are tabulated in Table 9.2.
4) See A.1.3 for 2 span floor joist requirements.

EXCERPTED BY PERMISSION OF AF&PA.

Addendum to Design Values for Joists and Rafters

Species[1]	Compression design value perpendicular to grain, psi "$F_{c\perp}$"
Douglas Fir-Larch	625
Eastern White Pine	350
Hem-Fir	405
Southern Pine	
Dense	660
Select Structural No. 1, No. 2, No. 3, Stud, Construction, Standard, Utility	565
Non-Dense	480
Spruce-Pine-Fir	425
Spruce-Pine-Fir (South)	335

[1] Design values apply to all grades for the species listed unless otherwise indicated in the table above.

EXCERPTED BY PERMISSION OF AF&PA.

Figure 6. *Check to see that the lumber species selected has the necessary compression strength perpendicular to the grain. Table 9.1 (left) in Span Tables for Joists and Rafters gives the required values for various design conditions; an addendum that comes with Design Values for Joists and Rafters (right) gives the values for specific species.*

E values for various species, sizes, and grades of dimensional lumber, and *Span Tables for Joists and Rafters*, which assigns allowable spans to various combinations of E and Fb. I find the AF&PA documents easy to follow. And if you get stuck, the association's technical staff can help you.

Western Wood Products Association (WWPA, Portland, OR; 503/224-3930; www.wwpa.org) also publishes span tables. WWPA's tables are more flexible than AF&PA's, so some designers and engineers prefer them for calculating loads on complex structures. However, they're also harder to use, because they require the correct use of numerical multipliers. The AF&PA publications, by contrast, use a simplified approach that's suitable for most wood-frame homes. This makes them a better tool for most architects and builders.

Sizing Floor Joists

Let's work through an example that illustrates the steps involved in using the tables. Let's say you're building a 16-foot addition and have to select the correct size and species of lumber for the floor joists. The joists will be 16 inches on-center. Their design span — the exact length from face to face of the supports — is 15 feet 1 inch (Figure 2).

Step 1: Check the Code

First, check the local code for allowable live load, dead load, and deflection (see Figure 3). For this example I'll use the *One and Two Family Dwelling Code*, which serves as the model for many state and local codes. This sets an allowable first-floor live load of 40 psf, a dead load of 10 psf, and a deflection of L/360.

Step 2: Span Table

Select the appropriate table in *Span Tables for Joists and Rafters*. The table of contents indicates that Table F-2 watches these loading conditions. Using Table F-2 (Figure 4), check each lumber size to see if a 16-inch spacing will permit a span of 15 feet 1 inch. Start with the "16.0" line in the "Spacing" column at the left of the table, then go to the right until you reach an appropriate span (at least 15 feet 1 inch in this case). Then drop down to find the appropriate Fb value for that span.

As the table shows, no 2x8s meet the span and spacing requirements, but a 2x10 with an E of 1,300,000 psi and an Fb of 1093 psi can span 15 feet 3 inches — more than enough. A 2x12 with an E of 800,000 psi and Fb of 790 psi also works, since it can span 15 feet 10 inches.

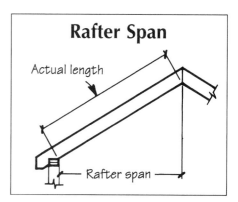

Figure 7. *Use the horizontal projection of a rafter, not its actual length, when figuring rafter span.*

Step 3: Wood Design Values

Now you must select a wood species and grade that meets the required Fb and E values, and that's available in your area. For this, use the tables in *Design Values for Joists and Rafters*. For this example, I've excerpted the relevant sections from tables for hem-fir, Douglas fir-larch, and spruce-pine-fir (Figure 5).

In hem-fir, either a No. 1 2x10 or a No. 2 2x12 would work. In Douglas fir-larch, either a No. 2 2x10 or a No. 2 2x12 works. In spruce-pine-fir, a No. 1 & 2 2x10 or 2x12 would do the job.

Step 4: Compression Check

The final step is to make sure the lum-

How To Read Span Tables

ber you've chosen meets the required design value for *compression perpendicular to the grain*. The loads carried by floor joists, ceiling joists, and rafters are transferred through their end points to supporting walls and beams. The ends of these members must be able to resist these loads without crushing.

Table 9.1 in *Span Tables for Joists and Rafters* (Figure 6, previous page) gives a required compression value of 237 psi for a span of 16 feet and a bearing length of 1.5 inches. (The tables permit a bearing length of up to 3.5 inches, but since 1.5 is probably the worst case that you'll encounter for joist or rafter bearing, it's a safe value.) You can get the compression design value for various species selected from the addendum that comes with *Design Values for Joists and Rafters*. For instance, hem-fir has an acceptable value of 405 psi, spruce-pine-fir of 425.

Ceiling Joists and Rafters

Ceiling joists are sized like floor joists except that deflection limits vary depending on whether the joists will be used for attic storage or will have a plaster or drywall finish. Check your code and follow the AF&PA tables accordingly.

When using the tables to size rafters, there are two points to keep in mind. First, remember that the rafter's span is not its actual length but its total horizontal projection (see Figure 7). Second, use the snow load value for your region in determining which rafter table to use. If your code book says your snow load is 40 psf, then you must use the 40 psf live load rafter table. The fact that snow loads only act part of the year has been taken into account in the rafter tables, but don't forget to use the "Snow Loading" column to get the Fb design value.

By Paul Fisette, a wood technologist and director of the Building Materials Technology and Management program at the University of Massachusetts in Amherst.

Beyond Code: Preventing Floor Vibration

Floor vibration, or bounce, is not a safety issue — it's a performance issue, and one that's likely to be important to homeowners. No one likes to hear the china rattling in the cabinet when they walk across the room. But at what point is the floor stiff enough, and how can a builder predict how the floor will perform?

Unfortunately, there's no clear-cut rule for a builder to follow, and the physics of vibration are so complicated that it's no easy matter to design a guaranteed bounce-free floor. Also, "acceptable" floor performance is highly subjective: What's good enough for one homeowner may not be good enough for another.

The building codes don't help much in this regard. They're primarily concerned with safety — in other words, the strength of the beam rather than its stiffness. The most stringent code limit for joist deflection is 1/360 of the span: For example, a joist with a clear span of 15 feet must not deflect more than 1/2 inch under live load (people and furniture). The dead load — the weight of the floor materials — is not typically included in calculating deflection.

And yet it has been known for decades that a $^{span}/_{360}$ live-load deflection limit will not necessarily yield floors that are acceptable to everyone when it comes to vibration.

By following these simple rules of thumb, however, you can take the annoying vibrations out of floor systems, whether you're framing with solid-sawn joists, metal-plate-connected floor trusses, or wood I-joists. There's no guarantee that every customer will be satisfied if you follow these guidelines, but they should prevent the vast majority of complaints.

Some Quick Rules of Thumb

Before looking at specific types of joists, here are some general guidelines for controlling bounce.

✔ **Shorten the span.** In general, shorter spans make for stiffer floors. For example, if the $^L/_{360}$ span table tells you a joist of a given size, grade, and species will just barely work for your span, shorten the span by adding a girder near the center of the original span. The resulting floor will vibrate less.

✔ **Increase the joist depth one size.** If the code requires a 2x8 at 16 inches on-center, then use a 2x10 of the same grade and species. Or use a 14-inch-deep floor truss when a 12-inch deep truss would meet code requirements. This may not be the most cost-effective solution in every case, but it's easy to remember and will save time and worry.

Probably the least efficient way to improve floor performance is to reduce the on-center spacing — 16 inches to 12 inches, for instance. Occupants feel "bounce" as a result of a foot impacting an individual joist. But even at 12 inches on-center, the joists are not close enough for the shock of a foot to be carried by two joists.

✔ **Glue and screw the sheathing.** Floor sheathing should always be glued down. Screws work better than nails for long-term bounce control.

Design for Solid-Sawn Joists

Our recommendation for stiffening solid-sawn floors is a simple modification of a rule that was published in 1964 by the FHA: For floors up to 15 feet, limit live-load deflection to $^{span}/_{360}$; for spans over 15 feet, limit the live-load deflection to 1/2 inch (see Table 1). In adopting this rule, we encourage builders and designers to ignore the reduced live load of 30 psf for sleeping areas, and instead use the standard 40 psf live load for all rooms. After all, a bedroom can become a study or home office, and the traffic may be heavier than in a living room.

Table 1. Maximum Clear Spans (Lmax) for Joists Longer than 15 Feet
(deflection limited to 1/2 inch)

LCODE	LMAX	LCODE	LMAX	LCODE	LMAX	LCODE	LMAX	LCODE	LMAX
Less than 15-0	Same as Lcode	16-0	15-9	17-0	16-5	18-0	17-2	19-0	17-11
15-1	15-0	16-1	15-9	17-1	16-6	18-1	17-3	19-1	17-11
15-2	15-1	16-2	15-10	17-2	16-7	18-2	17-3	19-2	18-0
15-3	15-2	16-3	15-11	17-3	16-7	18-3	17-4	19-3	18-1
15-4	15-3	16-4	15-11	17-4	16-8	18-4	17-5	19-4	18-1
15-5	15-3	16-5	16-0	17-5	16-9	18-5	17-6	19-5	18-2
15-6	15-4	16-6	16-1	17-6	16-10	18-6	17-6	19-6	18-3
15-7	15-5	16-7	16-2	17-7	16-10	18-7	17-7	19-7	18-3
15-8	15-6	16-8	16-2	17-8	16-11	18-8	17-8	19-8	18-4
15-9	15-6	16-9	16-3	17-6	17-0	18-9	17-8	19-9	18-5
15-10	15-7	16-10	16-4	17-10	17-1	18-10	17-9	19-10	18-6
15-11	15-8	16-11	16-5	17-11	17-1	18-11	17-10	19-11	18-6

*For code spans 20-0 and greater, Lmax=(180 L3code) 0.25, where Lcode is in inches.

Limiting joist deflection to 1/2 inch is an effective way to reduce annoying floor vibrations. For spans longer than 15 feet, the code L/360 maximum deflection limit results in actual deflections greater than 1/2 inch. In this chart, numbers in the "LCODE" columns represent code-allowable joist spans (in feet-inches) assuming a deflection limit of L/360. In the "LMAX" columns, those spans have been reduced so that the actual deflection is limited to 1/2 inch. To use this chart, locate your required clear span in the "LMAX" columns. Then, using a span table designed for L/360 maximum deflection, 40 psf live load, and the appropriate dead load, find a joist size and species that will work for the corresponding number in the "LCODE" column. Your joist will then be sized to limit live load deflection to 1 1/2 inch.

Table 2. Expected Vibrational Performance of Residential Floor Trusses (40 psf Live Load)

Live Load Deflection Limit	Strongback Installed	Truss Spacing (inches)	Vibration Rating
Span/360	No	24 or less	Code minimum; not rated
Span/360	Yes	24 or less	Good*
Span/480	No	24 or less	Good*
Span/480	Yes	24 or less	Very Good*

*Ratings require a minimum 23/32" APA-Rated sheathing, glued to truss chord and using nails or screws, and span-to-depth ratio of 20 or less. Ratings apply to maximum spans at the tabulated deflection limit. The ratings are based on specific input from experienced wood-truss design professionals, and our interpretation of opinions of experts and case studies.

Metal-Plate-Connected Floor Trusses

Floor trusses are a unique product in that they accommodate effective strongback bracing). The consensus among wood truss professionals is that strongbacks are effective in minimizing annoying vibrations, and that they are well worth the time and money it takes to install them.

Table 2 illustrates the expected performance of various floor truss designs, using a 40-psf live load. Table 3 gives guidelines for sizing and installing strongbacks. For best performance, strongbacks should be installed near the center of the span (versus two at the third points) in upright position and attached to a vertical web. The strongback should also be located at the bottom of a vertical web. To be effective, the strongback must be snugly attached to each web, as indicated by the nailing recommendations in Table 3.

When, for whatever reason, the vertical webs don't line up, you can attach a 2x4 or 2x6 scab to the top and bottom chords for attaching the strongback to the truss (Figure 8). The total number of nails used to attach the scabs to the truss chords should match the number used to attach the strongback to the vertical web.

Some of the truss professionals that we interviewed when developing Table 2 had more restrictive rules to offer, but none had less restrictive design advice. Again, no design criteria is guaranteed to totally eliminate vibrations, but we believe that following the recommendations in the table will minimize complaints.

Wood I-Joists

When using wood I-joists, a simple way to get good results is to always use the tables designed for span/480 deflection. Any I-joist stamped under the new APA standard for performance rated I-joists is automatically designed to meet the span/480 limit. The standard also uses 40 psf as the minimum live load for any floor. The APA standard is now being used by some I-joist manufacturers to make selection of I-joists easier. The allowable spans for various spacings are printed right on each joist.

Another design system for control vibration in wood I-joist floors is Trus

Table 3. Sizing and Attaching Strongbacks

Clear Span	Strongback Size	Connection Requirement at each Truss Web (minimum)
Greater than 15 feet, but less than or equal to 20 feet	2x6	4-16d Box (0.135"x3.5")
Greater than 20 feet	One 2x8 (or 2 2x6s)	8-16d Box (0.135"x3.5")

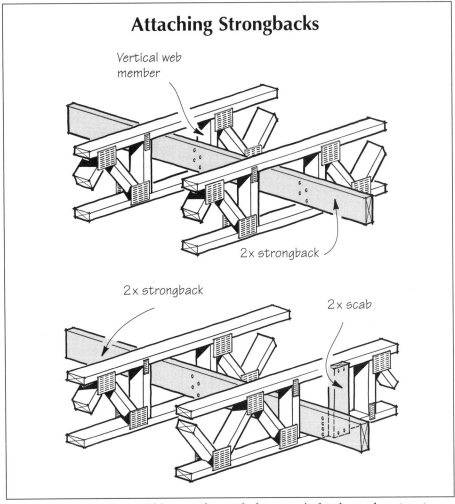

Figure 8. *Strongbacks should be securely attached to a vertical web member at center span next to the bottom chord (top). If the vertical web members don't line up properly, you can attach a 2x4 or 2x6 scab from chord to chord and nail the strongback to the scab (above). To transfer the load, use as many nails to attach the scab as you use to attach the strongback (see Table 3).*

Joist MacMillan's TJ-Beam software. Trus Joist has done extensive testing of floor performance and has developed its own rating system. Using the software, a user can select a number between 20 and 70, with 70 offering the greatest level of protection against potential floor problems as judged by an occupant. For example, a design that is rated at 55 is expected to be judged as "Good to Excellent" by 96% of the population, while 2% should judge such floors as "Marginal," and 2% should judge the floor to be "Unacceptable." This system allows the homeowners, through their contractors or architects, to select the level of floor performance to meet their expectations.

We tested the software for a 16-foot clear span supported by 2x4 walls (16 ft. 7 in. outside-to-outside), with I-joists 16 inches on-center and a residential load of 40/12 (live load/dead load). Using a 9.5-inch TJI Pro-250, the rating was 35. Increasing the depth to a 14-inch TJI Pro-250, the rating was a 53. Tightening up the spacing of the 9.5-inch I-joist to 12 inches on-center increased the rating only to 42 — illustrating that going to a deeper joist at the same spacing is a better solution.

The TJ-Beam software also provides a relative cost index that tells the user how much extra an improved floor will cost. Often an improved performance design can be obtained with the same or even lower cost than the original design.

By Frank Woeste, P.E. and Dan Dolan, P.E., professors at Virginia Tech University in Blacksburg, Va.

Calculating Loads on Beams and Headers

Most builders install built-up dimension lumber headers over standard window and door openings without worrying about the loads. You just put in the same-size double 2x10 or 2x12 headers throughout the house — not only will they carry standard residential loads, but they also automatically keep the window and door heads at a uniform height.

You can't beat sawn lumber for most small window headers, but as spans and loads increase, stronger materials are a better choice. Sawn lumber limits design potential and in some cases just doesn't work. LVL (laminated veneer lumber), Parallam, Timberstrand, and Anthony Power Beam are all good alternatives for longer-span and heavily loaded beams.

But regardless of the material you choose, you've got to accurately calculate the loads the beam will carry.

Following the Load Path

The job of headers and beams is simple: They transfer loads from above to the foundation below through a network of structural elements. The idea behind sizing headers and beams is straightforward: Add together all the live loads and dead loads that act on the member and then choose a material that can support the load. The beam must be strong enough in bending (its F_b value) so that it doesn't break and stiff enough (its E value) so that it doesn't deflect excessively under the load.

Engineered lumber beams are typically sized using span tables that match the span to pounds per linear foot (plf) of beam. So to use the tables in the manufacturers' design guides, you have to translate the load on a beam into pounds per linear foot.

Uniform vs. point loads. Loads are considered to be either uniformly distributed or point loads. A layer of sand spread evenly over a surface is an example of a pure uniformly distributed load. Each square foot of the surface feels the same load. Live and dead loads listed in the building code for roofs and floors are approximations of distributed loads, and are stated in pounds per square foot (psf).

Point loads occur when a weight bears on one spot in a structure, the way a column imposes its load at the base. The load is not shared equally by the supporting structure. Analysis of point loading is best left to engineers. Here, we'll consider only uniformly distributed loads.

Some Examples

Let's trace distributed loads for several different houses. Assume that all are located in the same climate, but have different loading paths because of the way they are built. These examples illustrate how distributed loads are assigned to structural elements. Our sample homes are in an area where the snow load is 50 pounds per square foot of roof area (treat snow as live load). In a warmer climate, of course, the snow load would be less, so you need to check your code book for live loads and dead loads in your region. All loads are listed as pounds per square foot of horizontal projection (footprint area), as in Figure 9.

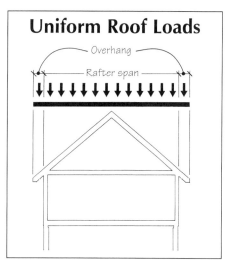

Figure 9. *Uniform roof loads are based upon the horizontal footprint area of the roof, not the sloped area of the roof.*

Header 1: One Story, Truss Roof

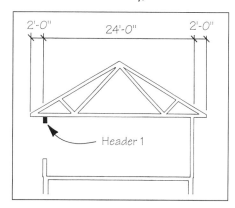

Here, each square foot of roof system delivers 50 pounds of live load and 15 pounds of dead load (an average weight for a roof built with trusses), or a total load of 65 psf, to the structural support system. Remember, these loads are distributed uniformly over the entire surface of the roof. Each exterior wall (and any headers within) will carry all loads from the center of the house to the outside, including the roof overhang. This distance, the "tributary" width, is 14 feet (12 feet + 2 feet). So each linear foot of wall must carry the loads imposed by a 1-foot-wide strip 14 feet long. Thus, each lineal foot of wall supports

live load (snow): 50 psf x 14 ft. = 700 plf
roof dead load: 15 psf x 14 ft. = 210 plf
total load: 910 plf

It is important to list live load, dead load, and total load separately because live load is used to compute stiffness and total load is used to calculate strength.

Header 2: One Story, Stick Roof

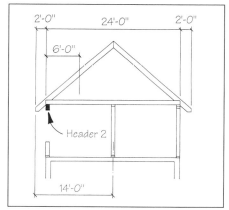

This house is identical to our first example except it is stick-built. As a result, the live load, dead load, and distribution of forces are different. Unlike the trussed roof, live load and dead load of the rafters and ceiling joists must be accounted for as separate systems. Since it is possible to use the attic for storage, the live load of the attic floor is set at 20 psf according to code. So each lineal foot of header supports

live load (snow): 50 psf x 14 ft. = 700 plf
roof dead load: 10 psf x 14 ft. = 140 plf
ceiling live load: 20 psf x 6 ft. = 120 plf
ceiling dead load: 10 psf x 6 ft. = 60 plf
total load: 1,020 plf

Header 3: Two Story, Truss Roof

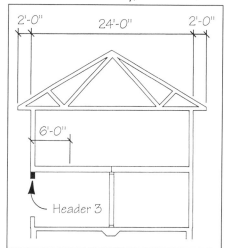

Again, this house has the same width dimension, but it has two levels. Loads are contributed to the lower header by

Calculating Loads on Beams and Headers

the roof, upper walls, and second floor. For second-story floor loads, code usually gives a live load of 30 psf and a dead load of 10 psf. *Architectural Graphic Standards* lists the weight of an exterior 2x6 wall as 16 pounds per square foot. So an 8-foot-tall wall weighs 128 plf (8 ft. x 16 psf). The loads delivered to the header are

live load (snow): 50 psf x 14 ft. = 700 plf
roof dead load: 15 psf x 14 ft. = 210 plf
upper level wall: 16 psf x 8 ft. = 128 plf
2nd floor live load: 30 psf x 6 ft. = 180 plf
2nd floor dead load: 10 psf x 6 ft. = 60 plf
total load: 1,278 plf

Ridge Beam & Center Girder

This figure illustrates two structural elements: a structural ridge beam and a center girder. Each has a tributary width of 12 feet. The load per foot of beam is determined the same way as for headers. For the ridge beam, the load is

live load (snow): 50 psf x 12 ft. = 600 plf
roof dead load: 10 psf x 12 ft. = 120 plf
total load: 720 plf

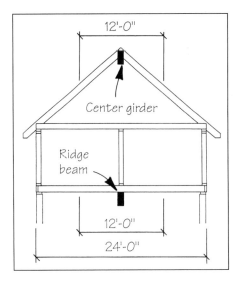

Center girder. The center beam carries half the floor load, the weight of the partition wall, and half the second-floor load (an attic treated as living space). Code load for first-story rooms is 40 psf live and 10 psf dead. *Graphic Standards* lists the weight of a 2x4 partition as 10 pounds per square foot. So the load on this girder is

1st floor live load: 40 psf x 12 ft. = 480 plf
1st floor dead load: 10 psf x 12 ft.= 120 plf
8-foot-tall partition: 10 psf x 8 ft. = 80 plf
2nd floor live load: 30 psf x 12 ft. = 360 plf
2nd floor dead load:10 psf x12 ft.= 120 plf
total load: 1,160 plf

In Summary

These examples are typical of the kind of calculations you will have to do to figure the uniform load on a beam. Armed with this information, you can size a dimension lumber beam, as long as you can confidently wade through the necessary formulas. Or you can go to the engineered lumber design guides and safely choose a beam that will carry the load for the span in question.

—*Paul Fisette*

Sizing Simple Beams

The fundamental structural element in a building is the beam, a horizontal member that supports something. It's used everywhere in wood-framed structures — joists, headers, girders. Here we'll find out how engineers size these beams (headers and girders at least; for joists, you just use the tables). We'll also go into the raw strength of a beam.

All beams, whether wood, steel, concrete, or composite, have five principal properties of interest to the structural engineer: support configuration, cross-section, material, length, and load. From these five properties, we can usually determine whether the beam will do the job.

All these properties apply to the wood members found in a frame house. To simplify things, let's restrict this discussion to "simply supported rectangular wooden beams with evenly distributed loads." What's all this mean? "Simply supported" means that the ends of the beam are resting on two props or supports (see Figure 10). These supports hold the ends up but allow the beam to bend and flex freely. (Other support configurations include cantilevers and beams fixed at one or both ends.) "Rectangular" refers to the section, "wooden" to the material. An "evenly distributed load" is placed along the entire beam.

In the case of floor systems, the "evenly distributed load" is not quite real; it is an ideal used for convenience. Real loads are spotty and moving; they are concentrated where furnishings and people are, and they can move around. Since designers have to set up the floor before knowing where things will be, they use a distributed load — so many pounds per square foot. This is usually a much higher load than would occur in a real building, but it serves for designing.

A Worked Example

Let's look at a specific beam and see if it does the job. Our arbitrary beam (Figure 11) will be an 18-foot-long girder down the center of a 20-foot-wide living room. The structural system is 10-foot-long 2x8 joists running from the walls to the girder at 16 inches on-center. The joists are #2 Doug fir and the girder is built up of select structural Doug fir 2x12s. According to the Uniform Building Code, the floor of a residential building should be designed to support a load of 40 pounds per square foot. We will use all of these specifications to calculate the *bending stress* — the raw strength of the beam.

Stress is the relative amount of load supported by a material. It is measured in pounds per square inch: load divided by area. It is *stress*, not load, that tears things apart. *Bending stress* (sometimes called *extreme fiber stress in bending*) is

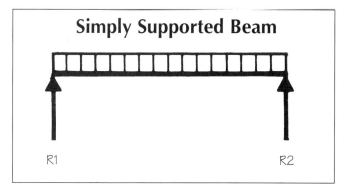

Figure 10. *A "simply supported" beam rests on its supports but is free to flex and bend. Most of the beams in a typical wood frame house — headers and girders, for example — are simply supported beams.*

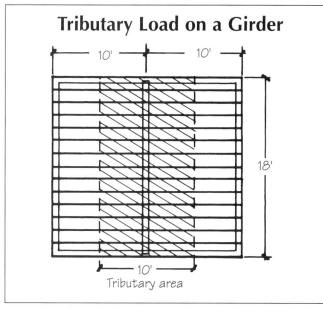

Figure 11. *A typical central girder carries half the load of the joists that rest on it in this case, an area 10 feet wide by the 18-foot length of the beam.*

the stress in the outermost (extreme) fibers of a wood beam, caused by loads perpendicular to the beam.

The symbol for bending stress is f_b. With a capital "F," F_b refers to the *maximum allowable bending stress*, a tested design value that depends on wood species. With a small "f," f_b refers to the *actual bending stress* induced in a beam by its load. In sizing wood beams, it is important to ensure that f_b doesn't exceed F_b.

Step 1. First, we calculate the load on the beam from the known 40-psf floor load. The floor space that directly loads the girder is called the *tributary area* for the girder — in this case, a 10-foot-wide area down the middle of the floor. At 40 psf, the beam load (w) in pounds per inch is:

$$w = \frac{10 \text{ ft.} \times 40 \text{ psf}}{12 \text{ in/ft}}$$

$$= 33.3 \text{ pounds per inch}$$

Because we're working in inches and pounds — the standard English unit of stress for centuries — there is a 12 thrown into the calculation to make the units come out right.

Step 2. Next we calculate a quantity called the *maximum bending moment* (M). This is the tendency for the beam load to stress the beam, and it depends on the beam load, the length of the beam, and the load configuration. For each load configuration there is a specific bending moment formula. These formulas are tabulated in various references; I personally use the AISC *Manual of Steel Construction* (American Institute of Steel Construction, One E. Wacker Dr., Suite 3100, Chicago, IL 60601; 312/670-2400). Another source is the AFPA *Wood Structural Design Data* book (American Forest and Paper Association, 1111 19th St., Suite 800, Washington, DC 20036; 202/463-2700).

The bending moment formula for a simply supported beam with an evenly distributed load is:

$$M = \frac{(w \times L^2)}{8}$$

where w is the unit load per inch on the beam and L represents its total length. For our problem beam:

$$M = \frac{33.3 \text{ lb./in.} \times (18 \text{ ft.} \times 12 \text{ in./ft.})^2}{8}$$

$$= 194{,}206 \text{ in. lb.}$$

So our 18-foot-long girder must be sized to resist a 194,206 in. lb. bending moment.

Step 3. Next we have to select a beam size that we think will be adequate. To check this, we calculate its *section modulus (S)*. Section modulus is a subordinate property of a beam, an expression of its strength in terms of its cross-section (Figure 12). For a rectangular beam, the formula for the section modulus is:

$$S = \frac{b \times h^2}{6}$$

where b is the width and h is the depth of the member. From the formula you can see what every framer knows intuitively: that a wood member increases in bending strength more in relation to its depth (which is squared) than its width.

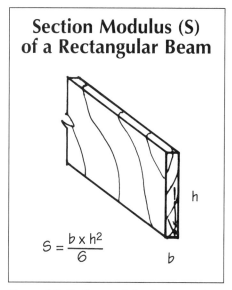

Figure 12. *Section modulus is a property of a beam that engineers use to calculate the bending stress of a given beam under specific loading conditions. The formula shows what every framer knows — that a stick of wood is stronger placed on edge than on its face.*

We'll need a pretty big girder to span 18 feet — say, a triple 2x12. A single 2x12 has a section modulus:

$$S = \frac{1.5 \times 11.25^2}{6}$$
$$= 31.6 \text{ in}^3$$

A triple 2x12 is three times this: 94.8 in^3.

No one actually calculates the section modulus for common lumber — we just look up the numbers in a table. You can find one in the National Design Specification supplement, Design Values for Wood Construction (also published by AFPA), which I'll excerpt:

Lumber Size	Section Modulus
2x4	3.1
2x6	7.6
2x8	13.1
2x10	21.4
2x12	31.6

Step 4. Now we use the section modulus to calculate the *actual bending stress* (fb) that the bending moment will create in the triple 2x12. The relationship between bending moment, section modulus, and bending stress can be expressed in a simple formula:

$$M = fb \times S$$

Since we know M and S, to solve for fb we set up the equation like this:

$$fb = \frac{M}{S}$$

$$fb = \frac{194{,}206 \text{ in. lb.}}{94.8 \text{ in}^3}$$

$$fb = 2{,}049 \text{ lb./in.}^2 \text{ (psi)}$$

Now we look in the tables of *allowable bending stress* (Fb) for various woods. In *Design Values for Wood Construction* we see that select structural Douglas fir-Larch used as a timber (beams 5x5-inches and larger) has an *allowable* bending stress of 1,600 psi. The *actual* bending stress of 2,049 in the triple 2x12 girder exceeds this by quite a bit, so our first guess was faulty. Let's try a quadruple 2x12, which has a section modulus of 126.4.

We recalculate, using the same bending moment:

$$fb = \frac{194{,}206 \text{ in. lb.}}{126.4}$$
$$= 1{,}536 \text{ psi}$$

So the actual bending stress (fb) in the quadruple 2x12 beam — 1,536 psi — is less than the maximum allowable bending stress for Doug fir-Larch lumber — 1,600. The beam is strong enough, but still must be checked for shear and deflection.

Checking for Shear

Shear is a word that is used a little casually. It actually has two meanings: One is "to cut off," as when you shear a branch with hedge trimmers. In our case, the relevant meaning is "to slide by in a parallel plane," as when a rock layer in the earth fractures during an earthquake and two strata slide past each other. To see shear in action, try a little experiment. Pick up a magazine and hold it flat in front of you, then fold it in half vertically. At the unbound edge, the pages slip by one another to form a knife edge — an example of shear. Now fold the magazine again, this time pinching the unbound edge. Now there is a bulge inside the magazine, as the pinch resists the shear.

Shear in wood. Shear is a natural consequence of bending. When a beam is flexed, it goes into a curve. The distance along the outside of the curve — the bottom of the beam — is longer than the distance around the inside — the top of the beam. As the bending stresses force the wood to take on different lengths along its two edges, the wood fibers begin to tear apart as they "slide by" one another parallel to the length of the beam.

Checking for shear is a three-part process: First, we find the maximum shearing force (Vmax) that the load will impose on the beam. Next, we use V to calculate the maximum shear stress (fv) that will develop in the beam as a result of the shearing force. Finally, we check fv against Fv, the maximum allowable shear stress for the species of wood we're working with — select Doug fir, in the case of our problem beam.

From our reference book, the formula for maximum shearing force (Vmax) for this type beam is:

$$Vmax = w \times \frac{L}{2}$$
$$= 33.3 \text{ lb./in.} \times \frac{(18 \text{ ft.} \times 12 \text{ in./ft.})}{2}$$
$$= 3{,}596 \text{ lb.}$$

where w is the unit load on the beam and L is its total length.

The maximum shear stress (fv) in the beam is calculated by dividing the shearing force (V) by the cross-sectional area (A) of the beam and multiplying by a factor of 1.5:

$$fv = 1.5 \times \frac{V}{A}$$
$$= 1.5 \times \frac{3{,}596 \text{ lb.}}{6 \text{ in.} \times 11.25 \text{ in.}}$$
$$= 80 \text{ lb./sq. in.}$$

The National Design Specification Supplement (published by American Forest & Paper Assn.; 202/463-2700) gives a maximum allowable shear stress (Fv) for Doug fir of 95 psi, so we are okay for shear.

Most of the time shear is not a problem, except when we have short, deep beams — that is, when the depth of the beam is greater than about one-sixth its length. Here, the greater distance between the inside and outside fibers causes a greater change in length and thus more shear. Regardless of the beam size, though, never skip the shear calculation.

Deflection Check

Let's move on to deflection. Above and beyond any calculations on the strength of a beam is the idea of "serviceability": Will the beam provide a useful horizontal support? "Useful" means that the beam will not only

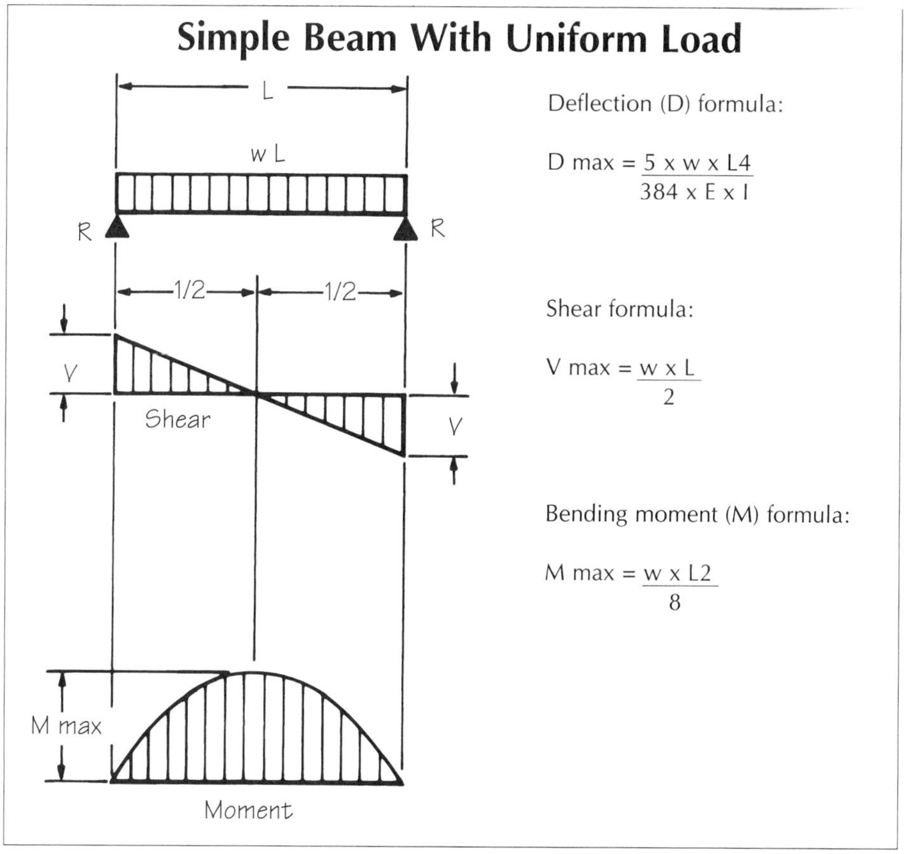

Simple Beam With Uniform Load

Deflection (D) formula:

$$D_{max} = \frac{5 \times w \times L^4}{384 \times E \times I}$$

Shear formula:

$$V_{max} = \frac{w \times L}{2}$$

Bending moment (M) formula:

$$M_{max} = \frac{w \times L^2}{8}$$

Figure 13. *Engineers use schematic drawings like this to model various loading conditions. The top of the sketch represents the beam's uniform load. The reactions (R) are at the support points. The middle of the sketch indicates that shear stress increases uniformly from none at center span to the greatest shear at the ends of the beam. The moment sketch, at the bottom, shows that bending forces increase toward the center of the beam, with the maximum bending at center span.*

resist breaking, but also will not flex so much that the floor structure is bouncy and unusable, transmitting shocks to other parts of the building. Most beam failures are not breakages, but serviceability failures. Ironically, most beam analyses are only for breakage. This is understandable when you consider that a breaking beam is a threat to life and safety. In residential construction, a breaking beam is extremely rare. The floors that sag and bounce are the ones that cost in repairs and hard feelings.

So we calculate *deflection* — the distance that the beam will sag when the design load is placed upon it. The deflection formula for our "simple beam with a uniformly distributed load" is:

$$D = \frac{(5 \times w \times L^4)}{(384 \times E \times I)}$$

In this formula, E is the *modulus of elasticity*, a property of the material from which the beam is made. E is derived by measuring the amount of stress required to stretch a specially sized sample of the material. Here are some typical values:

Material	E
S-P-F (constr. grade)	1,300,000 psi
Doug fir (select grade)	1,900,000 psi
Glulam	2,000,000 psi
Aluminum	12,000,000 psi
Steel	29,000,000 psi

The last two figures are included to give some sense of relative magnitude — steel is about 20 times stiffer than wood.

The other item in the deflection formula, *I*, is the *moment of inertia*. Like the *section modulus*, *I* is a property of the beam section. Calculated for a 2x12, where *b* and *h* are the width and depth of the member:

$$I = \frac{b \times h^3}{12}$$
$$= \frac{1.5 \times 11.25^3}{12}$$
$$= 178 \ in.^4$$

For the quad 2x12, *I* is four times as great, or 712 in.^{4}

We almost never calculate the moment of inertia — we just look it up in the tables. Here is the moment of inertia for some common lumber sections:

2x6	20.8 in.^{4}
2x8	47.6
2x10	96.9
2x12	178.0
G3x12	450.0
G5x15	1441.0

Plugging in the values for *E* and *I* we calculate the deflection (*D*) of the Douglas fir girder:

$$D = \frac{5 \times 33.3 \ lb./in. \times (18 \ ft. \times 12 \ in./ft.)^4}{384 \times 1,900,000 \ lb./in.^2 \times 712 \ in.^4}$$

$$= 0.70 \ in.$$

By code, the deflection of the girder must be less than 1/360 of the total length (*L*) of the girder:

$$\frac{L}{360} = \frac{18 \ ft. \times 12 \ in./ft.}{360}$$
$$= 0.60 \ in.$$

Our beam isn't quite stiff enough. The easy out is to add another stick, coming up to a quintuple 2x12. This beam has a moment of inertia of 890 in.4. Recalculating the deflection as above gives 0.56 in., less than the limit of 0.60 in.

But What About Bounce?

So now we're okay for deflection, at least by code. Your friendly building code enforcement officer will accept this structural system. But I'm not so sure that we have done a good design analysis. Most of the code-required analyses just consider the way buildings respond to static loadings. These are loadings that are applied and do not move. Real problems occur under dynamic, or moving, loads.

Sizing Simple Beams

The first time I encountered this condition was years ago in a house where the floor system met code for strength and deflection. Yet the floor was so springy that walking across the dining room whipped up a tsunami in the living room fishbowl. The serviceability just wasn't there.

To ensure serviceability, I consider the bounciness of the floor. I ask, "Will a 175-pound person falling 6 inches cause the floor to deflect more than 1/2 inch?" I realize that this is an arbitrary measure based on my own experience and intuition. It could certainly use some tuning up, but since I know of no other index, it's a useful working tool. The analysis involves an energy balance calculation, which I'll skip here and just give the result: The girder we're looking at will flex 0.51 inch under this measure — close enough.

However, when you look at the entire floor system — the 2x8 joists sitting on the girder — the bounciness is more than an inch. This excessive bounciness suggests that our code-acceptable floor won't really work very well in the real world. Here's what we can do about it: Make the system stiffer with 2x12 joists. If this seems a little excessive, consider that for an extra 200 board feet, you can build a much more solid floor system. Eighty to a hundred bucks will make the building a lot more comfortable. A little outside the textbook maybe, but worth thinking about.

By Harris Hyman, P.E., a civil engineer in Portland, Ore.

Simple Approach to Sizing Built-Up Headers

Selecting the right-size header is a common problem for contractors. Reference tables are available, but they can be long and confusing. Here I will share my simple reference for sizing standard headers.

There are many definitions of the word "header," but we are talking about main supporting beams that carry vertical loads around openings in walls. These can be door headers, window headers, or headers above portals between columns or posts. We will refer exclusively to wood headers in wood frame construction with wooden floor, ceiling, and roof framing running perpendicular to the direction of the header.

Using The Table

The load factors in Table 4 are grouped by lumber species. Load capacities are given per linear foot for headers of double and triple 2x4, 2x6, 2x8, 2x10, and 2x12 construction. To use the table:

1. Multiply the building width (eaves-to-eaves) by the *load factor* appropriate to the header location. The load factor has been calculated to take into account how much load is being carried by the wall and header of concern. When you multiply your building width by the appropriate load factor, the result is the uniform load in pounds per linear foot that the wall and header must be able to carry.

2. Look in the appropriate table for the header length you need (rounded up to the next whole foot). Look down that column for the first row with a load capacity equal to or greater than your calculated uniform load. At the left you'll find the size of framing lumber to use. It's as simple as that.

3. Notice the gray shaded area, which indicates increased jack stud requirements. For these loads, you will need two jack studs under the header.

Example

For example, say you are asked to replace the back door of a 28x42-foot colonial with double patio doors. Your new rough opening needs to be 5 feet 3 inches wide. How big should your header be?

First, find the right house type and load factor. In this case, the colonial has

Table 4. Header Load Factors

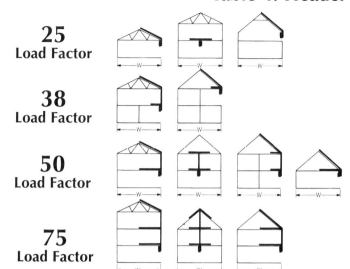

25 Load Factor
38 Load Factor
50 Load Factor
75 Load Factor

Load Factor x W = Header Load (in lbs / lin. ft.)

Calculate the header load, then find the size of lumber you will need for the span of the header and the species of lumber you are using.

			MAX. LOAD (LBS./LIN. FT.) FOR DOUBLE 2X HEADER										MAX. LOAD (LBS./LIN. FT.) FOR TRIPLE 2X HEADER									
			SPAN OF HEADER										SPAN OF HEADER									
			3'	4'	5'	6'	7'	8'	9'	10'	11'	12'	3'	4'	5'	6'	7'	8'	9'	10'	11'	12'
No. 2 S. Pine, Doug. Fir-Larch E=1,600,000 Fb=1,200 Fv=90	Nom. Lumber Size	2x4	521	306	195								781	459	293							
		2x6	950	642	483	336	246						1425	963	725	504	370	283				
		2x8	1458	935	688	544	429	328	259	210			2187	1403	1032	817	643	492	389	315	254	196
		2x10	2280	1353	962	746	609	515	422	342	282	237		2030	1443	1119	914	773	633	513	424	356
		2x12		1905	1296	981	790	661	568	498	418	351			1944	1472	1185	991	852	747	627	527
No. 2 Hem-Fir E=1,400,000 Fb=1,000 Fv=75	Nom. Lumber Size	2x4	434	255									651	382	244							
		2x6	792	535	403	280	205						1188	802	604	420	308	236				
		2x8	1215	779	573	454	357	273	216				1822	1169	860	681	536	410	324	262	217	
		2x10	1900	1128	802	622	508	430	352	285	235	198		1692	1203	933	762	645	528	427	353	297
		2x12		1588	1080	818	658	551	473	415	348	292		2382	1620	1227	987	826	710	623	522	439
Spruce-Pine-Fir E=1,300,000 Fb=875 Fv=70	Nom. Lumber Size	2x4	396	223	142								595	334	214							
		2x6	739	499	353	245	180						1108	749	529	367	270	206				
		2x8	1134	727	535	423	312	239	189				1701	1091	803	635	469	359	283	229		
		2x10	1773	1052	748	580	474	389	308	249	206			1579	1122	871	711	584	462	374	309	259
		2x12		1482	1008	763	614	514	442	369	305	256		2223	1512	1145	921	771	663	553	457	384

Additional Notes on the Table:
- The "load factors" are based on a total design load (live plus dead load) of 50 pounds per square foot. This yields conservative header design in most residential situations where uniform loading is involved. If you are working in a building where floor or roof live loads exceed 40 psf, or where point loads or other unusual loading is involved, do not use this table.
- Loads of greater than 2,000 pounds per linear foot will very rarely be encountered. They are tabulated here for reference, but if you come up with such large numbers, check your arithmetic.
- Note that for certain building types the required load capacity can be as much as three times greater than that of others.
- The guidance provided in this article is intended to apply to a wide variety of standard header situations. When the words "standard header" don't apply, it is time to consult the experts: Call a licensed engineer. You may be surprised how helpful he or she can be.

Simple Approach to Sizing Built-Up Headers

two stories, with a bearing wall running down the center of the first and second stories, a framed roof, and a full-width attic used for storage (Figure 14).

According to Table 4, this header location has a load factor of 50. Multiplying the load factor (50) by the width of the house (28) gives the pounds per linear foot the header must carry: 1,400.

Now determine the species of wood you'll be building with — in this case, Doug Fir-Larch. Round up the rough opening to 6 feet. Looking in the 6-foot column under "Triple 2x Headers," you find that a triple 2x12 header in Doug Fir-Larch will carry 1,472 pounds per linear foot. Note that the number is in the gray area, which indicates a second jack stud is needed.

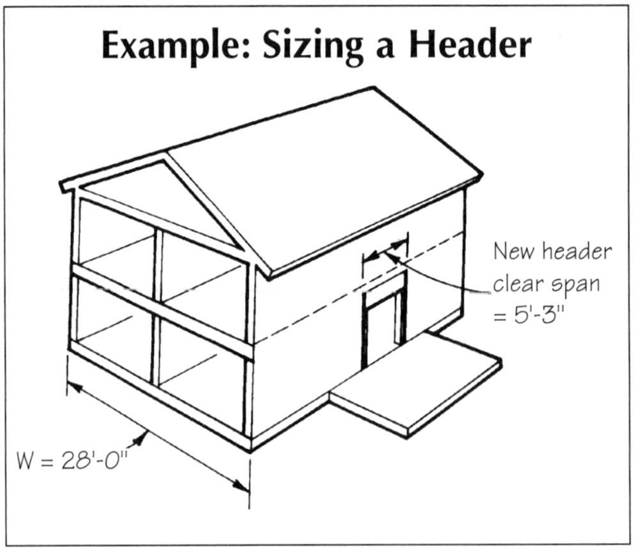

Choosing Header Stock

When making headers, particularly deeper sizes such as 2x10 and 2x12, use the driest lumber you can find.

The use of dry lumber for headers is very important because of how dramatically new "kiln-dried" framing lumber can shrink in a short time. With a 2x12 header you can often get over $1/4$-inch shrinkage across its $11 1/4$-inch width, resulting in a corresponding sag in the top plates. Sometimes the plywood sheathing may carry the loads during the construction period, leaving the header carrying no load at all, and a $1/4$-inch gap above. Then during an unusual loading condition, such as heavy snow, sagging may occur, resulting in cracked interior wall finish and trim.

When you don't have dry header stock available, it becomes even more important to use the smallest header that will carry the load. For this reason, I rarely specify headers larger than 2x8s except above garage doors. If 2x6s are adequate, use them: The shrinkage in a 2x6 will be less than half what you'll get with 2x12s. For normal-width doors, you'll usually find the 2x6 headers meet the calculated load capacity requirement, but check to be sure.

Wood defects. Select wood carefully for your headers. Plan your cuts to avoid including knots or grain defects at center-span bottom or near ends at

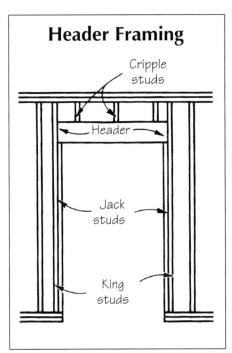

Figure 14. *First, multiply load factor (50) by building width (28 ft.) to get load: 50 x 28 = 1,400 lbs./lin. ft.*

Then, round up header span (5'-3") to the next whole foot: 6'.

Finally, look in the 6' column for the size header you need. In this case a triple 2x12 in Doug Fir-Larch is the only header that will do the job; it will need double jack studs.

Figure 15. *Fasten headers by end-nailing through the king studs using 20d nails. Jack studs, which must be nailed securely to the king studs, carry header loads to the floor. With longer spans or particularly heavy loads, double jack studs may be required.*

mid-height. Do not include cracked pieces or ones with even small splits in the ends. In my inspection work I have seen several cases where a single defective piece of wood necessitated costly ripout and rebuilding at a later date.

Making The Header

The use of plywood spacers or fillers is a common and effective way to bring the thickness of the header up to the dimension of the wall thickness, and it is quite acceptable. For a double 2x header in a wall with 2x4 studs, this requires a $1/2$-inch plywood spacer, which can be placed between the 2xs or on the outside. A frequently encountered myth is that the plywood is a major factor in increasing the strength of the header. This is really not true. For the standard header just discussed, a good grade plywood, installed with the face grain parallel to the header's grain, would add at most only about 8% to the strength — hardly major.

For northern climates, it's just as wise to include rigid foam insulation on the outside between the header and the exterior sheathing and to forget about the plywood.

To nail the individual 2xs to each other, you should use no less than two 16d nails every 16 inches. For 2x12s, use rows of three nails. With triple 2x construction, use 20d nails.

At the ends, the header should be attached with 20d nails spiked through the first full-height stud (Figure 15). Use at least one 20d nail for every 2x3 equivalent: Thus, for a double 2x6 there would be the equivalent of four 2x3s, and you would use four nails at each end. Where proper nailing cannot be accomplished for some reason, use galvanized steel header hangers or framing clips. Don't leave a header secured with only a couple of toenails.

When Wood Won't Work

There are some cases where built-up

wood headers are not appropriate. They typically include situations with unusually long spans, heavy loadings, or both. Also, a transom light, palladian window, or other feature may occasionally make the depth of a standard wood header unacceptable. In such cases, consider using glulams, laminated veneer lumber, steel flitchplates, or structural steel beams. Choosing the right option will involve evaluation of the strength and stiffness requirements and is generally best left to the professional engineer.

By Robert Randall, P.E., a professional engineer licensed to practice in New York, New Jersey, Connecticut, and Wisconsin.

Sizing Stiff Floor Girders

We received a call a while back from a homeowner who had a problem he wanted us to solve. He had just built a garage with living space above, and the upstairs floor was too bouncy. The odd thing was that the homeowner had specifically asked the building designer to ensure that the floor was very stiff, so the designer had specified floor joists (wood I-joists, in this case) with an L/720 live-load deflection limit. So why was the floor so springy?

The problem, as it turned out, was the midspan girder that the joists crossed. Although the girder had been sized correctly by "code" for strength and stiffness, its deflection limit was much greater than that of the joists. As a result, the whole floor system felt bouncy.

This is not an unusual condition. A group of professors and graduate students at Virginia Tech conducted five years of research on residential floor vibration. We analyzed more than 200 actual wood-frame (dimensional lumber, floor truss, and wood I-joist) floor systems. As a result of this research activity, homeowners and engineers frequently call us for help with residential floor vibration problems. We've found that most of the problem floors — floors with excessive vibration — are framed with joists running to a girder, and that the girders are typically too long between supports and therefore too flexible.

Bad Vibes

Traditionally, the stiffness of wood-frame floor systems has been controlled by limiting live load deflection to L/360 (where L is the joist span). But while this deflection limit may prevent plaster from cracking (the reason it was developed), it may not prevent annoying vibration of the floor, especially as the span gets longer.

All floors vibrate when someone walks across them; the question is whether you feel it or not. The perception of "excessive" vibration is subjective, so complaints by homeowners will vary. However, research done in the past suggests that most people are sensitive to vibrations in the 8 to 10 Hertz range. (A Hertz is a measure of vibration frequency; one Hertz equals one cycle per second). This sensitivity is believed to be because the human organs have a natural frequency of 8 to 10 Hertz; floor vibration in the same range is therefore perceived to be uncomfortable.

Sizing Built-Up 2x10 Girders
(L/600 deflection, 40 psf live load, 10 psf dead load)

To use this chart, first figure out the tributary width (TW) that the girder must support (see illustrations). Then find the required Fb and E values for the 2x10 lumber. Note that the required values increase as girder span increases, but decrease when you upgrade from a three-ply to a four-ply girder. In the box on the next page, there is a partial list of design values from some commonly available species and grades. Select a species and grade that meets or exceeds the required design values for your conditions.

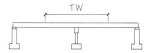
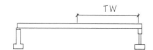

Three-Ply 2x10 Girder

	TW (Feet)	Required Fb	Required E (million)
8-Foot Pier Spacing	10	650	0.78
	12	781	0.93
	14	911	1.09
	16	1041	1.24
	18	1171	1.4
	20	1301	1.55
9-Foot Pier Spacing	8	659	0.88
	10	823	1.11
	12	988	1.33
	14	1153	1.55
	16	1317	1.77
10-Foot Pier Spacing	6	610	0.91
	8	813	1.21
	10	1016	1.52
	12	1220	1.82

Four-Ply 2x10 Girder

	TW (Feet)	Required Fb	Required E (million)
8-Foot Pier Spacing	14	683	0.82
	16	780	0.93
	18	878	1.05
	20	976	1.16
	22	1073	1.28
	24	1171	1.4
9-Foot Pier Spacing	12	741	0.99
	14	864	1.16
	16	988	1.33
	18	1111	1.49
	20	1235	1.66
	22	1358	1.82
10-Foot Pier Spacing	8	610	0.91
	10	762	1.14
	12	915	1.36
	14	1067	1.60
	16	1220	1.82

Crunching the Numbers

Joist Vibration

For a simple-span joist supported by rigid supports, such as a foundation wall, the *fundamental frequency* can be calculated by

$$f = 1.57 \sqrt{\frac{386 EI}{WL^3}}$$

where f is the calculated frequency (Hz); E is the modulus of elasticity of the lumber (psi); I is the moment of inertia of the lumber (in.^{4}); W is the actual dead weight of the joist and floor material (lb.); and L is the clear span (in.).

Floor System Vibration

When floor joists are supported by a girder, the vibration frequency of the floor system can be calculated by

$$f_{system} = \sqrt{\frac{f^2_{girder} \times f^2_{joist}}{f^2_{girder} + f^2_{joist}}}$$

Note that the floor system frequency is less than that of the component parts.

Lumber Design Values*

Species	Grade	Fb (psi)	E (psi)
Douglas Fir–Larch	Select Structural	1,450	1,900,000
	No. 1 & Better	1,150	1,800,000
	No. 1	1,000	1,700,000
	No. 2	875	1,600,000
Hem - Fir	Select Structural	1,400	1,600,000
	No. 1 & Better	1,050	1,500,000
	No. 1	950	1,500,000
	No. 2	850	1,300,000
Spruce-Pine-Fir (South)	Select Structural	1,300	1,300,000
	No. 1	850	1,200,000
	No. 2	750	1,100,000
Southern Pine	Select Structural	2,050	1,800,000
	No. 1	1,300	1,700,000
	No. 2	1,050	1,600,000

Source: NDS Supplement, 1991.

*Applicable adjustment factors have already been included in the "Required F_b" in the sizing chart on the previous page. When selecting a lumber species, use Base Design Values without any adjustments.

The Research

Because of the subjective nature of the problem, we combined a subjective measure of floor vibration with a scientific measurement. One person would do a "heel drop" — that is, stand on his toes and drop at once on his heels. Another person standing a few feet away would assess the vibration produced as either "acceptable," "marginal," or "unacceptable."

The researchers would then measure the floor's actual vibration with an accelerometer. What we found was that floors judged to be "acceptable" generally vibrated at 15 Hertz or higher. The floors judged "marginal" tended to be in the 11- to 13-Hertz range, while the "unacceptable" floors had vibration rates below 11 Hertz.

Recommendations

It makes sense to try to build floor systems that have a fundamental frequency well above 10 Hertz. Based on our findings, a target frequency of 15 Hertz will result in an "acceptable" floor most of the time.

For simple joist spans running from foundation wall to foundation wall, the calculation is fairly straightforward (see "Crunching the Numbers," above). For those who want to avoid the math, a simpler recommendation is to size floor joists for L/480 deflection instead of L/360. An easy way to do this is to take the allowable joist span from an L/360 span table and multiply by .91.

When the floor joists are supported by a girder, however, the calculations get a little more complicated. If you work through the formula for "system" frequency, you'll see that the frequency for the floor system is less than that of the component parts — the joists and girder. For example, when both the joists and girder have a frequency of 15 Hertz, the frequency of the floor system is only 10.6 Hertz.

To avoid the complication of having to calculate the vibration frequency of every floor system, we recommend designing built-up wood girders for floor systems with a live load deflection limit of L/600, and limiting girder span to 10-feet for dimension-lumber girders.

The sizing chart, (page 17), gives the required bending strength (F_b) and stiffness (modulus of elasticity, or E) for 2x10 lumber used in built-up girders for normal residential floor loads. In the box at left, the structural design values of several common wood species are given. These values were taken from the National Design Specification Supplement, 1993 edition, published by the American Forest & Paper Association (202/463-2700).

—*Frank Woeste and Dan Dolan*

When Columns Buckle

Most framers install thousands of wooden studs and built-up posts every year without ever questioning their ability to carry vertical loads. Conventional job-site wisdom says that, in most cases, 2x4s are plenty strong and that 2x6s, used primarily for the extra room they give for insulation, are overkill — even on 24-inch centers.

For the most part, this conventional wisdom is right. Studs and columns rarely fail from compression — that is, being crushed under load. But there are situations where extra wall height or greater than normal loading might cause a stud or a built-up post to fail by *buckling*, or suddenly bowing out to one side.

In theory, a slender column made from elastic material such as wood or metal will buckle into a curved shape at some predictable loading, then snap back to its original shape when the load is removed. You can demonstrate this by standing a yardstick on end. Press down the top and the yardstick bends into a curve parallel to its thin dimension; let up, and it springs back. You can do this over and over with no permanent change in the yardstick.

In a building, however, buckling usually leads to a devastating collapse because the load that creates the buckling continues to press down until the buckled member fails. Buckling failure is prevented by properly designing columns and tall studs.

Take a look at the photo in Figure 16. This shot was taken on the job site of a high-end house in New England. The rafters of the 16-foot-high great room ceiling are resting on a strong LVL ridge beam. But note the post that's holding up the end of the beam — a couple of 16-foot-long 2x4s spiked together. Note, too, that the sheathing behind the post is an insulative fiberboard — not a structural sheathing.

Is this post strong enough to resist buckling? Let's find out.

Preliminaries: Slenderness Ratio

The 1992 *National Design Specification*,

Figure 16. *A double 2x4 post supports the structural ridge beam of this 16-foot-high cathedral ceiling. Is the post strong enough?*

or *NDS* (American Forest and Paper Association; 202/463-2700), widely accepted by building codes for the design of wood members, states that the ratio of the *unbraced* length of a stud or column divided by its width in the direction of buckling cannot exceed 50 (except during construction, when it is temporarily allowed to be as high as 75). This index is called the *slenderness ratio*.

The *NDS* (Appendix A.11.3) also states, in effect, that stud walls with adequately fastened structural sheathing on at least one side are protected from sideways buckling and thus only need to be evaluated for buckling in their stronger direction — the 3 1/2-inch dimension of a 2x4, for instance (see Figure 17). The sheathings the *NDS* refers to are the common structural sheathings such as plywood and OSB, but not drywall, thin paneling, foam sheathing board, or other non-structural insulative sheathings.

Now apply this to the double-stud column in Figure 16, which is 16 feet tall. Let's evaluate it in its weaker direction first — the 3-inch dimension:

$$\frac{16 \text{ ft.} \times 12 \text{ in./ft.}}{3 \text{ in.}} = 64$$

The post definitely fails the test in its

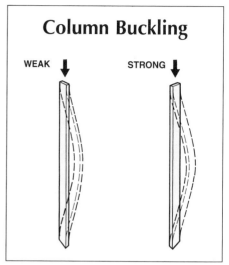

Figure 17. *A stud or post must resist buckling in two directions — sideways (the lumber's weaker direction) and perpendicular to the wall plane (the stronger direction). Structural sheathing or horizontal blocking will prevent sideways buckling, so often the "stronger" direction is the problem.*

weaker direction, meaning that it might have a tendency to buckle sideways under load. Even evaluated for the 3 1/2-inch dimension, it has a slenderness ratio of 55 — still too high. Solid horizontal blocking or structural sheathing could fix the problem for the weaker direction by shortening the unbraced length of the post. Or you could add an additional stud lamination, making the post 4 1/2 inches wide in its sideways dimension. Calculating the slenderness ratio for a triple 2x4 post:

$$\frac{16 \text{ ft.} \times 12 \text{ in./ft.}}{4.5 \text{ in.}} = 42.7$$

Now the post is okay in the sideways direction. But there's no easy way to brace the post in its 3 1/2-inch direction, unless we use a larger stud size. With a 2x6 post, the slenderness ratio is 35, but now the builder has a post sticking out of the wall.

Figuring the Actual Compressive Stress

The high slenderness ratio gives us an idea that the column may not perform well under load. But we still haven't looked at the actual load the column is supposed to carry. Column design gets a little more difficult at this point. The process involves determining the *allowable compressive stress parallel to the grain* (F_c), reduced according to *NDS* design criteria for columns, and comparing this reduced F_c to the *actual* compressive stress (f_c) the column or stud will experience under maximum loading, to make sure the actual stress is less than the allowable.

Determining the actual compressive stress is a matter of basic load calculation: figuring out how many pounds per square inch of load are pressing down on the top of the stud or column. Let's find the stress on the post in Figure 16. The ridge beam the post is carrying is 16 feet long and supports the rafters for a room 16 feet wide. So the tributary area for the end of the beam supported by the post is 8x8 feet, or 64 square feet (Figure 18). Since the house is in an area that gets a lot of snow, we'll assume a total roof design load of 50 psf. That means that under maximum loading, 3,200 pounds are bearing down on the top of that post.

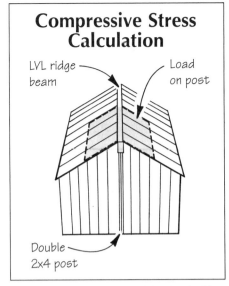

Compressive Stress Calculation

Figure 18. *The load carried by the double 2x4 post is one-fourth the total roof load for the space.*

The top of the post is 3x3 1/2 inches, or 10 1/2 square inches. So the actual compressive stress is:

$$f_c = \frac{3{,}200 \text{ lb.}}{10.5 \text{ in.}^2} = 305 \text{ psi}$$

That's the actual maximum stress; now we have to figure out the *reduced* allowable stress and compare them.

Comparing With the Allowable Stress

The *NDS* procedure for determining the stability of columns requires you to adjust the listed allowable compressive stress, F_c, for the species and grade of lumber you are using by multiplying by a *column stability factor* (C_p). Sounds simple enough, except that the formula for calculating C_p is quite complex. Rather than plow through the algebra, I have simplified the C_p calculation into a graph that applies to sawn wood studs and columns used *under normal service conditions*. (For unusual conditions, such as high temperatures or moisture levels, refer to the *NDS* for additional correction factors.)

To determine allowable stress for a column or stud, use the following procedure:

Step 1: Look up the values for the modulus of elasticity (E) and compressive stress parallel to the grain (F_c) for the lumber you are using. These are found in the *NDS Supplement* and other lumber references.

Step 2: Calculate the value of x using the following formula:

$$x = \frac{0.3(E)}{(F_c)(L/d)^2}$$

where

L = the length of the stud, or length between lateral bracing points, *in inches*

d = the width of the stud in the direction of potential buckling *in inches*

E = the modulus of elasticity for the species and grade of lumber (psi)

F_c = the allowable compressive stress of the lumber (psi), modified by other correction factors from the *NDS* as necessary for nontypical applications

I derived this formula from more complicated formulas found on pages 14 and 104 of the *NDS*. (Note that we are talking about visually graded sawn lumber used in normal service conditions.)

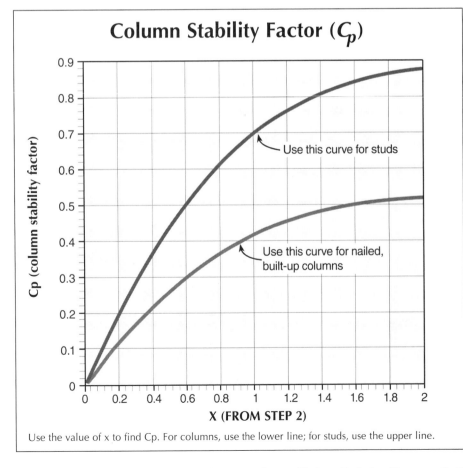

Use the value of x to find Cp. For columns, use the lower line; for studs, use the upper line.

Step 3: Use the value of x to look up the value of C_p from the graph, page 62. Make sure you distinguish between columns and studs when using the graph.

Step 4: Multiply F_c by C_p to get the reduced allowable compressive stress for the stud or column. Compare this with the actual compressive stress; if the actual stress is less than the allowable stress, the stud or column is adequate.

Let's do this for the post in Figure 16.

Step 1: The lumber is No. 1&2 Spruce-Pine-Fir, which has an E value of 1,400,000 psi and a listed F_c value of 1,265 psi (including the 1.15 size factor increase from *NDS* Table 4A).

Step 2: Plug these values into the formula to solve for x:

$$x = \frac{0.3(1,400,000 \text{ psi})}{(1,265 \text{ psi})(192 \text{ in.}/3.5 \text{ in.})^2} = .11$$

(Note that we're evaluating the post in its stronger orientation, the 3½-inch dimension. If it fails in its stronger direction, it will also fail in the weaker direction if it's not braced.)

Step 3: Using the blue, or lower, line on the graph for nailed built-up columns, $C_p = .06$

Step 4: Multiplying F_c by C_p, we get the reduced allowable compressive stress on the post:

1,265 psi x .06 = 76 psi

Stressed Out

Compared with the actual stress of 305 psi, the post is overstressed by a factor of 4! This post should not be considered stable against buckling. Remember, too, that we looked at the post's stronger dimension — the 3½-inch dimension. Since the insulating sheathing should not be given structural credit, the post should be looked at in its weaker direction, too — the 3-inch dimension.

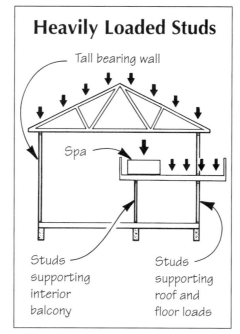

Figure 19. *This building section illustrates three areas where you might need to check for stud buckling: tall bearing walls, exterior walls carrying unusually high combined loads, and unsheathed interior bearing walls carrying heavy loads.*

Without running through all the numbers again, the allowable stress, evaluating the post in its weaker direction, is 50.6 psi — only about 16.6% of the actual maximum compressive stress.

What's the Best Fix?

We've already seen above that the 3½-inch dimension of a 2x4 post is too weak for such a tall post. The builder could have framed that end wall with 2x6s. Using the previous four-step procedure, the reduced allowable stress for a 2x6 post is approximately 206 psi (x = 0.29, C_p = 0.17. If you check this number, remember that the *NDS* F_c value for a 2x6 — 1,210 psi — is slightly less than the F_c for a 2x4, because the size factor increase gets smaller as the lumber gets wider.) The actual compressive stress on the top of a double 2x6 post is

$$f_c = \frac{3,200 \text{ lb.}}{(3 \text{ in.} \times 5.5 \text{ in.})} = 194 \text{ psi}$$

This is less than the reduced allowable stress, so it's okay. Remember, too, that with nonstructural sheathing, the builder should also add horizontal blocking at midspan all the way out to the corner of the room. The blocking cuts the effective length of the post in half, and brings the slenderness ratio down to 32 — well below the 50 maximum.

Other Places to Check

In addition to structural posts, you may need to check studs in tall bearing walls, as well as bearing walls that carry unusually high loads. Figure 19 gives some examples of places to check for stud buckling. Keep in mind that interior finishes like drywall and ¼-inch paneling don't offer much restraint against sideways buckling, so heavily loaded interior bearing walls may need blocking or structural sheathing.

Temporary bracing. A couple of years ago I was called in to design a floor heating system for a stone house that was being gutted and retrofit with a new roof. Since a new slab was to be poured, the carpenter had left out all the interior walls and temporarily supported the massive new roof framing with a few long 2x6s spiked together as columns under the long LVL ridge beam.

Construction slowed as the weather got colder, and then we were hit with one of the worst winters of the century in this area. Several feet of snow piled up on that roof and eventually took its toll as the temporary posts buckled and the roof collapsed. Fortunately no one was hurt; the last I heard, the contractor's liability insurance was about to take a hit.

Would the roof have collapsed under normal circumstances? Probably not. But it demonstrates the kind of temporary conditions to watch out for. Anytime, for example, that you're having a pallet of plywood or roof shingles lifted up to the top of an unsheathed structure, watch out. Such a concentrated load can cause a wall to bow.

By John Siegenthaler, P.E., a building systems designer in Holland Patent, N.Y.

Misplaced Load Paths

Not long ago, Mr. & Mrs. P. called me in to help them figure out why their kitchen counter was pulling away from the wall. The problem was bad enough that you could fit your fingertips into the gap behind the backsplash. They also noticed cracked tiles, sloping floors, and on the second floor, doors that were sticking or not closing correctly.

A Classic Case

What I found was a classic case of misplaced load paths. The construction was inadequate, traced to a deficient design that had been followed in all innocence by the well-meaning contractor. It seems he bought some plans from a plan mill in a distant state and hired a local engineer to "stamp" them. Unfortunately, neither the designer nor the local engineer nor the building official spent much time reviewing the design. As a result, there were some seriously misaligned load paths carrying loads from the upper floors and roof down onto floor framing that wasn't strong enough.

Figure 20 shows a section through the middle of the house. Rafters, joists, and studs are all set at 16-inch centers, and the forces indicated are the loads at various points carried by each rafter, joist, or stud. Load design criteria for southern New York are 30 psf snow load, 30 psf live load on upper floors, and 40 psf live load on the main floor. I used 10 psf for the dead load of floor and roof assemblies, a common conservative value.

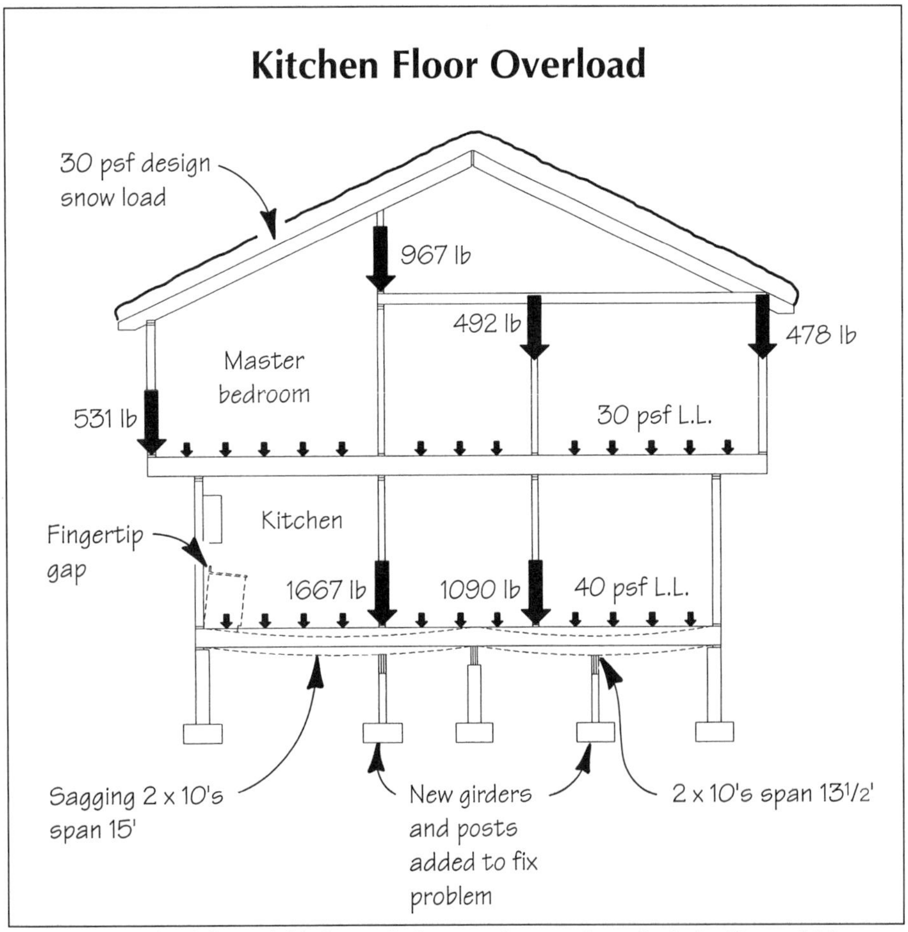

Significantly, the distress experienced by Mr. and Mrs. P. was primarily due to dead loads, as they had little furniture and a small family. The sagging took place in the first three years following construction.

The Biggest Mistake

The worst mistake by the designer was to overlook the fact that the cathedral ceiling in the master bedroom eliminated the truss effect that results from attic floor joists working with the rafters to form a triangle. This common roof-triangle "truss" normally transfers all the roof loads to the exterior walls. But with the addition of the cathedral ceiling, more than half of the roof loads were being carried by the interior wall of the master bedroom.

Additional loads from the attic floor, the dead loads of walls, and second-floor loads were all added in, creating a concentrated, or point, load on the first-floor framing. The 2x10 floor joists had probably been sized using a uniform load span table; the spans would have been pretty much maxed out even without this added point load.

Repairs

The calculated bending stress ($f_b = 3,370$ psi) was more than three times the allowable bending stress in the kitchen floor framing, and almost as bad at the front of the house. While a structural collapse probably would not have occurred, the excessive sagging would have continued to cause problems with sticking doors and cracking tile and drywall, not to mention the gaping space behind the kitchen counter.

The builder, a man of integrity, agreed to jack the floors level, install additional girders and piers, and repair all the secondary damages at his own expense. He probably won't overlook the matter of proper load paths again.

—*Robert Randall*

Figure 20. *A fatal flaw of this house design was the upstairs cathedral ceiling, which interrupted the second-story ceiling joists and placed more than half of the roof load on the interior wall of the master bedroom. This load, along with floor loads from the second story and attic, greatly overstressed the 2x10 kitchen floor joists. The numbers represent the total concentrated loads in pounds carried by each individual stud onto the joists below.*

Chapter Two
MATERIALS AND CONNECTIONS

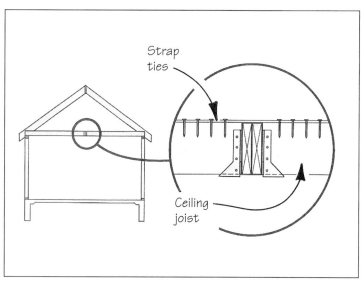

Structural Sheathing: Plywood vs. OSB 24

Strength of Nails 26

Box vs. Common Nails 28

Using Metal Connectors 29

Frequently Asked Framing Questions 32

Structural Sheathing: Plywood vs. OSB

The issue for most builders who choose between plywood and OSB is durability. OSB looks like a bunch of wood chips glued together (that's what it is). Its detractors are quick to say that OSB might fall apart if it gets wet. This opinion has a familiar ring — plywood suffered the same criticism not too long ago. Delamination of early plywood sheathings gave plywood a bad name. Many "old-timers" swore by solid board sheathing until the day they hung up their aprons. But not many builders today share that view of plywood.

Similarities of OSB and Plywood

The model building codes all use the phrase "wood structural panel" to describe plywood and OSB, basically recognizing these two materials as equivalents. Likewise, APA — The Engineered Wood Association (the group responsible for approving more than 75% of the structural panels used in residential construction) treats OSB and plywood as equals in their published performance guidelines. And wood scientists agree that there is no significant difference in structural performance between the two materials.

OSB and plywood share the same exposure durability classifications: Interior, Exposure 1 (95% of all structural panels), Exposure 2, and Exterior. They also share the same set of performance standards and span ratings. Both materials are installed on roofs, walls, and floors using the same installation recommendations. Requirements about the use of H-clips on roofs and blocking on floors are identical.

The weights of OSB and plywood are similar: 7/16-inch OSB and 1/2-inch plywood weigh in at 46 and 48 pounds per sheet, respectively.

Researchers at the University of Illinois studied the withdrawal and head pull-through performance of nails and staples in plywood, waferboard, and OSB. They found that in both dry and six-cycle aged tests, OSB and waferboard performed equal to or better than C/D-grade plywood. The results of another independent study conducted by the Weyerhauser Technology Center in Tacoma, Wash., also showed that withdrawal strengths in OSB and plywood are the same.

The Differences

But while the two products may perform the same structurally, they are undeniably different materials. Plywood is made by hot-gluing thin sheets of veneer that are peeled from a spinning log. Resulting veneers have pure tangential grain orientation, since the slicing follows the growth rings of the log. Throughout the thickness of the panel, the grain of each layer is positioned in a perpendicular direction to the adjacent layer. There is always an odd number of layers in plywood panels so that the panel is balanced around its central axis. This strategy makes plywood stable and less likely to shrink, swell, cup, or warp.

To make OSB, logs are ground into thin wood strands. The dried strands are mixed with wax and adhesive, formed into thick mats, then hot-pressed into panels. But don't mistake OSB for chipboard or waferboard: The strands in OSB are aligned, or "oriented." Strand plies are positioned as alternating layers that run perpendicular to each other. This structure mimics plywood; OSB is engineered to have strength and stiffness equivalent to plywood. Waferboard, a weaker and less stiff cousin of OSB, is a homogeneous, random composition.

On Site With OSB

Under ideal conditions, the performance of the two materials is similar, but in the real world there are differences in service.

Irreversible swelling. All wood products expand when they get wet. When OSB is exposed to wet conditions, it expands faster around the perimeter of the panel than it does in the middle. The swollen edges of OSB panels can telegraph through thin coverings like asphalt roof shingles — so-called "ghost lines" or "roof ridging." The Structural Board Association (SBA, 45 Sheppard Ave. East, Suite 412 Willowdale, ON M2N 5W9, Canada; 416/730-9090), a trade association that represents OSB manufacturers in North America, has issued a technical bulletin outlining a plan to prevent this phenomenon. SBA correctly indicates that dry storage, proper installation, adequate roof ventilation, and application of a warm-side vapor barrier will help prevent roof ridging.

Irreversible edge swelling has been the biggest problem with OSB. Manufacturers have done a good job of addressing this issue at the manufacturing facility and during transportation by coating panel edges. But the reality is that builders don't limit OSB use to full-sized sheets. The edges of cut sheets are seldom, if ever, treated in the field. Houses under construction get rained on. And if you use OSB in an area of very high humidity, such as above an improperly vented attic or over a poorly constructed crawlspace, you're asking for trouble.

OSB responds slowly to changes in relative humidity and exposure to liquid water. It takes a long time for water to soak OSB, and conversely, once water gets into OSB, it is very slow to leave.

The longer that water remains within OSB, the more likely it is to rot. The panel's wood species has a significant impact: OSB made from aspen or poplar has practically no natural decay resistance. Many of the western woods used to manufacture plywood have at least moderate decay resistance.

EIFS problems. In the mid 1990's, the walls in many Southeastern homes covered with the Exterior Finish and Insulation System (EIFS) were found to be rotting. In these cases, rigid foam insulation was applied over OSB and coated with a stuccolike covering. When the exterior foam boards were removed, wet, rotted, crumbling OSB was exposed. OSB was criticized in the press, but the real problem wasn't with OSB. In all the cases I'm familiar with, improper installation of flashing or protective coverings was the culprit.

Louisiana-Pacific's OSB-based Inner-Seal siding also made the news in 1996 when LP settled a class-action suit for $350 million. The claims were that OSB siding was rotting on the walls of many homes in the South and Pacific Northwest, both moist climates. LP said the problems were caused by improper installation. But builders and consultants involved in this case think the material simply doesn't work in permanently exposed applications. To my knowledge, there has not been a problem of similar scale associated with plywood siding.

Clearly, OSB, in its current state of development, is sensitive to moist conditions. Plywood, although not immune, is somewhat forgiving. Plywood actually gets saturated much faster than OSB but is not prone to edge-swelling and dries out more quickly.

OSB Pluses

On the positive side, OSB is a more consistent product. It's truly an engineered material. You'll never have a soft spot in an OSB panel because of overlapping knot holes, as you can with plywood. Nor do you have to worry about knot holes at the edge of an OSB panel where you are nailing. Delaminations are virtually nonexistent.

OSB is approximately 50 strands thick, so its characteristics are averaged out over many more "layers" than is the case with plywood. OSB is consistently stiff, whereas plywood has a broader range of variability. During the manufacturing process, plywood veneers are randomly selected and stacked up into panels. You may get four veneers of early-growth wood stacked above one veneer of old-growth wood. Most plywood panels are "overbuilt" to cover the statistical range that guarantees each sheet of plywood will meet the minimum standard. OSB, on average, is 7% less stiff because it stays closer to its target spec. However, OSB sometimes feels stiffer on a floor because there are no occasional weak panels as there are with plywood.

OSB is stronger than plywood in shear. Shear values through its thickness are about two times greater than those of plywood. This is one of the reasons OSB is often used for the webs of wood I-joists. However, nail-holding ability controls performance in shear wall applications, so plywood and OSB perform equally well as structural sheathing.

Subflooring and Underlayment

While OSB and plywood are equal structurally, flooring manufacturers make different recommendations regarding their use as a substrate.

Hardwood flooring. The National Oak Flooring Association (NOFA), in Memphis, recommends either 5/8-inch or thicker plywood, 3/4-inch OSB, or 1x6 dense Group 1 softwood boards installed at a diagonal under hardwood flooring. The NOFA recommendation is based on research conducted at Virginia Polytechnic Institute in Blacksburg, Va. Researchers there simulated what happens on a real construction site. They built several full-sized floors out of diagonal boards, plywood, and OSB, and weathered them for five weeks before installing hardwood flooring. Finished floor systems were cycled in an environmental chamber to simulate the changes that occur in summer and winter months.

The study showed that diagonal-board subflooring was far and away the best system. Statistically, 5/8-inch plywood and 3/4-inch OSB performed the same. But two significant observations were made during the study: Some of the plywood delaminated during the weathering experiment and new patches had to be spliced into the subfloor system. Also, researchers learned that the best floors of all were the control specimens in each group, which had been protected from any weathering. This speaks volumes for the importance of protecting materials during transport, storage, and the early stages of construction.

Tile. According to Joe Tarver, Executive Director of the National Tile Contractors Association (NTCA), in Jackson, Miss., "OSB is not an acceptable substrate to receive ceramic tile." NTCA lists OSB, along with pressboard and lauan plywood, as "not acceptable" in its reference manual's section on substrate materials. This has to do with thickness swelling: If OSB gets wet, it swells, transferring stress and causing the tile to fail.

Resilient flooring. The Resilient Floor Covering Institute (RFCI), a trade association that represents manufacturers of vinyl sheet-flooring and tiles, also favors plywood. RFCI installation specifications recommend plywood as an underlayment material, although OSB is allowed as a subfloor material. While manufacturers have not seen a deluge of failures due to the use of OSB under resilient flooring, they have received complaints of edge swelling that has telegraphed through their flooring products. Manufacturers feel more comfortable guaranteeing their products when they are installed over plywood.

Roof and Wall Sheathing

All manufacturers of siding products contacted for this report agree that OSB and plywood are equals when it comes to wall sheathing. An engineer with the Western Wood Products Association in Seattle, Wash., assures us, "There have been no problems reported from the field. Nail-holding and racking resistance are the same."

Roof sheathing draws mixed opinions. The National Roofing Contractors Association (NRCA) in Rosemont, Ill., and the Asphalt Roofing Manufacturers

Association (ARMA) in Rockville, Md., both recommend the use of APA performance-rated OSB and plywood panels. However, ARMA, NRCA, and representatives from at least two roofing manufacturers, Celotex and TAMKO, prefer plywood roof decks. Warranties on shingles are extended to both substrates, but manufacturers feel more comfortable with plywood. Mark Graham, NRCA's associate director of technical services explains, "We hear a lot of complaints related to dimensional stability. And a disproportionate number are related to OSB. So we are a little bit cautious." Graham also acknowledges that APA, an organization he clearly respects, is standing firmly behind the OSB product.

Florida's Dade County is the only building code district in the country that prohibits the use of OSB as a roof deck. Damage to roofs during Hurricane Andrew was originally blamed on OSB's poor nail-holding power. Dade's banning of OSB spawned several research initiatives to explore the suitability of OSB as a structural sheathing. This research has conclusively proven OSB seaworthy. Many experts think the ban makes no sense. Dade's position is perceived by many industry insiders to be an overreaction aimed at satisfying public concern.

Future Watch

OSB is pushing plywood aside as the structural panel of choice. APA's market data indicates that more than half the structural panels sold in residential construction in 1995 were OSB. And OSB plants are being built while plywood plants are closing down. The good news for builders is that the increase in OSB production is expected to depress the price of all structural panels. Also, strong supplies help reduce price volatility.

Two things are certain: OSB is in our future, and it will improve. Production will reflect market needs. Perhaps thickness swelling will be included in future performance standards (it should be). OSB manufacturers can formulate their process to provide virtually any property they want. They can build panels to resist high relative humidity, deliver more strength, or provide a harder surface. It becomes a question of cost vs. performance; builders will dictate the performance of the final product.

—*Paul Fisette*

Strength of Nails

Typical porch deck construction is joists running perpendicular to the side of the house. The joists are then supported by a ledger strip nailed to the side of the house. Is this strong enough? Let's see.

Suppose our deck is 6 feet deep by 12 feet wide, with joists placed 16 inches on-center. The inside ends of the joists sit on the ledger or in joist hangers attached to the ledger. A box joist is nailed into the outboard ends, and this box joist is supported by posts. Building codes specify a working design load of 60 pounds per square foot on small exterior residential balconies. Because the joists are placed 16 inches (1.33 feet) on-center, the load per running foot on each joist is:

$w = 1.33 \text{ sq ft/ft} \times 60 \text{ lb/sq ft} = 80 \text{ lb/ft}$

The total load on each 6-foot joist is:

$P = 6 \text{ ft} \times 80 \text{ lb/ft} = 480 \text{ lb}$

This load is shared, half held up by the ledger, and half held up by the box joist. The force of the half load that is felt by the ledger is called the "end reaction," which is:

$R = \frac{P}{2} = \frac{480 \text{ lb}}{2} = 240 \text{ lb}$

Therefore, each 16-inch section of the ledger must have sufficient nails to support a lateral load of 240 pounds.

The Humble Nail

It is almost too obvious to say that nails are the basic fasteners in wood frame construction. However, few engineers and fewer builders have considered the strength of nails. We just bang 'em in, and where we get worried, we bang in a few extra ones, running pretty much on feel and experience. Meanwhile, a lot of research has actually been done on nailing. The holding power of a nail depends on the diameter of the nail, the distance it penetrates into the wood and the type of wood used. The research also looks at two different loading situations: lateral loading and withdrawal.

Lateral loading is a load that is perpendicular to the nail; the ledger nails are laterally loaded. Most builders never use nails loaded in withdrawal situations, where the load tends to pull the nail out. It just isn't done, except for nailing up ceiling drywall or the like.

The research has come up with nailing tables and rules. The table in Figure 1, for example, is from the Uniform Building Code.

Let's get back to the deck ledger. Each 16 inches of ledger supports 240 pounds. Here in the West, most of the framing is Doug fir, so we need three 10d or 12d nails for each 16-inch section of ledger. I have never seen nailing tables for eastern spruce, but I have seen research reports. These suggest working values of about two-thirds the safe loads of nails driven into Doug fir. So in New England, you'll need four nails in each 16 inches of ledger. At the other end of the joist, three nails are needed to support the (Doug fir) joists into the box joist; 4 nails in spruce.

Now take a closer look at the table: There really isn't any difference in holding strength between 10d and 12d nails. This is because they both have a diameter of 0.148 inch, but the 12d nails are 1/4 inch longer. And there really isn't too much decrease in strength by dropping to 8d nails. Also, you don't gain much of an increase in strength by using 16d nails. The sharper carpenters have known this for some time and a lot of excellent framing is done with 8d nails.

I'll suggest a rule of thumb: 80 pounds per nail — any nail — in Doug fir, 60 pounds per nail for spruce, and never use nails in withdrawal situations. This is pretty workable, and if strength is really a serious consideration, use some fastener other than a nail — like a screw or lag bolt. Screws are about twice as strong in lateral loading and about four times as strong in withdrawal.

There are some suggested rules for nailing.
- Never rely on a single nail — use a minimum of two. Wood is not homogeneous and a single nail could be driven into a soft spot.
- Space nails no closer than half their length, and do not place nails closer than one-fourth of their length to the edge of the wood — this avoids the risk of splitting. For example, 10d nails are 3 inches long. By this rule, don't drive them into the ledger within $3/4$ inch of the edge and don't place them any closer than $1^{1}/_{2}$ inches apart.
- Nails need 11 diameters of penetration to hold at full strength. This is about $1^{1}/_{2}$ inches for 8d to 16d nails. If the penetration is less than 4 diameters, about $1/2$ inch, just forget about any possible contribution of that nail. So when you nail a two-by ledger onto framing through $1/2$-inch sheathing, the ledger and the sheathing take up 2 inches of the nail. To penetrate $1^{1}/_{2}$ inches, you'll need a nail at least $3^{1}/_{2}$ inches long — a 16d. But there's a problem: The 16d nail is a little fatter (0.162 inches) and 11 diameters is about $1^{3}/_{4}$ inches. So instead, use a 20d nail — it's 4 inches long and has a diameter of 0.192 inches. Even this is a little shy of the 11 diameters needed, but it's close enough.
- In unseasoned wood, nails have only about three-quarters of their holding power.
- Metal side plates increase the holding power of nails by about 25%, if they are used according to the manufacturer's instructions. Joist hangers are typical holding plates, attached with lots of small nails. Keep a Teco or Simpson catalog on hand to help with messy framing situations.
- Do not toenail for strength. Toenailing should be used only to hold things in place until stronger fastenings can be applied. The bottom of a toenailed stud is a model of this — the stud is really held in place with sheathing nails.
- In exterior applications, galvanized nails usually have the zinc pounded off the head during driving, so that they rust and stain the wood. Their real value is increased holding power from the rougher shank. If staining is a problem, spring for stainless steel nails.

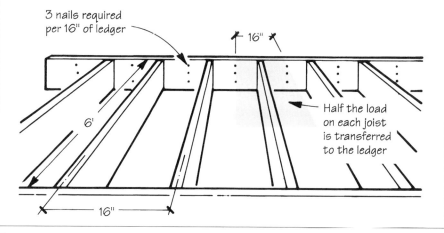

Strength of Nails

Nail Size	6d	8d	10d	12d	16d
Safe lateral load, Doug fir, So. pine	63lb	78lb	94lb	94lb	107lb
Safe withdrawal load, Doug fir	29lb	34lb	38lb	38lb	42lb
Safe withdrawal load, So. pine	42lb	48lb	55lb	55lb	60lb

Figure 1. *For a 6-foot-wide deck with a 60 lb./sq. ft. design load, each joist framed 16 in. on-center must carry 480 lb. Half the load is transferred to the ledger, so each 16-in. section of ledger must support a 240-lb. lateral load. According to the chart, for Doug fir or Southern pine lumber, you'll need at least three 10d nails for every 16 inches of ledger (three 10d nails x 94 lb. = 282 lb.).*

In rural areas, spiral-shank, or "pole barn," nails are used a lot. These nails are rated for the same strength as common nails, but they maintain full rated strength in unseasoned wood. (Personally, I do not agree with this rating. I have done some measured tests and found that pole barn nails have one-third to one-half more holding power than common nails. This is speculative on my part, as my tests were not extensive and systematic.)

There are other types of nails that are very strong, particularly ring-shank nails used in marine applications. For strength values, look up the manufacturer.

In the past few years, air nailers have become very popular. They are extremely quick and effective, as any user well knows. Air-driven nails have been tested and do have the same strength as hammer-driven nails even though the nail magazine manufacturers are using lighter wire for the nails.

Back to the porch where we started. The joists that rest on the nailed up ledger will certainly keep from falling, but the ledger will not keep the joists from pulling away from the building (that's one advantage of joist hangers). Each particular design should be examined for this possibility and appropriate measures should be taken.

Subtle creatures, these nails.

—*Harris Hyman*

Box vs. Common Nails

A contractor recently told me about a case involving a building with a leaky roof. In the attempt to figure out why the roof was leaking, it was discovered that the "plans" called for eight nails at a certain connection, but that the contractor had installed only six. The contractor had to fix the problem, at great expense. The lack of two nails per connection became a big issue. By the time the builder paid his legal expenses, he figured the extra five pounds of nails cost him over $1,000 per pound!

While something like that may never have happened to you, improper or inadequate nailing can be a serious matter in wood construction. Perhaps one of the most common problems I see is the substitution of box nails for common nails, with no allowance for their difference in strength.

Skinny Commons

In simple terms, a box nail can be thought of as a skinny common nail. A 16-penny common nail and a 16-penny box nail are both 3 1/2 inches long, but a common 16-penny nail has a diameter of 0.162 inch, while the 16-penny box nail is 0.135 inch in diameter. The question is, so what?

Nails are rated for "lateral capacity" and "withdrawal capacity." Both ratings are based on nail diameter and the density of the lumber being used. The lateral and the withdrawal strength are greatly affected by the nail's diameter (see Table A). For this reason, you can't just make a one-to-one substitution of common nails specified in a plan with the same number of box nails. Engineers and architects routinely specify common nails, and their designs are based on rated capacities of commons. But since box nails are popular with many carpenters and are widely available in builder supply stores, I've offered a formula (based on Spruce-Pine-Fir lumber) for making a safe conversion from common to box nails (Table B).

An Example

Frequently, home builders encounter a specification for attaching a "strong-

A. Lateral Strength of Common vs. Box Nails (in Spruce-Pine-Fir)*

Penny Wgt.	8d		10d		12d		16d		20d	
Nail Type	Common	Box	Common	Box	Common	Box	Common	Box	Common	Box
Diameter (in.)	0.131	0.113	0.148	0.128	0.148	0.128	0.162	0.135	0.192	0.148
Lateral Rating (lb.)	70	57	83	68	83	68	120	88	144	100
Side Member Thickness (in.)	3/4	3/4	3/4	3/4	3/4	3/4	1 1/2	1 1/2	1 1/2	1 1/2

*Nail penetration of at least 12 diameters in the main member is assumed. Based on National Design Specification for Wood Construction (1997).

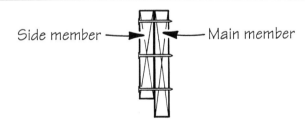
Side member — Main member

B. Conversion Ratio for Common to Box Nails (Lateral Capacity)*

To convert required number of common nails of a given size to box nails, multiply by the appropriate ratio and round up.

Penny Wgt.	8d	10d	12d	16d	20d
Ratio	1.23	1.22	1.22	1.36	1.44

*Based on values in Table A. S-P-F lumber assumed.

C. Lateral Strength of Gun Nails*

Length (in.)	3	3 1/4	3 1/2	3 1/2
Diameter (in.)	0.131	0.131	0.131	0.161
Lateral Rating (lb.)	64	64	64	88

*Based on NER-272. S-P-F lumber assumed.

back" to a metal-plate-connected floor truss. The specification might read, "Attach strongback at bottom of vertical web next to centerline of truss with minimum 3-10d common nails." Assuming the contractor wants to use box nails instead, the rule-of-thumb calculation using Table B would be

1.22 x 3= 3.66, or 4

So instead of three 10-penny common nails, you would use four 10-penny box nails. Always round up.

Check With the Designer

For the comparison presented in Table A (which applies only to Spruce-Pine-Fir framing lumber), the common nail typically has a rated lateral capacity 22% to 44% higher than that of a box nail of the same penny weight. Note that the ratios presented in Table B will vary depending upon species of the lumber used and the thickness of the wood members being joined. The multipliers in Table B are good approximations for other species besides Spruce-Pine-Fir. If your plans specify an architect or engineer of record, always contact this design professional for an equivalent nailing schedule.

Pneumatic Nails

Pneumatically driven nails are yet another complication for the diligent contractor. For the most part, gun nails of the same length as common nails have a smaller diameter (Table C), although many nail gun manufacturers are now making a full-size 16-penny nail (0.161 inch diameter). The manufacturers will provide technical data that state the lateral capacity of their various nails. Some manufacturers will also provide a free copy of the code report, *NER-272*, that serves as the minimum guideline for the pneumatic fastener industry. Before substituting a pneumatic nail for a common nail in a critical connection, always check with the building designer.

—*Frank Woeste, P.E.*

Using Metal Connectors

Many structural failures in wood-frame buildings occur not because a member breaks but because a connection comes apart. Yet, in residential work, making strong, engineered connections in the field has never been easier. There are many different metal connectors available today, and manufacturers have been innovative as well as responsive to industry needs.

Every designer and builder should have some manufacturers' product catalogs on hand for quick help in finding solutions to connection problems. I keep these catalogs in a ring binder on a reference shelf at arm's reach from my work station, and refer to them often.

In this section, I'll look at some of my favorite connectors, as well as a few that I would avoid.

Double-Shear Joist Hangers

High on my list of favorites is the newer series of face-mount joist connectors featuring double-shear nailing (see Figure 2), such as the Simpson LUS and HUS series (Simpson Strong-Tie, 800/999-5099) and the USP SU series (USP, 800/328-5934). Installation of these hangers involves cross-toe-nailing through the hanger and the heel of the joist into the supporting header. This nailing pattern increases load capacity, uplift resistance, and resistance to separation between joist and header. The connection is also more resistant to shrinkage sagging of the hung joist and, perhaps most important, is easier to install in most cases because there is more room to swing the hammer on the 45-degree angle.

The downside to these (and most) hangers is that drying shrinkage of the hung joist can either result in settlement at the top of the joist or leave a gap between the hanger's saddle and the bottom of the joist. If such shrinking does occur, you can backfit wedges or shims to support the bottom of the joist and keep the top flush.

Strap Ties

This generic term applies to a variety of connectors that are simple straps of galvanized sheet stock with lots of holes for the number of nails required to develop the full rated strength. Pay attention to the nailing requirements listed in the catalog. It is often surprising how many nails it takes to reach the rated strength.

Strap ties can be used for many applications, several of which have been discussed here. These include rafter-to-joist connections in raised eaves construction, and resisting wind uplift, especially where there are sheathing joints at or near the plate lines. Straps can also be used to provide tension capacity

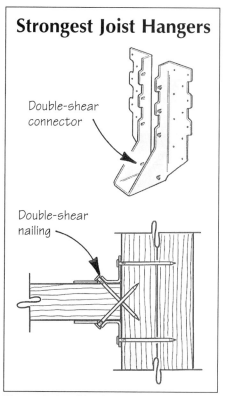

Figure 2. *Double-shear connectors provide greater strength with fewer nails than ordinary hangers. The angled nail configuration also makes it easier to drive the nails in a joist bay.*

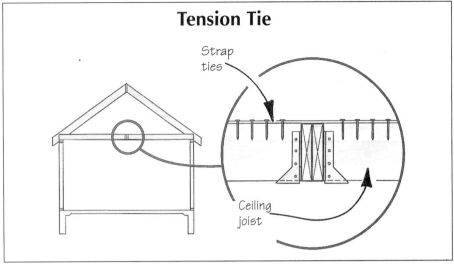

Figure 3. *In the absence of a structural ridge or center bearing wall, ceiling joists are critical for preventing rafters from spreading. Where the joists are interrupted by a flush-framed beam, strap ties make an ideal tension connection.*

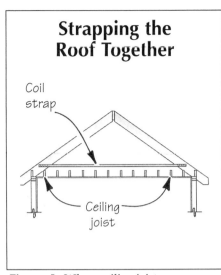

Figure 5. *Where ceiling joists run perpendicular to the rafters, coil strap makes an excellent rafter tie.*

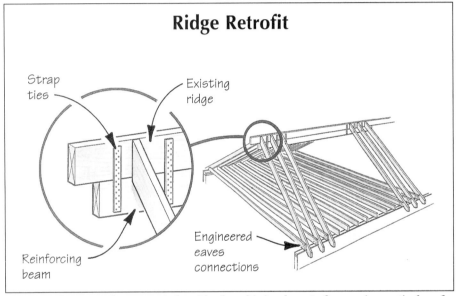

Figure 4. *In cases where an existing ridge board is inadequate for carrying vertical roof loads, as in a shed dormer retrofit, you can use strap ties to suspend a reinforcing beam below the original ridge.*

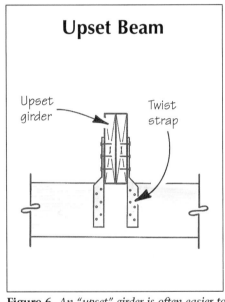

Figure 6. *An "upset" girder is often easier to frame than a flush-framed beam. The joists hang by twist straps, installed in pairs.*

through ceiling joists interrupted by a flush header (Figure 3). Joist hangers are never to be depended on for tension perpendicular to a header.

Reinforced ridge beam. On occasion, you'll be asked to remove ceiling joists to create either a tray ceiling or a full cathedral. Or you may be faced with a shed dormer addition, where you have to remove the rafters themselves. In these cases, it may be necessary to double the existing ridge, which is easier if the new beam is added below. If the rafter-ceiling joist connections in the undisturbed portion of the roof are strengthened, the "ridge doubler" can be hung from the existing ridge (Figure 4).

Strap ties are a simple way to provide a high-strength connection at the ends of such a reinforcing beam, where simply nailing a 2x4 on each side would probably not be adequate.

Coil strap. Simpson offers CS and CMST series coil strap, which is simply a long coil of strap tie in a convenient flat box. Many contractors I know keep a box in the cab of their truck to avoid having to run off to the supply house for the right length of strap tie. Just snip the coil strap to the right length and nail it up.

Coil strap is a simple solution when added tension ties are needed between the eaves, whether in new construction or remodeling. When the ceiling joists run the wrong way (Figure 5) or access is difficult to connect to them, you can run coil strap the width of the attic and nail it directly to the rafters at opposite eaves. Coil strap can also be twisted to lie flat on joists or plywood subfloor and requires no nailing except at end connections.

Twist straps. These handy connectors are simply strap ties with a twist. The 90-degree twist allows them to connect to faces of members that are per-

The Right Angle

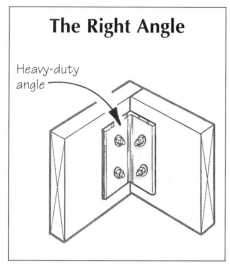

Figure 7. *Heavy-duty angles like Simpson's HL or USP's KHL are a good solution for attaching that last deck joist to the ledger board.*

pendicular. Usually used in pairs, they serve well to support ceiling joists beneath an upset girder (Figure 6). As always, fill all the holes with nails.

Heavy-Duty Angles

Simpson's HL series or USP's KHL angles come in handy for structural connection of perpendicular members in a variety of applications. For example, these heavy angles are often a good answer to the tricky connection of the last deck joist to a flush ledger where a regular joist hanger wouldn't work (Figure 7). Heavy angles can also be used to support connections to the faces of masonry foundations.

Least Favorite Connectors

There are a few products in the

Hangers to Avoid

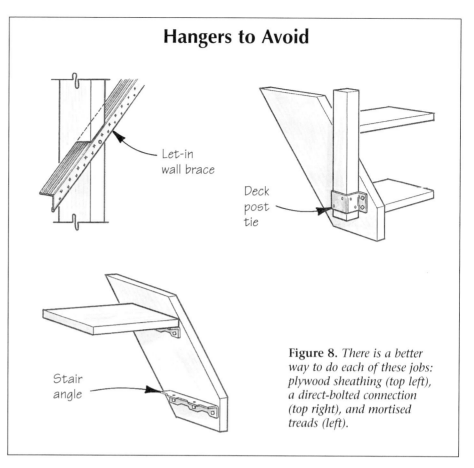

Figure 8. *There is a better way to do each of these jobs: plywood sheathing (top left), a direct-bolted connection (top right), and mortised treads (left).*

connector catalogs that make me wary (Figure 8). While they may have their place, I think they may lead to unsatisfactory results and advise against their use.

Wall bracing. These metal-strap braces are "designed to fulfill the same code bracing requirements as 1x4 let-in bracing," which isn't saying much. These let-in braces are woefully inadequate for full-blown wind loading and in many cases probably leave it to the plaster or drywall to keep the walls from racking. I strongly advise against depending on this type of bracing.

Deck post ties. This is an interesting idea that I find lacking in merit. Direct-bolted connections are more rigid.

Stair angles. I am a firm advocate of let-in stair treads. Either cut traditional stair carriages, or mortise the stringers (this can be done quickly with a router template) and let the treads bear directly on the shoulder of the mortises for a stronger, stiffer stair.

—*Robert Randall*

Frequently Asked Framing Questions

When to Double Rafters

Q *When you remove one rafter to install a skylight, do you have to double the two rafters at the sides of the opening? What about the headers?*

A When you're framing a skylight that requires you to remove no more than two rafters, a common rule of thumb says to double the rafters on either side of the opening and use double headers. This is often required by building inspectors, but in fact, it's a very conservative guideline that often results in unnecessary framing. Only in rare cases are the doubled headers required. And in many cases, particularly when 2x10 or 2x12 rafters are chosen for insulation thickness rather than for strength, doubled rafters may not be required.

In my experience, the inspectors also often insist on joist hangers at the headers, but these are also rarely needed. Usually, an adequate number of 16d nails, as many as 8 nails for a 2x12 connection, depending on the loads) can handle the reaction forces.

Watch out, though: As the opening size increases to the point where you're removing three or more rafters, even doubling the perimeter framing may not be sufficient.

—Robert Randall

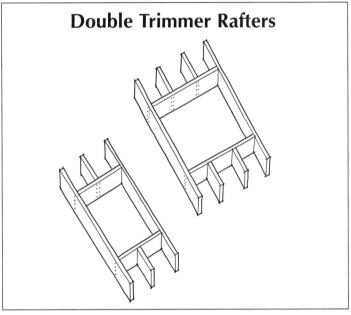

Figure 9. *If deep rafters have been selected to accommodate insulation requirements rather than structural requirements, it may be unnecessary to double trimmer rafters. It would rarely be necessary to double the opening headers.*

Purpose of T&G Plywood

Q *Does tongue-and-groove plywood add extra strength or stiffness to a floor system, or does it just help prevent floor squeaks?*

A The tongue-and-groove joint doesn't add strength, but it does help to distribute loads to adjacent panels, improving the perceived stiffness of the floor. T&G plywood was developed as a labor-saving alternative to installing solid wood blocking at unsupported panel edges.

Without the tongue and groove, a load on one panel edge causes that panel to deflect relative to the adjacent sheet. A wood floor that spans across the joint would experience a wedging action, causing a floor squeak. Tongue-and-groove plywood is actually more effective than solid blocking at preventing squeaks, because over time the blocking will shrink, leaving unsupported edges.

—Scott McVicker

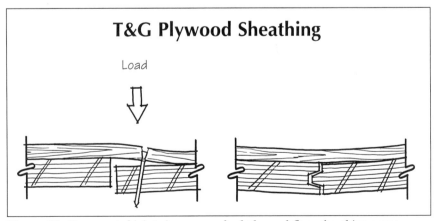

Figure 10. *Unsupported joints in square-edged plywood floor sheathing cause squeaks when someone steps on them. T&G plywood sheathing prevents squeaks and makes the floor feel stiffer, though it doesn't actually increase the design strength of the floor system.*

Horizontal vs. Vertical Sheathing

Q *Is it true that installing plywood sheathing horizontally (perpendicular to the direction of the studs), with joints staggered, is stronger than installing it vertically? Is this true of roof sheathing too?*

A This is true for wall sheathing in many instances, but not for roof sheathing. To understand why, we need to look at how the grain of the plies is oriented relative to the direction of the applied force. Each layer of wood in plywood is oriented either parallel or perpendicular to the long direction of the sheet. Most of the shear force is resisted by those plies whose grain runs parallel to the direction of the applied force. So for 3-ply plywood, for instance, which has two face plies running parallel with the long dimension of the sheet, and a single central ply running perpendicular, most of the wood fibers are oriented parallel to the length of the sheet, so that is the plywood's stronger direction.

This fact is reflected in the Uniform Building Code's nailing schedule for structural panel shear walls (1997 UBC, Table 23-II-I-1), which permits the allowable shear for $3/8$-inch and $7/16$-inch panels, if oriented horizontally across the wall studs, to be increased to that of corresponding $15/32$-inch panels. As plywood gets thicker, this rule is less important because the overall percentage of fibers running parallel with the long dimension decreases as the number of plies increases.

Note that the UBC table applies only to fully blocked shear walls; in other words, all the plywood edges have to be supported by a minimum of 2-by framing. Regardless of plywood orientation, a plywood panel fully supported at all edges is always stronger than a panel with some edges unsupported.

So far we've talked only about wall sheathing, which mainly resists lateral loads from high wind or earthquakes. Roof sheathing is another matter, since roofs experience forces applied both parallel and perpendicular to the long direction

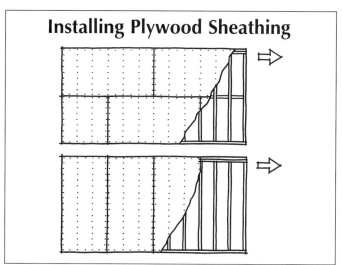

Figure 11. *When all the panel edges are supported by solid framing, $1/2$-inch plywood sheathing is stronger against racking forces when installed horizontally (at top). If there's no blocking at the 4-foot mark, the sheathing is stronger installed vertically, as long as all edges are supported (at bottom).*

of the plywood. We could, in theory, credit a plywood panel installed perpendicular to the rafters with the higher shear force in that one direction, but we would be forced to accept the basic code value in the opposite direction. In such a case, the designer generally assigns the lower shear value to the plywood in both directions. If a greater shear value is needed, the designer may specify increased nailing or thicker plywood.

—*Scott McVicker*

Plywood in Built-Up Headers

Q *Do the layers of plywood in typical built-up headers add significant strength to the header?*

A The most important thing the plywood adds is thickness. Of course, the plywood does add some strength, but for several reasons engineers almost never count on this strength in their designs.

Only the layers of plywood with the grain oriented horizontally (parallel with the direction of the header) are really adding any strength. A quick look at the thicknesses involved shows that the additional strength is small. If half the layers in $1/2$-inch plywood are horizontal, that's $1/4$ inch of extra material. Compared with 3 inches of 2x10, that's an increase of only 8%. What's more, you only get the full effect of this extra thickness if there are no splices in the plywood near the middle of the span, or better yet, no splices at all. For headers at openings wider than 8 feet, that's not often the case. But it's these longer headers that will most likely need some extra strength.

Combine these drawbacks with size limitations and the plywood almost never makes a critical difference in safety. What I mean by size limitations is that when I design a header, the numbers may tell me I need two 2x9s. Since two 2x9s are about 30% stronger than two 2x8s, the 2x8s plus 8% from 1/2 inch of plywood wouldn't be strong enough. And I wouldn't ask the framer to rip some 2x9s, I'd simply call for 2x10s. What's more, he'll probably use double 2x10s for all his headers, big and small. Because headers only come in certain depths, there's usually extra strength in the 2x10s to begin with. And that extra strength in the 2x10s means that the small extra strength from the plywood is rarely important. But the thickness is helpful.

—*Christopher DeBlois*

Strength of Toe-Nails vs. End-Nails

Q *What is the strength of toe-nails compared with end-nails when nailing studs or joists?*

A The answer to this question is found in the National Design Specification for Wood Construction (NDS), published by the American Forest and Paper Association (202/463-2700). The NDS gives the design capacities for nails loaded laterally and in withdrawal. The nail-strength tables are further categorized by wood species and the size of the members being joined. The nail capacities given in the tables assume that the two members are being joined side by side, the way you nail overlapping joists to one another above a girder or bearing wall (single shear connections). Toe-nailed and end-nailed applications require strength reductions.

For toe-nails loaded laterally (for example, the nail loading that results from wind pressure on a stud wall), the reduction factor is 0.83. For toe-nails loaded in withdrawal (for example, uplift on a stud wall due to wind suction on the roof), the reduction factor is 0.67.

For laterally loaded end-nails, the reduction factor is 0.67 (called the "end grain factor"). The NDS (widely used as the basis for code requirements) doesn't allow nails driven into end grain to be loaded in withdrawal.

As you can see, a correctly installed toe-nail (Figure 12) is stronger than an end nail of the same size. Of course most carpenters use smaller nails when toe-nailing to avoid splitting, so this also has to be taken into account. For example, when attaching a 2x4 stud to the sole plate, the BOCA code prescribes two 16d nails driven through the plate into the stud. If toe-nailing the stud in place, the code prescribes four 8d commons. In this case, the four toe-nails would be stronger in lateral loading than the two end-nails. Get a copy of the 1997 NDS for specifics.

—*Frank Woeste*

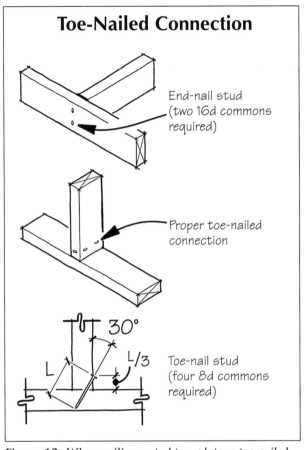

Figure 12. *When nailing a stud to a plate, a toe-nailed connection is typically stronger against lateral forces than an end-nailed connection.*

Safety Factor in Wood Construction

Q *Doesn't the safety factor in wood construction mean that most wood structures are way overbuilt?*

A The idea that "wood structures are way overbuilt" may be the greatest myth in the wood construction field. It is possible that at one time in history wood structures were overbuilt, but it is certainly not true today. The safety factor for bending strength for visually graded dimension lumber is 1.3; by contrast, the safety factor for structural steel, which has much less variability from piece to piece, is as much as 2.

So how are safety factors applied? To arrive at the design values used in wood design, thousands of pieces of lumber of representative sizes, grades, and species have been tested. These tests are run for about ten minutes to determine the stress that will cause a piece of lumber to fail. The test data for every piece of lumber of a given grade, size, and species is recorded. In a test of bending strength, for example, the values from a batch of lumber might range from 3,000 to 15,000 psi. By convention, the value of the 5th percentile is calculated (in other words, 95% of the pieces tested fall above this number, 5% fall below). Choosing a value at the 5th percentile is a way of accounting for the wide variability in the strength of pieces of visually graded lumber (due to knots, slope of grain, etc.).

This number — let's say it's 4,000 psi — is then divided by 1.62 to convert it to a ten-year duration value, which is the load duration that is used in the design of wood floor systems. (Remember, the test lasts only ten minutes; lumber can resist more stress for short periods of time.) Finally, the ten-year value is divided by a safety factor of 1.3. So a 5th percentile value of 4,000 psi would become 1,899 psi. This is the number that is published in the allowable design stress tables.

It's a grave mistake to make design decisions based on an assumption that the wood safety factor is excessive.

—*Frank Woeste*

Deflection of Plywood vs. OSB

Q *Does OSB sag more than plywood when installed horizontally over 24-inch-center rafters?*

A It depends on the materials used to make the OSB, which can be manufactured from a variety of species. These include aspen, southern pine, sweet-gum, yellow poplar, and birch. The Modulus of Elasticity (MOE) of the wood used will determine the relative flexibility of an OSB panel. The list at right shows the MOE values for some of the woods used to make OSB and plywood (from NDS Supplement).

On the same roof with rafters spaced at 24-inch centers, a plywood panel made from high-grade Douglas Fir-Larch veneers is going to deflect less than an OSB panel made from aspen. Run the same test using an OSB panel made from similar materials and you will most likely find no difference in their deflections.

Probably the reason that OSB has the reputation for flexing more than plywood is that much of the OSB sold is manufactured from lower grade fibers. This is why OSB typically costs a lot less than good plywood. —*Scott McVicker*

Wood Species	MOE
Aspen	800,000 to 1,100,000
Yellow Poplar	1,100,000 to 1,500,000
Beech-Birch-Hickory	1,200,000 to 1,700,000
Douglas Fir-Larch	1,300,000 to 1,900,000
Southern Pine	1,200,000 to 1,900,000

Calculating Loads on Sloped Roofs

Q *I often see roof loads calculated based on the horizontal run of the roof. But isn't it more accurate to figure the weight of snow and roofing materials by measuring along the actual length of the rafter? Thus, as the roof gets steeper, the rafter gets longer, and the weight of roofing materials increases.*

A Actually, with snow, you don't get more load along the slope of the rafter; you get the same load as a flat roof with the same run would get. This is because as snow falls vertically it spreads itself further along a sloping rafter and so accumulates less depth. The BOCA code recognizes this and allows you to use the horizontal projection of the roof when calculating snow loads. BOCA also allows you to reduce the snow load for roofs with slopes greater than 30 degrees, presumably because snow will slide or blow off steeply pitched roofs.

For dead loads, you are correct. Technically you should use the actual rafter length when adding up the weight of roofing materials. However, in my practice, I typically use the horizontal run of the roof for both types of load. To do this, I use conservative (too heavy) dead loads and full snow loads regardless of pitch. I ignore the slope factor altogether for snow load reduction which adds another measure of conservatism. (Slope length cannot be ignored for wind load analysis, though.)
—*Robert Randall*

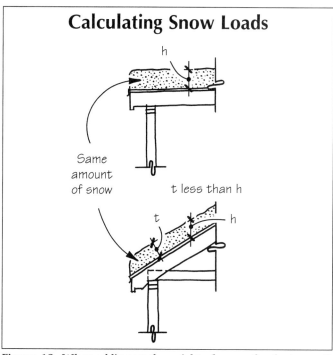

Figure 13. *When adding up the weight of snow of a sloping roof, use the horizontal run of the roof, since the same amount of snow would accumulate on a perfectly flat roof. The code also allows you to apply a slope reduction factor, to account for snow blowing or sliding off a sloping surface.*

Nailing Patterns for Built-Up Beams

Q *What's the best nailing pattern for built-up beams?*

A The critical issue with built-up beams is that all the layers must deflect together and by the same distance in order to be properly sharing the load. For beams where the load comes down evenly on top of the beam, such as drop beams or beams directly under bearing walls, the nailing pattern is not all that critical. All you need are enough nails to hold the layers together and keep them from twisting. For beams loaded from the side, however, and especially for beams loaded from one side only, the nailing pattern is critical.

When beams are loaded from the side, there must be enough nails to transfer the load through the loaded member and into the attached members. For example, if a beam consists of three 2x10s loaded from one side only, the loaded member should only carry one-third of the weight. To transfer the rest of the load into the attached members there must be enough nails from the loaded 2x10 into the center 2x10 to transfer two-thirds of the load, and enough nails from the far side 2x10 into the center 2x10 to transfer the final one-third of the load into that outer member. These numbers assume that all three 2x10s rest fully on the supports; the situation gets more complicated when the members are not all the same size or material. The bottom line, though, is that if all the pieces deflect together and equally, the beam should perform as designed.

At a minimum, I recommend pairs of 16d nails every 12 inches along the beam, with the top row of nails 1 1/2 inches or so from the top of the beam, and the bottom row 1 1/2 inches or so up from the bottom. Use the same nailing pattern on both sides for triple beams, and check with an engineer whenever you think the loads involved might be unusually heavy.

—Christopher DeBlois

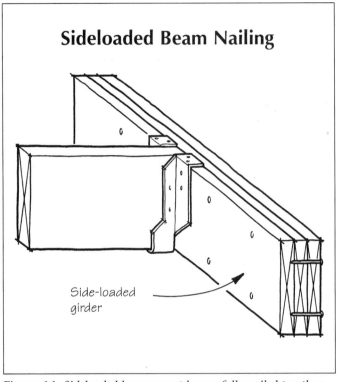

Figure 14. *Sideloaded beams must be carefully nailed together, to ensure that all the beam members share the load.*

Rafter Thrust in a Shed Roof

Q *Does a cathedral shed roof addition need collar ties to restrain the outward thrust of the rafters?*

A A shed roof with a proper shear connection at the ridge has no lateral thrust. Think of a ladder leaning against a building. Imagine the ladder has wheels at the bottom. With no restraint at the top, the ladder will roll away from the building. Attach the ladder to the building at the top, however, and it stays put. A shed roof is similar; as long as it's properly attached at the top, the bottom can't move. "Collar ties" are an exercise in futility. Use them as ceiling joists if needed; otherwise, leave them out.

—Robert Randall

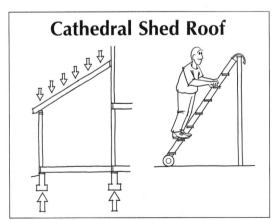

Figure 15. *Collar ties are not necessary in a cathedral shed roof, since there's no outward thrust to restrain. The situation is analogous to a ladder leaning against a wall, and attached at the top. Even if it's on wheels, the ladder can't move away from the wall.*

When Are Collar Ties Needed?

Q *Collar ties don't seem necessary in attics where the rafters come all the way down to the ceiling joists. Can you remove some of them to create headroom?*

A The most common reason for installing collar ties is to prevent rafters from spreading apart under load. However, in a conventionally framed peaked roof, like the kind you describe, collar ties would probably serve little or no function, since the attic floor joists serve as ties to prevent the rafters from spreading. Note that the connections between the rafters and the joists must be adequate, and that the overlapping joists at midspan must also be properly nailed.

There are some exceptions, however, when collar ties might be useful even in a conventional attic roof. For example, very long rafters in a relatively steeply pitched roof (slopes above 6/12, for instance) may benefit from a stabilizing effect if adequately connected collar ties are installed on every rafter pair. In this case, the collars serve not as ties but as spreaders. Also, in high wind situations with lower pitched roofs, collar ties may help hold the ridge assembly together, although steel strap ties installed just below the ridge board would probably work better.

My call is that in the vast majority of such cases, collar ties can be removed with no detrimental effect. In most of the cases I have observed, the existing connections between the collar ties and the rafters are inadequate to provide any meaningful beneficial effect anyway. —*Robert Randall*

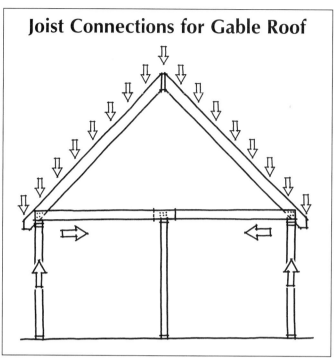

Figure 16. *Collar ties are usually not needed in conventional gable roof attics, as long as the floor/ceiling joists are properly connected to handle the tension forces created by the outward thrust of the roof.*

Shear Strength of Gypboard

Q *I've heard that engineers give no structural "credit" to gypboard, but I know it greatly stiffens partitions when I nail it up. How much shear strength does drywall really have, and why not credit it in the design?*

A As you suspect, properly fastened gypboard does have the capacity to resist racking and/or lateral forces. The 1997 Uniform Building Code (Table 25-1) gives shear values for both gypsum wallboard and gypsum sheathing. In fact, the allowable lateral force on a wall with fully blocked 5/8-inch gypboard on both sides nailed at 4-inch centers (350 plf) actually exceeds that of a wall with 1/2-inch Structural I plywood fastened with 10-penny nails at 6-inch centers (340 plf). Be careful, though: If you are working in seismic Zones 3 or 4, note that even with fully blocked edges you must reduce the allowable lateral load on gypboard by 50%.

As to crediting the design for the strength of the gypboard, this decision is based on the materials selected for the particular structure. If you build a house with rigid-foam insulation panels on the exterior (under finishes) and gypboard on the interior, then the gypboard is the lateral force-resisting material. However, if the interior gypboard is combined with plywood sheathing on the exterior (or with diagonally braced structural steel studs), then the strength of the gypboard is discounted. In the latter case, the plywood is considered the primary lateral-force-resisting material because of its greater strength and stiffness. In both instances, the designer must make certain that the primary lateral-force-resisting material is sufficiently fastened to the framing to resist the total lateral load despite the presence of other secondary materials.

In reality, it is the combination of all the primary and secondary materials that will resist the applied lateral loads. However, should the loading persist, the repetitive cycles of load/release will cause fatigue of the weaker materials (like gypboard) until essentially only the primary lateral material remains functional. If we were to credit the strength of the gypboard towards the total lateral load (and reduce the plywood nailing accordingly), our structure would lack critical capacity after the time when the gypboard had yielded. This is the reason gypboard receives no credit for its strength.

—*Scott McVicker*

Using Rafter Purlins

Q *Does using purlins and struts at midspan allow you to cut the roof span in half compared with what's given in a rafter table?*

A Yes. Actually, in theory, a rafter with purlin support at midspan could be a little longer than twice the maximum allowable single span length. This is due to the effect of moment continuity across the support. This means that the roof load on one side of the purlin has a slight lifting effect on the other side of the purlin.

When using purlins, you must be careful that the struts are properly supported. Always carefully trace the load path down to the ground, verifying the adequacy of each element.

To be most effective, struts should be installed as close to vertical as possible so as not to create lateral forces that have to be dealt with. This will depend, of course, on the location of the bearing wall below. And keep in mind that when the struts get longer than 6 feet, they may require lateral bracing.
—*Robert Randall*

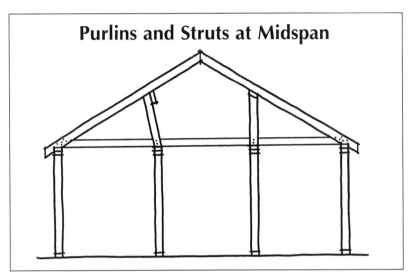

Figure 17. *Purlins and struts reduce rafter spans just as bearing walls do. The struts, or a kneewall, should be installed as close to vertical as possible, and must be properly supported below.*

Splices in Built-Up Beams

Q *Is it necessary to place splices in built-up lumber beams directly above the support posts?*

A The easy answer to this question is Yes, but it's not entirely true. What is true is that you can't run into trouble locating all splices directly over support posts.

In reality, the most efficient location for splices is at points of inflection. Figure 18 shows the expected deflection of a uniformly loaded beam without any splices spanning from wall to wall across a center post. Note how the beam sags near the centers of the spans, while the deflection curve turns upward over the post. The points where the curvature of the beam transitions from concave down over the post to concave up between the posts are the inflection points. At those points, stresses in the wood due to bending are lowest — in fact, they are zero. Unfortunately, shear stresses won't be zero at these points, so if you spliced all the members of a built-up lumber beam at inflection points, you would still need some type of steel or wood shear plates nailed or bolted across the splice to transfer the loads from one section of beam to the next. That's a trick that's common in commercial steel construction, but that becomes a pain for wood framing.

A second problem is that wood beams aren't flexible enough to see the shape of the curvature and reveal the inflection points; their locations must be calculated. Since the location of

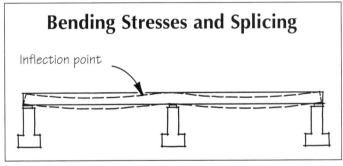

Figure 18. *Bending stresses disappear at a beam's inflection points, making this a good place for splices in a built-up beam, as long as metal shear plates are used to handle shear stresses. But because it's difficult to figure out exactly where the inflection points are, it's always a safe bet to place splices directly over posts.*

each inflection point depends on the relative length of adjacent spans, the number of spans, and the variations in load along the beam, there is no easy rule of thumb for locating the inflection points and hence the best location for splices.

So my suggestion is to take the safe route and set all your splices in multiple span built-up beams directly over the posts.
—*Christopher DeBlois*

Weight of Steel vs. Lumber Beams

Q *For the same loads, which is heavier, structural steel beams or lumber beams?*

A When depth is not a restriction, it is almost always possible to design a steel I-beam that is lighter than the lightest structurally acceptable wood beam design, including glulams, LVL, and Parallam beams. And no matter how hard you try, solid timbers, built-up 2x beams, and flitch beams are almost always heavier than the lightest steel I-beam option — usually a lot heavier. Yes, it's true that steel as a material is heavier than wood given two chunks of the same size. That's because the density of steel is 12 times or so higher than the density of Southern Yellow Pine, for example. One cubic foot of steel weighs about 490 pounds, while the same size chunk of kiln-dried SYP wouldn't top 40 pounds. But because the steel can be formed into very efficient shapes, like I beams, the overall weight of a steel beam is often lower than the lightest wood option.

In some cases, steel may be the only type of beam that will solve a problem. A good example is that common remodeling problem of removing a loadbearing wall without having the new support beam project below the ceiling. For long spans in a 2x10 floor, you can't get enough stiffness from $9^1/_4$-inch LVLs or 9-inch glulams, but 8-inch steel I-beams come in a variety of widths and weights to handle almost any situation like this. In such a case, the framer may complain that the steel beam is very heavy, but it's not heavier than the alternatives when there are none. There are also times when steel is ideal not because it can hold up a lot of weight, but because it can be welded into rigid frames. The modern two-story window wall leaves little room for plywood shear panels, but in high wind and seismic areas you can't ignore the potential for racking that accompanies these lateral loads. A stiff moment frame of steel tubes or I-beams can often solve this problem when wood just won't do the job.

—*Christopher DeBlois*

Strength of PT Lumber

Q *Does CCA pressure treatment adversely affect the strength or durability of framing lumber?*

A According to the National Design Specification for Wood Construction (1997), pressure preservative treatment with CCA (chromated copper arsenate) does not affect the strength of lumber except in the case of impact loads (loads that last about a second). Fortunately, impact loads are not typical in residential construction.

The bigger concern with PT lumber is that in use it is typically exposed to the weather. Thus, its design strength is subject to a "wet service" reduction factor, and thus most fasteners have weaker values. Therefore, most lumber properties are lower and most connections are weaker. One way to avoid a moisture penalty for connections in PT lumber is to use threaded hardened-steel nails, which have been shown in testing to have full rated strength even in wet lumber. When working with pressure-treated lumber, choose fasteners that resist corrosion from CCA, such as hot-dipped galvanized.

—*Frank Woeste*

Thanks to the structural engineers who contributed to this section: Christopher DeBlois, P.E., Palmer Engineering, Tucker, Ga. Scott McVicker, N.E., McVicker & Associates, Palo Alto, Calif. Robert Randall, P.E., Randall Engineering Mohegan Lake, N.Y. Frank Woeste, P.E., Professor of Wood Construction, Virginia Tech, Blacksburg, Va.

Chapter Three
ROOF STRUCTURE

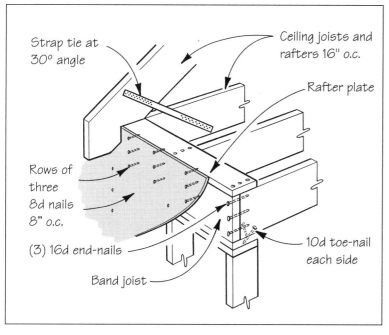

Holding the Roof Up 42

Framing with a Raised Rafter Plate 43

Resisting Wind Uplift 45

Straight Talk About Hips and Valleys . . .47

Holding the Roof Up

One morning following a heavy snowstorm, I got an emergency call about a roof collapse. It was, in fact, an example of what we engineers call "catastrophic failure" — a classic case of underdesigned (or undesigned?) eaves tie connections. Although these things don't occur often, such a collapse serves as a reminder of what might happen if connections are not properly designed and executed in the field.

This old, solidly built structural-brick building had full-size 3x8 attic floor joists at 16 inches on-center, with full 1-inch-thick plank flooring. (As it turned out, the sturdy attic floor saved the day, halting the fall of the roof assembly and preventing damage to the walls and ceilings below.) The roof was framed with full-cut 3x6 rafters at 24 inches on-center, which meant that only every fourth floor joist aligned with a rafter and could serve as an eaves tie.

The calculated load at this connection is 1,680 pounds, obviously more than the four or five toe-nails provided were capable of carrying. The building had stood for over a hundred years with no apparent problem until this sudden and very dramatic failure.

Eaves Ties Explained

Any peaked roof that does not have a structural ridge capable of carrying the vertical loads of the entire center half of the roof will inevitably develop lateral thrust, usually at the eaves (see Figure 1). This thrust is a horizontal force tending to spread the top of supporting walls and is usually resisted by eaves ties, also known as ceiling joists or attic floor joists. Regardless of what you call them, these members carry a lot of force in tension and must be designed accordingly. The weak link is usually in the connections, both at the eaves and at center splices, which may be overlooked as significant structural connections.

Table 1 shows how dramatic the variation of this thrust can be with varying roof pitch. The low-pitched roofs, 3:12 and 4:12, need ten and more nails per connection, a nailing schedule seldom seen in practice. When the number of nails required reaches ten or more, I usually specify bolts instead and hope that there will be two or three nails as well.

In most parts of the country, you don't need to look very far to see a roof with noticeable sagging of the ridge line. When you get into the attic and investigate, which is often part of my job, you frequently find eaves tie connections that have too few fasteners. Over the years, the result is a sagging ridge and cracked ceilings below, especially the flush ceilings between a living room and dining room common in ranch-style homes.

Weak Center Splices

Ceiling cracks near the center of the building occur when ceiling joists are not adequately spliced. The center splices should have the same number of nails or bolts as the connection to the rafters. When a flush-framed header or girder is installed to support ceiling joists, you should use at least 16-gauge steel strap ties to bridge across the header and connect the joists, which are interrupted by the header (Figure 2). Use the same number of nails in the strap (usually 10-penny) as for the rafter-joist connection. I typically recommend Simpson-brand strap ties, because they're load-rated and the catalog specifies the required nailing.

Another common eaves tie deficiency I see is in two-story homes where the master bedroom extends from the front to the rear of the house and the ceiling joists have been rotated to run parallel with the ridge to create a flush ceiling (Figure 3). Without eaves ties or a structural ridge, the only thing left to keep this roof standing is the shear diaphragm action of plywood sheathing, which is often not adequate.

Not a Collar Tie

Eaves ties, as discussed here, are not the same as collar ties, which may be used to stiffen roof rafters. In some cases, eaves

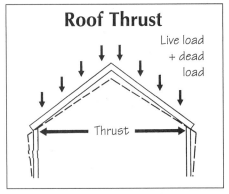

Figure 1. *In a typical roof with no structural ridge, the loads push out on the exterior walls. These forces must be resisted by nails or bolts at the joist-rafter connections and at the midspan splices in the joists.*

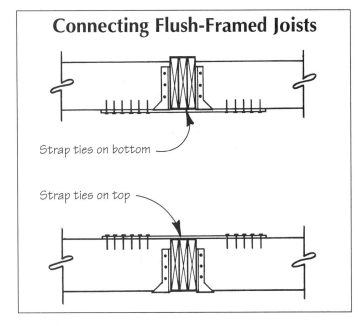

Figure 2. *When a flush-framed header or girder interrupts the ceiling joists at midspan, use strap ties to connect the joists and carry the tension created by the roof load.*

Table 1. Nailing & Bolting Schedule for Eaves Tie Connections

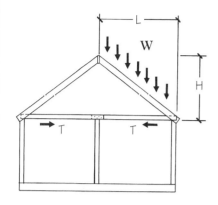

$W = L \text{ (ft)} \times \text{rafter spacing (ft)} \times \text{roof design load (L.L. + D.L. in psf)}$

$T = \dfrac{W \times L}{2H}$

	Building Width (2L)			
Roof Pitch	20'	24'	30'	36'
3/12	T = 1,064 lb. 11 nails or 3 bolts	T = 1,276 lb. 13 nails or 4 bolts	T = 1,596 lb. 16 nails or 4 bolts	T = 1,915 lb. 20 nails or 5 bolts
4/12	T = 798 lb. 8 nails or 2 bolts	T = 958 lb. 10 nails or 3 bolts	T = 1,197 lb. 12 nails or 3 bolts	T = 1,437 lb. 15 nails or 4 bolts
6/12	T = 532 lb. 6 nails or 2 bolts	T = 638 lb. 7 nails or 2 bolts	T = 798 lb. 8 nails or 2 bolts	T = 957 lb. 10 nails or 3 bolts
9/12	T = 355 lb. 4 nails or 1 bolt	T = 425 lb. 5 nails or 2 bolts	T = 532 lb. 6 nails or 2 bolts	T = 638 lb. 7 nails or 2 bolts
12/12	T = 266 lb. 3 nails or 1 bolt	T = 319 lb. 4 nails or 1 bolt	T = 399 lb. 4 nails or 1 bolt	T = 479 lb. 5 nails or 2 bolts

Note: This table is based on 30-pound psf live load plus 10-pound psf dead load. Nails are assumed to be 16d, at 100-pound capacity each; bolts are 1/2 inch, at 400-pound capacity each. Rafters are 16 inches on-center.

No Eaves Ties

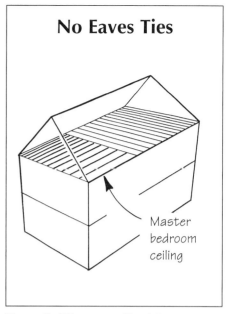

Figure 3. *Whenever ceiling joists are turned parallel to the ridge, as in this second-story master bedroom design, there is no eaves tie effect. A structural ridge must be used for this section of the roof to prevent sagging.*

ties may be raised above the wall plates to resemble collar ties, but when this is done, careful analysis is needed to avoid overloading the rafters with bending moment.

Fortunately, most roof failures occur progressively, and visible cracks and sagging provide advance warning of trouble developing. The best strategy for a builder, of course, is to take care to prevent problems by using a few extra nails or, for low-pitched roofs, using bolts. And when in doubt, consult an engineer.

—Robert Randall

Framing with a Raised Rafter Plate

The last section ("Holding the Roof Up") looked at the issue of lateral thrust in roofs — the tendency for rafters to spread apart under load. Conventionally framed roofs (Figure 4) are often strong enough to resist this lateral thrust as long as the carpenters use enough nails or bolts where the rafter laps the attic joist (though this can be a challenge in shallow-pitched roofs).

A problem with this framing detail, especially in cold climates, is that the attic insulation tends to get compressed above the wall plate. Even with ventilation chutes, the increased heat loss at the plates may contribute to ice dam troubles. So many energy-conscious builders use a "raised rafter plate" detail (Figure 5) which lifts the rafter end above the attic joist, allowing for more insulation over the wall plates. This helps to solve the heat loss problem, and also has other advantages: It creates more headroom and floor space in the attic, it allows more sunlight to come in below the roof overhang, and it gives a greater reveal outside for a wide frieze board above the windows.

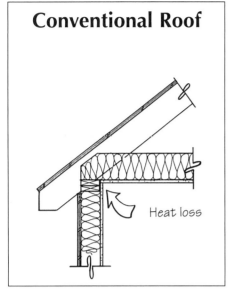

Figure 4. *Conventional attic framing provides a strong eaves tie connection where the rafter laps the joist, but it doesn't allow much space for insulation above the wall plates.*

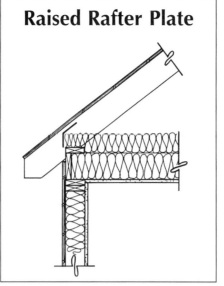

Figure 5. *The raised rafter plate allows room for insulation above the wall plate, but the lateral thrust of the roof must be accounted for in the design.*

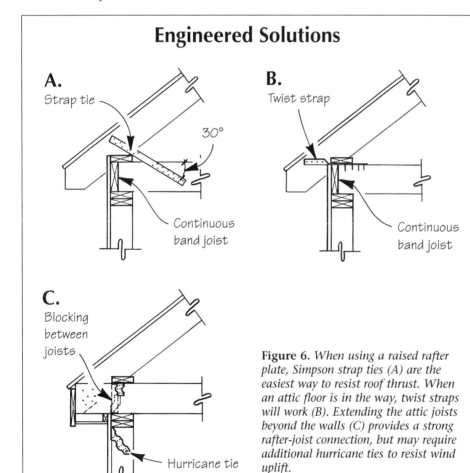

Figure 6. *When using a raised rafter plate, Simpson strap ties (A) are the easiest way to resist roof thrust. When an attic floor is in the way, twist straps will work (B). Extending the attic joists beyond the walls (C) provides a strong rafter-joist connection, but may require additional hurricane ties to resist wind uplift.*

Unfortunately, the structural requirements of this detail are often overlooked. The results can be disastrous, as the roof collapse cited in the previous section illustrates. Toenailing the rafters to the raised plate — the most common job-site solution — will rarely provide a strong enough connection. A really industrious carpenter might get six or eight toenails into each rafter, but if the raised plate itself has been nailed off with one or two nails per joist, it won't matter.

Raised Plate Details That Work

Here are three simple details that work well (Figure 6). All rely on metal connectors available from Simpson Strong-Tie Co. (Dublin, CA; 925/560-9000; www.strongtie.com) and other manufacturers.

Detail A, the simplest, uses either the Simpson ST strap tie or the CS coil strap — galvanized steel straps prepunched for the nails needed to carry a rated load. The Simpson catalog lists the capacity of each tie. Use the load calculation procedure from the previous section, increase the load by about 20% to allow for the slope of the strap tie, and just nail it in place with the nails specified in the catalog. For long spans, low roof pitches, or heavy snow loads, it may be easier to use two ties of lesser capacity rather than a single tie with 20 nails.

Note the placement of the exterior wall sheathing over the attic band joist, which provides a wind tie for the roof assembly. It also helps prevent ice dam leaks from getting into the wall framing.

Detail B allows for use of a plywood subfloor in the attic, which restricts the placement of the metal ties. In this case, you can use the Simpson TS series twist straps. Calculate the load as for Detail A, except there's no 20% increase because there's no angle involved. Because the twist straps are set perpendicular to the uplift force of the wind, Detail B has relatively poor wind lift resistance and may require supplementary wind ties in high-wind locations. By the way, remember to nail the twist straps to the joists before the plywood goes down.

Detail C provides excellent rafter thrust resistance because joists extend past wall plates and can be directly nailed or bolted to rafters. The addition of a steel wind tie at alternate rafters, such as the Simpson H7 shown here, makes this a dependable structural design. Note that the blocking above the plate is required by code to prevent the joists from rolling, but it also serves to prevent "wind washing" of the attic insulation. Dropping the soffit 2 inches helps prevent possible ice dam leaks.

—*Robert Randall*

Resisting Wind Uplift

Many contractors build roofs using a raised rafter plate detail to allow extra room for insulation at the eaves (see Figure 7). There are benefits with this detail, but my recommendation for using strap connectors confuses some people and merits further explanation.

In particular, builders have asked me whether the one or two toe-nails used to hold the joists to the top plate now become the weak link in the system (when it comes to uplift, not spreading).

Uplift vs. Spreading

This question highlights an important point about the eaves connections. Not only do these connections have to resist the tendency of the roof to spread apart, but they also have to resist uplift forces from the wind (and earthquakes in seismic zones). The previous section, "Framing With a Raised Rafter Plate," dealt mainly with rafter spread. But the detail in question also addresses uplift, because the 30-degree angle of the strap allows it to work effectively as both a horizontal tie and a vertical tie. (The pitch of the angle should equal the ratio of the horizontal force to the anticipated vertical force.)

What follows is an analysis of this detail to find out whether the construction can withstand a typical severe wind load.

Wind Load Design for Eaves

The design of structures for wind loads is a very complex subject and is covered in incredible detail (yet differently) by many codes. My purpose here is not to explain how to design for wind loads, but to analyze how well the detail in Figure 7 can resist typical uplift loads in a high-wind region. To do this, I'll assume that the BOCA code applies and we are in a 110-mph wind region in an exposed location. The design wind pressure will be 37.2 psi on walls. For a 4/12-pitch roof, an uplift factor of .75 applies, so the design wind lift is 28 psf. Figure 8 shows a typical wind lift calculation using this value.

Holding Capacity of Nails

The connections in Figure 7 depend entirely on nails, so the first step in determining the strength of the assembly is to figure out the holding capacity of each individual nail. Design values for nails vary widely, depending on the size and type of nail and the species of lum-

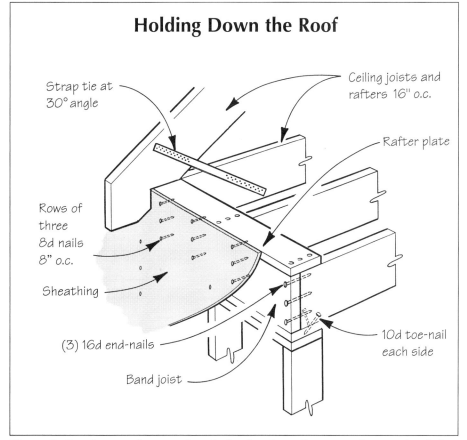

Figure 7. *Framing a roof on top of a raised rafter plate allows room for extra insulation above the top plates, but it may weaken the important connection between the ceiling joists and the rafter ends. Using a rated strap tie at every rafter maintains this connection, which helps prevent the roof from spreading. The strap tie also transfers uplift forces from the roof to the ceiling joists.*

Table 2. Common Nail Design Values in Hem-Fir Framing
(based on NDS 1991 tables & adjustment factors)

Nail Size:	8d	10d	16d
Direction of Load:	Lateral	Withdrawal	Lateral
Application:	Through 1/2-in. sheathing into band joist and rafter plate	Toe-nail ceiling joists into top plate, 2 inches penetration	End-nail band joist into ceiling joist ends
Adjustment Factors:	Wind loading: 1.6	Wind loading: 1.6 Toe-nail in withdrawal: .67	Wind loading: 1.6 Nailing into end-grain: .67
Design Value:	63 lb. x 1.6 = **100.8 lb.**	25 lb./in. x 2 in. x .67 x 1.6 = **53.6 lb.**	122 x .67 x 1.6 = **130.7 lb.**

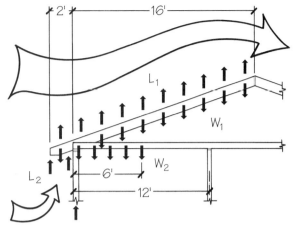

Figure 8. *Wind creates negative pressure, which exerts a lifting force on the roof. The eaves overhang is also subject to direct positive wind pressure, which must be added into the total design wind load. The dead loads from the roof and ceiling assembly are subtracted from the wind uplift for a total net uplift.*

L_1 = Roof Lift = 18 ft. × 1.33 sq. ft.* × 28 psf = 671 lb. per rafter
L_2 = Eaves Lift = 2 ft. × 1.33 sq. ft.* × 28 psf = 75 lb. per rafter
Total Lift on Roof = 746 lb. per rafter

W_1 = Roof Dead Load = 18 ft. × 1.33 sq. ft.* × 7 psf = 167 lb. per rafter
W_2 = Ceiling Dead Load = 6 ft. × 1.33 sq. ft.* × 5 psf = 40 lb. per rafter

Net Lift at Top Plate = $L_1 + L_2 - [W_1 + W_2]$ = 539 lb. per rafter

*Number of square feet of roof per linear foot of rafter at 16 inches on-center

ber used. There are also adjustment factors for such things as load duration, excessive moisture, high temperature, nailing into end-grain, and toe-nailing. The National Design Specification for Wood Construction (NDS, 1991 edition) explains these factors in detail, and gives tabulated design values for many nail types and wood species. Table 2, derived from the NDS, gives capacities for the nails shown in Figure 7, with the appropriate adjustment factors applied, and based on Hem-Fir framing lumber.

Note that a nail can be loaded either laterally (load perpendicular to the shank) or in withdrawal (load parallel to the shank). The NDS specifies different values and correction factors depending on the direction of load; obviously, a nail loaded in withdrawal has much less holding capacity than a laterally loaded nail. I advise against ever depending solely on nails loaded in withdrawal for a framing connection.

Plywood–Band Joist Connection

With that background, we can answer the question posed: Isn't the weak link the connection of the ceiling joists to the top plate?

In this case, the plywood lapping the band joist is a lot stronger than you might think. Note from that chart that every 8d nail into the band joist can withstand a little more than 100 pounds of lateral (in this case, wind lift) load. So if the plywood is nailed with two 8d nails into the band joist and one into the rafter plate every 8 inches (or three nails altogether every 8 inches), as in Figure 7, the band joist is held down with six nails per rafter, or 600 pounds of resistance. That's more than enough to handle the uplift loads in the sample calculation in Figure 8.

What if the builder used 6d nails to nail off the sheathing? The design value for 6d nails in this situation would be 78 pounds per nail, so it would take more nails to achieve the same result.

Ceiling Joist to Band Joist Connection

The plywood does a good job of holding down the band joist. But note that in Figure 7, the strap ties are attached to the ceiling joists, not to the band joist. If the wind tries to pick up the roof, the straps will transfer the force directly to the ceiling joists. So a careful analysis needs to look at the nails holding the joists in place. First, there are the three 16d end-nails through the band joist. At 130.7 pounds each, these three nails can resist 392 pounds total — not enough to satisfy our sample design requirement. But there are also a couple of 10d toenails through each ceiling joist into the top plate, which gives another 107 pounds of resistance. Because the rafter plate clamps the joist ends in place, we can also count the 8d nails securing the sheathing to the rafter plate — one every 8 inches, or two per joist. These nails are also laterally loaded, which gives another 100.8 pounds per nail, or 201 pounds per joist. So the sum total of the nails holding the ceiling joists to the band joist is

392 lb. + 107 lb. + 201 lb. = 700 lb. resistance

The strap tie detail, with the plywood properly nailed, should be plenty strong enough for most wind load conditions. For extremely severe circumstances — coastal or other high-wind zones, highly exposed locations, tall buildings, and so forth — engineered design is always in order. There are any number of rated metal connectors available from Simpson or USP that can easily solve wind load design problems.

Good Practice

My suggestion to a cautious builder is to spend a few minutes and a few dollars to incorporate these simple recommendations:

- Nail sheathing with 8d nails at 6 inches on-center along both wall plates, all band joists (in vertical rows of three nails), and all studs. Be sure that plywood seams do not fall near ends of studs at floor platforms. In two-story houses, install the plywood so that it

laps the band joist and ties the first-story studs to the second-story studs. The sheathing is what holds the house together in a wind.
- Be wary of tall structures, large overhangs, and low-pitched roofs. Wind lift on a large overhang, including porch roofs, can peel a roof right up. Low-pitched roofs are particularly subject to aerodynamic lift.
- Give some thought to the eaves connections. Minimum code nailing practices cannot hold a candle to a strong wind. If you have any doubts, add a few hurricane ties; they're not that expensive.
- Make make sure the rafter-to-eaves tie connection is adequate, especially where you have a long span, low pitch, or heavy snow loads.
- Never set rafters 24 inches on-center if the ceiling joists are 16 inches on-center. The rafter-to-ceiling joist connection is extremely important.

Some readers may notice that there is a potential for conflict between these recommendations addressing wind uplift and the recommendations addressing lateral wind loads in Chapter 4, "Plywood vs. Let-In Bracing." If you meet lateral bracing requirements by placing the plywood edges to land on the top plate, then you should use steel straps across the band joist and rafters to address wind uplift. However, if you install the plywood so that it bridges from the wall plates across the band joist, as in my detail, you may need to add blocking along the bottom edge of the plywood, if it breaks in the middle of a stud, to handle horizontal wind shear forces.

—*Robert Randall*

Straight Talk About Hips and Valleys

Hip and valley construction, including the proper sizing of hip and valley rafters, is an aspect of wood frame construction that is not generally well understood and is often poorly executed. There are two important structural aspects of hip and valley rafters: sizing the member and providing support at the ends.

In plan view, hip and valley framing are similar: In either case, a main structural member — the hip rafter or valley rafter — runs diagonally between the high (ridge) corner and the low (eaves) corner of an area of intersection of two sloping roofs. At outside corners, this is a hip rafter; at inside corners, it's a valley rafter. Because their geometry is similar, the same calculations can be used to analyze either hip or valley rafters.

Determining Hip and Valley Loads

The *tributary area* is the portion of the roof from which loads are transferred onto the hip or valley rafter. The tributary area includes half the span of each jack rafter, which results in a kite-shaped figure (see Figure 9). These dimensions are measured in the horizontal plane only — you can usually ignore the slope when calculating roof loads.

As shown in Figure 9, the tributary area includes half the area of the entire hip or valley intersection. Note that I'm showing the hip or valley area as a square,

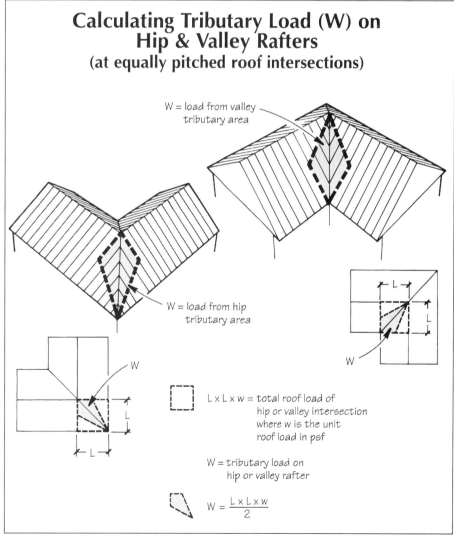

Figure 9. *A hip or valley rafter picks up half the load of all the jack rafters feeding into it — its so-called tributary load. In plan view, this tributary load is kite-shaped. Note that the hip's tributary area is wider at the top, the valley's is wider at the bottom.*

Supporting Hips & Valleys: Two Case Studies

Here are two examples that illustrate two different approaches to apex support of hip and valley rafters. In both cases, the hip or valley rafters have been selected by using the Table 3 found on page 51. The roof loading calculations in each case demonstrate how the apex loads are arrived at. These are not intended as general solutions for all similar roof designs. Each roof structure must be individually examined for the best support solution. This is especially important in large-span roofs, which can place very large tributary loads on hip and valley rafters, and in cases where all posts and horizontal members such as collar ties have been deleted to create a cathedral ceiling.

Case 1: Support Post

This is a typical L-shaped roof. The apex load comes from the hip rafter on one side and the valley rafter on the other. A center support post (a triple 2x6 or the equivalent) carries the 3,600-pound point load vertically down through the walls below to a foundation footing.

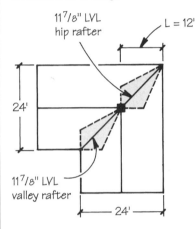

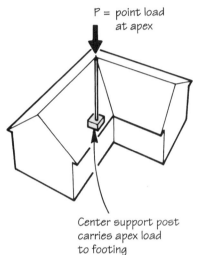

Design load (w) = 40 psf snow load
+ 10 psf dead load

To calculate total load (W) on hip or valley rafter:

$$W = \frac{w \times L^2}{2}$$

$$= \frac{(40 \text{ psf} + 10 \text{ psf}) \times 12^2}{2}$$

$$= 3{,}600 \text{ lb. for each hip and valley rafter}$$

To calculate total apex load (P):

$$P = \text{hip apex load} + \text{valley apex load}$$
$$= {}^2/_3 (W) + {}^1/_3 (W)$$
$$= {}^2/_3 (3{,}600 \text{ lb.}) + {}^1/_3 (3{,}600 \text{ lb.})$$
$$= 3{,}600 \text{ lb.}$$

P = point load at apex

Center support post carries apex load to footing

Case 2: Rafter Truss

In this typical shallow-pitched hip roof, the apex load is carried by a pair of rafters to the top plates of the exterior walls. Since the tension increases as the roof pitch gets lower, this creates a large tension load (4,900 pounds) in the ceiling joist. The rafter/joist connection requires six 1/2-inch carriage bolts at each end to handle this tension. This case is based on southern loading conditions; in northern latitudes, where heavy snow loads are common, the bolts can be applied to two sets of rafters, with the ridge beam sized accordingly to distribute the load.

The connection of the hip rafters to the ridge must be able to accommodate the total apex load (3,267 pounds, or twice that up North). The author often specifies 3/16-inch bent plate-steel connectors with carriage bolts for these connections.

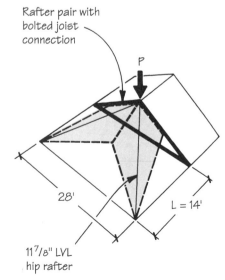

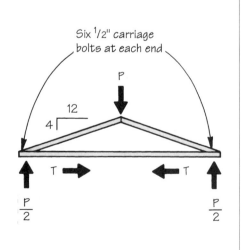

Design load (w) = 15 psf snow load
+ 10 psf dead load

To calculate total load (W) on each hip rafter:

$$W = \frac{w \times L^2}{2}$$

$$= \frac{(15 \text{ psf} + 10 \text{ psf}) \times 14^2}{2}$$

$$= 2{,}450 \text{ lb. for each hip rafter}$$

To calculate total apex load (P):

$$P = 2 \times \text{hip apex load}$$
$$= 2 \times {}^2/_3 W$$
$$= 2 \times {}^2/_3 (2{,}450 \text{ lb.})$$
$$= 3{,}267 \text{ lb.}$$

$$T = {}^1/_2 P \times {}^{12}/_4$$
$$= 4{,}900 \text{ lb.}$$

Figure 10. *The sagging hip rafter in this roof is more than 20 feet long. Because it was undersized, it has permanently deflected under the tributary load from the jack rafters.*

which is the case for the intersection of roofs that have equal slopes. This is the most common type of hip or valley construction, and the subject of this article.

Sizing Hip and Valley Rafters

I'll skip a lot of heavy calculations and jump to one obvious result of this applied load, namely a predictable pattern of deflection. In many engineering analyses, deflection is as much a concern as bending stress, or the tendency to break. In other words, the hip or valley may sag (Figure 10) and cause cracking walls and ceilings or leaking roofs (more likely in valleys) before the danger of outright collapse occurs. This is of particular concern where cathedral ceilings are involved, because ceiling finishes are more likely to crack when they're applied directly to rafters.

Sizing rules of thumb. There are several rules of thumb floating around for sizing hip and valley rafters made from dimensional lumber. One of these, for example, says to take the size of the jack rafters, increase it by one lumber size (to allow room for the long bevel cut at the top end), and double it up. Thus, 2x8 jack rafters would feed into a doubled 2x10 hip rafter. For the most part, such rules of thumb have worked reasonably well for two reasons:

First, there are secondary effects, such as diaphragm loads borne by plywood sheathing, tension loads carried by wall top plates, and truss effects of adjacent rafters and collars, that may combine to provide a significant portion of the required support for the hip or valley rafter. Most frame buildings benefit from such structural "redundancies," although these effects are difficult to quantify dependably and are usually ignored by engineers for structural design purposes.

Second, hip or valley rafters sized by one of the rules of thumb will typically be fairly close to correct in smaller structures. But the error increases dramatically as the size of the structure increases. This probably accounts for the pronounced sag in the roof in Figure 10, which is a relatively large hip roof.

The formula used to calculate deflection for hip and valley rafters takes the length of the hip or valley rafter and increases it to the fifth power. This means that if you double the length of the hip or valley rafter, you don't just double its tendency for deflection; you increase deflection by a factor of 25, or 32 times as much as the deflection in the shorter length!

Table 3 "Sizing Hip & Valley Rafters," page 55, summarizes a lengthy series of sizing calculations for various roof design loads. But choosing the right size hip or valley rafter is only half the battle; you've still got to properly support the apex connection.

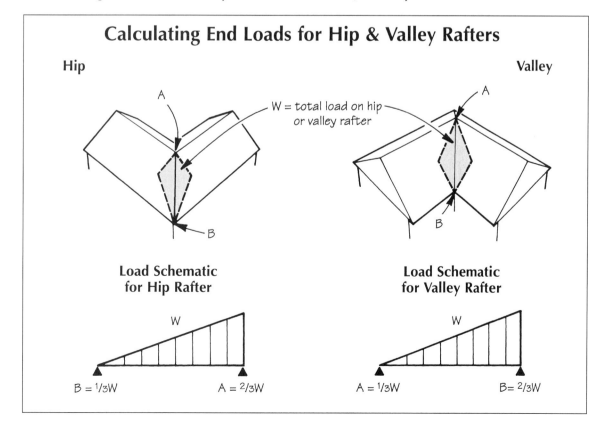

Figure 11. *Because the loads on hip and valley rafters increase from one end to the other, the loads distributed to the end supports are not equal: one-third of the load goes to one end, two-thirds to the other. For valleys, the larger share of the load is transferred to the lower end — the exterior wall in the drawing at left. For hips, the two-thirds load is carried at the top, or ridge, end.*

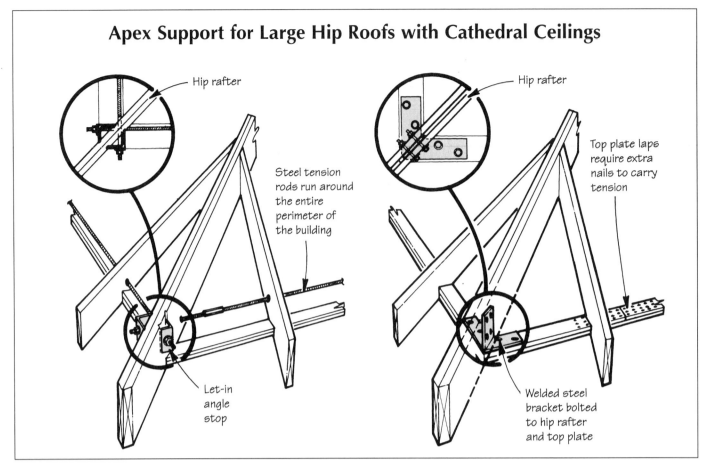

Figure 12. *In large hip roofs with full cathedral ceilings – typical of contemporary open-space design – there are no support posts or restraining horizontal members, such as collar ties, to resist the spread of the rafters. In such cases, the author often uses steel rod tension ties (left) around the perimeter of the building. The tension ties prevent the bottom ends of the hip rafters from spreading and thereby support the apex. In an alternate detail (right), the double 2x6 top plates act as the tension members, and welded steel brackets secure the hip rafters at the corners.*

The Apex Load

Standard hip and valley rafters are treated by engineers as "simply supported beams with the load increasing uniformly toward one end." The load "increases uniformly" as the jack rafters grow longer — toward the ridge in the case of a hip and toward the top plate in the case of a valley. By definition, in beams with uniformly increasing loads, one-third of the total load on the beam is transferred to one end, two-thirds to the other end. We engineers depict the loading on such beams with schematic sketches like the ones in Figure 11.

The high (ridge) end connection of a hip or valley rafter must support a major concentrated load — 2/3W for hips, 1/3W for valleys, where W is the total load on the hip or valley rafter. I refer to this as the apex load. In the most common hip construction, two hip rafters meet at an apex with a ridge running off in a different direction. The vertical load from the two hip rafters is then equal to 4/3W — a big load! This point load must be provided for in the building design.

Providing End Support at the Apex

During the house inspections that I do as part of my engineering consulting work, I have seen situations where the framer used a temporary 2x4 or 2x6 strut to hold up the apex during construction of a hip roof, then left the strut in place after the roof was finished. Over time, the eaves connections had slipped, allowing the structure to settle and bowing the "temporary" strut by as much as 3 inches. When the strut snaps, the progressive sagging will accelerate.

Another common sight is a strut about halfway down the length of the hip or valley rafter. The intention is good, but usually the strut transfers major roof loads onto one or two attic floor joists, resulting in sagging and/or cracking of the ceiling below.

So how should you support these large apex loads? Depending upon the design of the house, I usually recommend one of three different approaches: a center post, a rafter "truss," or tension ties.

The center post is often the simplest solution (see Case Study 1), although post sizing can get complex. I typically use a triple 2x6 or 6x6 post in the attic, and quadruple 2x4s, laterally braced with 2x blocks or skinned with plywood, in the walls below. It's important to align the load path of the post from floor to floor. Don't bring the post down onto a header or girder unless the girder has been designed to handle the load — the large point load could easily overload a typical built-up wood beam.

Table 3. Sizing Hip & Valley Rafters
(for equally pitched roof intersections only)

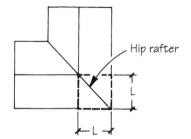
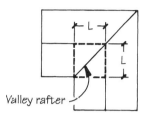

L in feet	15 psf live load	30 psf live load	45 psf live load	60 psf live load
4	2x6	2x6	2x6	2x8*
5	2x6	2x6	2x6	2x8
6	2x6	2x8	2x8	2x10
7	2x8	2x10	2x10	2x12
8	2x10	2x12	2x12	(2) 2x12
9	2x10	(2) 2x12	(2) 2x12	(2) 2x12
10	2x12	(2) 2x12	(2) 2x12	LVL 11 7/8
11	(2) 2x12	(2) 2x12	LVL 11 7/8	LVL 11 7/8
12	(2) 2x12	LVL 11 7/8	LVL 11 7/8	LVL 14
13	(2) 2x12	LVL 11 7/8	LVL 14	(2) LVL 14
14	LVL 11 7/8	LVL 14	(2) LVL 14	(2) LVL 14
15	LVL 11 7/8	LVL 14	(2) LVL 14	(2) LVL 14
16	LVL 11 7/8	(2) LVL 14	(2) LVL 14	(2) LVL 16*
17	LVL 14	(2) LVL 14	(2) LVL 16	(2) LVL 16
18	LVL 14	(2) LVL 14	(2) LVL 16	*
19	(2) LVL 14	(2) LVL 16*	*	
20	(2) LVL 14	(2) LVL 16		

*Deflection limited
Note: The calculations behind this table are based on the following:

10 psf dead load
L/240 deflection
No slope adjustments

Fb wood = 1,000 psi
E wood = 1,000,000 psi
Fb LVL = 2,800 psi
E LVL = 2,000,000 psi

The rafter truss (Case Study 2) will work with either hips or valleys or a combination of the two. The apex load is carried by a pair of rafters to the top plates of the exterior walls, creating a large tension load in the attached ceiling joist, which must resist the spreading tendency of the rafters. The rafter/joist connection typically requires several 1/2-inch carriage bolts at each end. In northern latitudes, where severe snow loads prevail, the bolts might be applied to two sets of rafters, with the ridge beam sized accordingly to distribute the load.

The tension tie is a solution I have used with large hip roofs with cathedral ceilings (Figure 12). The steel rod provides enough horizontal restraint at the eaves to prevent the hip rafters from spreading. With smaller hip roofs it is sometimes possible to provide the needed horizontal restraint by adding structural wood members or steel strapping, but this should only be undertaken on a case-by-case basis with the guidance of a licensed engineer.

—*Robert Randall*

Chapter Four
SEISMIC AND WIND BRACING

Plywood vs. Let-In Bracing 54

Stiffening Garage Door Openings 56

Earthquake Design 58

Shear Wall Construction Basics 60

Installing Seismic Framing Connectors . .64

Hold Down Problems and Solutions 70

Fixing Shear Wall Nailing Errors 73

Plywood vs. Let-In Bracing

Occasionally a builder will ask whether let-in bracing is as good as plywood sheathing for resisting lateral loads on buildings. The answer is simple: Let-in bracing does not come close to plywood. In structural design calculations, engineers usually ignore any contribution that let-in braces may make.

Code-Minimum Bracing

Building codes generally prescribe that exterior stud walls be braced, at each end, with let-in 1x4s, metal strap devices, diagonal wood sheathing, or plywood or OSB sheathing. Figure 1 shows two common ways of complying.

Let-in bracing. Building codes require you to nail let-in bracing with two 8d nails (or 1 3/4-inch staples) at every stud and each plate. Yet despite all these nails, the design lateral load capacity is limited by the capacity of the two nails in each plate, where the brace force is concentrated. The nails in the plates will always fail first, and once they fail, the brace is no longer very effective.

Plywood at corners. Some builders use vertical plywood sheets at wall corners, often in conjunction with nonstructural insulative sheathing on the rest of the wall. When the plywood is nailed according to code (6 inches on-center on supported panel edges and 12 inches on-center on intermediate supports), the total design shear load capacity is more than five times greater than for let in 1x4s.

Structural Sheathing

Figure 2 shows the two most common ways to sheathe walls with plywood or OSB. Often, the carpenters run the plywood horizontally. For residential construction, builders often use 7/16-inch plywood nailed with 6d common nails. If the plywood is nailed according to minimum code requirements, a 20-foot wall with horizontally applied plywood has a total design shear capacity of 2,460 pounds. That means the wall can resist over a ton of force applied laterally at the top plate.

When the same plywood is installed vertically, the shear capacity goes up to 3,280 pounds. The strength of the vertically sheathed wall is greater because all the plywood panel edges are fastened to solid framing and there are no plywood joints parallel to the shear force. The weak spot in the horizontally sheathed wall is the continuous plywood joint at the 4-foot height where the panel edges are least supported. (When plywood shear walls fail under load, the failure begins at the panel edges, with the plywood pulling away from the framing and pulling out the nails or tearing through the nail heads.) If you add horizontal 2-by blocking and nail the plywood at the horizontal plywood joints, the wall with horizontal panels would have the same shear strength as the wall with panels applied vertically.

Lateral Wind Loads

So what are the forces that create lateral loads on buildings? The two most common are wind and earthquake forces. Since seismic forces rarely control design in the area where I work, I'll concentrate this discussion on wind loading.

A typical building in a suburban neighborhood within 25 miles of the Massachusetts coast must be designed to resist a lateral wind load of 21 psf. Figure 3 shows a typical small garage or shop building in a coastal zone with wind blowing against the eaves side. The lateral wind load on the bottom half of the first story is transferred by the walls directly to the foundation (this assumes the walls are properly anchored to the foundation). The rest of the lateral wind load, from the top half of the first story up to the peak of the roof, must be collected by the walls and roof and transferred through the gable end walls to the foundation. Since there are two gable ends, each gable end carries half of the wind load.

The ability of the gable end walls to resist the resulting wind force, applied at

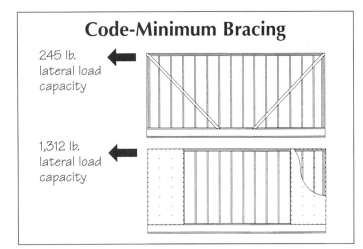

Figure 1. When nailed with 6d common nails 6 inches on-center on the edges and 12 inches in the field, a pair of 7/16-plywood braces has more than five times the strength of let-in 1x4s (assumes S-P-F faming lumber).

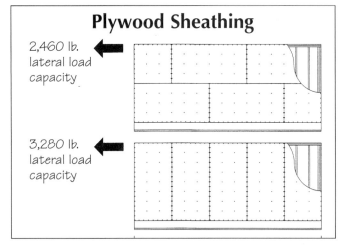

Figure 2. These lateral capacities assume 7/16 sheathing, S-P-F framing, and 6d nails 6 inches o.c. on the edges and 12 inches in the field. The vertically installed plywood is stronger because all panel edges fall on framing.

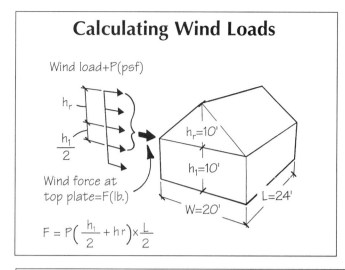

Figure 3. *In this example, the wind exerts a load in pounds per square feet against the side and roof of the building. Assuming the walls are securely bolted to the foundation, the load from the bottom half of the wall is transferred directly into the foundation. The load from the top half of the wall all the way to the roof peak is applied as a force at the top plates of the gable end walls.*

Table 1. Recommended Shear (lbs. per lin. ft.) for APA-Rated Sheathing Nailed to S-P-F Framing

Nominal Panel Thickness	Nail Size (Common or Galv. Box)	Nail Spacing at Panel Edges[1] 6"	4"
FULLY BLOCKED SHEAR WALLS			
3/8"	6d	164	246
3/8"	8d	188.6	295.2
7/16"	8d	209.1	323.9
15/32"	8d	229.6	352.6
UNBLOCKED SHEAR WALLS			
3/8"	6d	123[2]	
3/8"	8d	147.6[2]	
7/16"	8d	162[2]	
15/32"	8d	176.3[2]	

[1] Assumes 12-inch spacing on intermediate supports
[2] 16-in. max stud spacing

Adapted from page 42 of APA's Residential & Commercial Design Guide (available from APA-The Engineered Wood Association, 206/565-6600), and page 7 of APA Research Report 154, Structural Panel Shear Walls.

the top plate, is largely dependent on the lateral shear strength provided by the plywood sheathing. The lateral force to be resisted at the top plate of each gable end is:

$$F = 21 \text{ psf} \times \left(\frac{10 \text{ ft.}}{2} + 10 \text{ ft.}\right) \times \frac{24 \text{ ft.}}{2}$$
$$= 3{,}780 \text{ lb.}$$

Notice that neither of the walls shown in Figure 2 would have a design shear capacity large enough to resist this force, so a stronger wall is required. Table 1 is drawn from APA research on shear walls and diaphragms and shows how increasing the number or size of nails, or the plywood thickness, can increase the lateral design strength of the wall. (Note that the table is simplified and applies only to 2-by S-P-F lumber and other lumber of similar density).

Using the table, you can see how the design strength of the plywood sheathed walls is calculated. For horizontally applied plywood, without blocking at the horizontal plywood joints ("unblocked" shear walls), the shear strength is calculated by multiplying the length of the wall by the shear value for 6d nails:

20 ft. x 123 lb./ft. = 2,460 lb. capacity

Note that the shear values are limited by nail size, not by plywood thickness. With 6d nails, 3/8-inch plywood values apply even if you use 7/16-inch plywood. But if you use 8d nails with 7/16-inch plywood, the lateral capacity goes up to 3,240 pounds (20 ft. x 162 lb./ft.). This still isn't strong enough. The easiest solutions are to apply the plywood vertically (for a 10-foot-high wall, you'll need special-order 10-foot-long plywood) or to provide horizontal blocking at the joints and increase the nail size to 8d, which results in a 4,182-pound shear capacity (20 ft. x 209.1 lb./ft.). Or you can add blocking and use 6d nails spaced 4 inches on-center on the edges to get a 4,920-pound capacity (20 ft. x 246 lb./ft.).

This is a simplified example, meant only to give an idea of how plywood sheathed walls can be used to resist lateral loads. Most real world designs are more complicated and can involve varied building shapes and wall openings along with uplifting and/or overturning forces. Imagine putting a 16-foot-wide garage door in the gable end of the example above; there would only be 2-foot long plywood shear walls left to resist the lateral loads, and they would be subject to large overturning forces. Or suppose the building had a second floor with another 10 feet of height; that would add another 2,520 pounds to the lateral force at the first-story top plate, for a total of 6,300 pounds. These and other common situations might call for a more sophisticated structural design.

Don't underestimate the strength of plywood sheathing for resisting lateral loads. But remember, the strength is dependent on good nailing, especially at the panel edges. If you are trusting your least experienced carpenter to install sheathing, make sure you inspect the nailing before roofing and housewraps are applied. Otherwise, you may be sacrificing building strength without knowing it. And even if the plywood is properly nailed, the entire wall (or diaphragm) must be adequately anchored to the supporting structures. Often the code-minimum anchors will be adequate, but more complicated geometries and loadings may require specially designed components and connections, particularly when shear-wall or diaphragm action is being relied upon.

By Philip Westover, P.E., a structural engineer in Winchester, Mass.

Stiffening Garage Door Openings

A popular addition in some parts of the country is a two-car garage with an apartment over it. While this project looks like a complete winner, it has a number of engineering difficulties. The size of the garage footprint is pretty well dictated by the size of vehicles and available garage doors — usually 24 feet wide by 24 to 30 feet long. It can vary a little, but it's a pretty simple building.

Your major design concerns are probably the need to keep an open floor plan in the garage, and sizing the headers over the wide doors. These are usually solved by a main girder down the center and by big headers. You may need a couple of pieces of steel, but there's really nothing serious to worry about here. If you check the tables or do a few simple calculations, resolving all of the gravity loads is pretty straightforward. Then you're ready to build a strong, serviceable, economical building, right?

Not so fast. The obvious, self-evident structural solution won't always do the job. We've been working with only one dimension of the building. But there are loads other than gravity, in other directions beside down. One of these other loads is wind, which acts sideways and sometimes up.

Wind Loading

Wind loading can be a big problem with a garage. As an example, I'll use the conditions in eastern Oregon, where 80-mph gusts are not uncommon. A wind like that exerts a load of 15 psf on the sides of the building. When the wind blows parallel to the ridge, it strikes the gable end. A typical garage with living space above rises up about 18 feet to the eaves, and then another 8 feet to the ridge. This gives a gable end area of 528 square feet, for a total wind load of 7,920 pounds (see Figure 4).

Fortunately, on the gable end, there are no large unsupported areas. The bottom is fixed to the ground, the sides are supported by the eaves walls, and the top is supported by the roof pitches. In addition, the second-story floor deck acts as a large stiffener across the middle.

We have a different situation with the wind blowing across the ridge. The load is a little bigger because there is more total surface facing the wind: the side walls plus half of the roof area. Calculating the load as we did for the gable gives a 9,360-pound total load exerted by wind blowing against the side of the building.

But the real design difficulty comes from there being so little support in the gable end at the garage door openings. The exterior sheathing, which gives a frame wall its strength against racking forces, is interrupted and greatly reduced around the doors. The small area at the top of the door is simply not strong enough to resist a sideways force. This gable gives no support for the eaves side, and if a high gust strikes the side of the building, the whole building will twist and move (Figure 5). The occupants of the apartment will get a very uneasy feeling as the building jumps around an apparent 1/2 inch or so during severe gusts of wind. (Actually, the movement will be less, but it feels like a whole lot more.)

Three Solutions

Well, what do we do about it? Engineering is not just calculations, it's also finding a solution. I have worked with three designs that stiffen the building in the racking dimension: angle braces, steel stiffeners, and diaphragm sheathing (Figure 6).

Braces. Angle braces cut across the corners of the doors at or close to a 45-degree angle. When the door is framed and sheathed and trimmed out it creates an appearance that a lot of people like, and it is easy to build. I designed this once for a barn in Vermont. About a year later, a neighbor who saw the

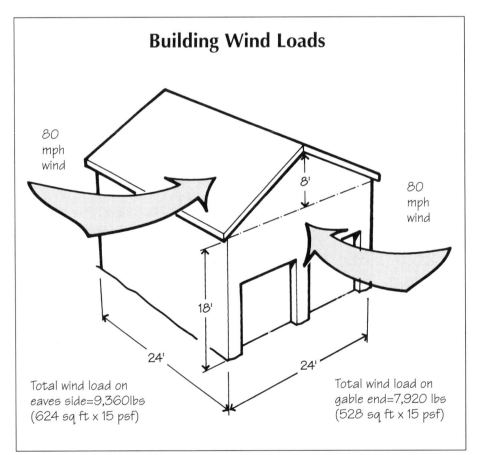

Figure 4. *An 80-mph wind exerts a 15-psf load against a building. For the building shown, this would mean a total load of 7,920 lb. for a wind blowing against the gable, or 9,360 lb. against the eaves side.*

Racking Under Wind Load

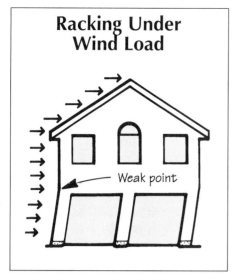

Figure 5. *A typical garage with living space above will rack under heavy wind loads if the gable end is not adequately braced.*

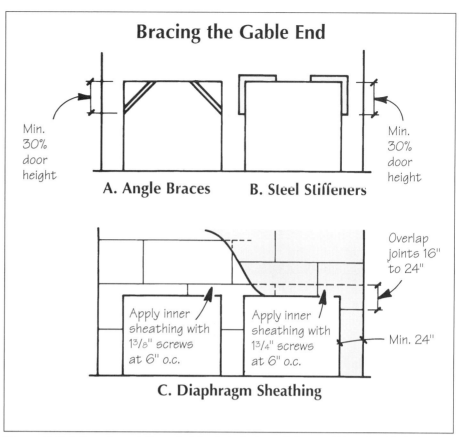

Figure 6. *Solutions for resisting wind loads include angle braces (A) and welded plate-steel stiffeners (B). Diaphragm sheathing (C), consisting of two layers of 3/8- or 1/2-inch plywood or OSB, provides the strongest wall.*

chamfered corners thought it was neat and wanted chamfered corners on his building. It became sort of a fad and a bunch of buildings in central Vermont all have angle braces on doors and windows, whether they need them or not.

A messy calculation that I won't repeat here (really, I looked it up in a table) shows that for adequate strength, the brace should be about 30% of the way down the upright. On a standard 7-foot-high door this is 25 1/4 inches. The centerline length of the brace is 35 5/8 inches. This says cut the braces out of 41-inch-long sticks. I suggest a double 2x6, spiked together with plenty of 10d nails. A 4x4 or a 4x6 might be tempting, but the brace must be free from checks and splits. This might be hard to find in a 4x6; 4x4s are usually better, but cull for a good piece.

Because this detail steals some of the clear height from the door openings, it might be preferable to use 8-foot-high doors for the extra headroom they provide.

Steel. Ninety-degree steel stiffeners preserve the square appearance of the doorway. They should be about 3/8 inch thick by 3 inches wide by 2 feet long on each leg. Bolt through the jamb and header framing with at least four 3/4-inch bolts in each leg. The framing details have to be worked out for each situation. This looks like a huge piece of steel, but it's really not: To have the same effect as the angle brace, it has to cover the same 30% of the door height.

Diaphragm sheathing. For those builders who want the square openings and prefer not to work with steel, there is diaphragm sheathing. This consists of two layers of plywood or OSB that overlap 16 to 24 inches, screwed to the framing. This is my favorite; the wall is exceptionally stiff against racking, and I don't particularly like the chamfered corners of the bracing. The paneling should be 3/8 or 1/2 inch thick. The panels should also be relatively free from delaminations and voids. This is sometimes hard to find in CDX plywood; most OSB is usually okay.

The inner layer goes up with 1 3/8-inch drywall screws on 6-inch centers; the outer layer needs 1 3/4-inch screws, also on 6-inch centers. The screws should go solidly into the framing.

For the diaphragm to work properly, the exterior jambs of the garage doors must be at least 24 inches wide, as indicated in Figure 6. The diaphragm is actually a whole lot stronger than necessary, but thinner plywood is inadequate for sheathing.

I find this problem interesting because it vividly illustrates that structural design is not simply resistance to vertical loads. Over the years, I've gone through this explanation a dozen times with real structures, both at the design stage and after the fact when the building had problems. One couple who came to me had actually had their entire house built over the garage — which made for a literally shaky foundation. Fixing it was an expensive mess on all fronts: contractual, emotional, and legal.

Try not to forget that there are several dimensions to building loading conditions. They all must be checked out for a safe and serviceable structure.

—*Harris Hyman*

Earthquake Design

On January 17, 1994, the Northridge quake struck in California and caused estimated damage of at least $30 billion, according to the Engineering News-Record. What caused so much damage, and what kind of building codes are supposed to ensure structures that can withstand earthquakes? Is it realistic to expect a building or a major structure, such as a highway bridge, to survive an earthquake? When can we say that a structure has survived or failed?

Most of us do not understand the discussions of earthquake construction because the engineers talk in terms of horizontal shear from lateral loading, or vertical shaking, or shear damage and flexural damage. What are they talking about? And more important, what can they do about it?

It can be assumed that any existing structure in a major metropolitan area in southern California will someday be subjected to an earthquake. The design codes are not meant to protect them from minor damage. Windows may break. Walls may deform. Buildings may shake. The codes are meant to ensure the survival of a building's occupants.

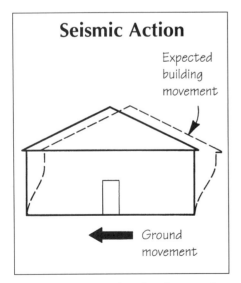

Figure 7. *In an earthquake, the ground under a building moves sideways, exerting potentially huge lateral forces on the structure. Assuming the building is properly anchored to its foundation, it will rack as shown unless properly braced.*

Even if a building must be rebuilt, if the occupants survived, the building code did its job. In terms of dollars and the financial disaster to landlords, this success might fall a little short of the mark. However, please remember that building codes specify only the minimum acceptable requirements.

The major problems are easy enough to describe but very difficult to solve. We want the foundations to hold together. In a building, that translates into more reinforcing and ties in the foundation. We want buildings to resist any form of deformation that would allow them to collapse. That means square walls should remain square and not distort (see Figure 7) and building stories should not collapse.

The three most common and economical ways of bracing building walls for earthquake protection are shear walls, diagonal bracing systems, and moment-frame systems (Figure 8).

Shear Walls

To understand how a shear wall works, stand a book on your desk in front of you. Set it so you can read the title on the front of the book. Try to push the top of the book to your left, but do not allow the bottom of the book to move. If you used a good book, you have a good shear wall; if you used a small magazine, you made a low-quality shear wall. If you place another book at right angles to the first one, you can see how a shear wall system in a building works to protect the occupants.

Until recently, drywall has been an acceptable material for interior shear walls in apartments. But because several people in California died in apartments with drywall walls, the code is apparently about to be changed to require plywood or OSB interior walls. The intent is to make a stronger shear wall that will offer more resistance to deformation. Exterior shear walls are also designed to withstand distortion. And because exterior walls need to have more resistance to shear deformation than the interior walls, materials such as concrete or reinforced masonry are often used in their construction.

Diagonal Bracing Systems

A square is not a very stable geometric shape. It is easy to distort. Think of a picture frame with no glass. The only force available to keep it square comes from the little nails in the corners, and even they don't do the job. Pick up the frame, and its own weight will distort it. On the other hand, it is difficult to distort a triangle, which is the most stable of all geometric shapes. Place two diagonal braces across a square, tying the four corners together diagonally. The square has now been made into four triangles and has become much more stable because of the diagonal bracing. That is exactly how a wall should be braced to protect it from an earthquake.

Unfortunately, the diagonal let-in braces (1x4s or metal straps) that residential builders are accustomed to using are not effective in earthquakes. Wood-frame builders are better off using plywood shear-wall construction.

Moment Frames

Think about the picture frame with no glass again. If the corner connections are fixed, say by placing a small angle iron in each corner, it will be more difficult for each corner connection to move out of its right-angle shape. Moment-frame systems are built in this fashion, with a strong connection at each joint in the frame to make it difficult for the angle between the frame pieces to change. If the angle of the connections cannot change, the overall shape of the frame cannot change, and the building will be protected. This type of construction requires placing steel angles at the corners of door and window framing and bolted steel plates at post-beam connections, or even embedding a welded steel frame within the framing.

Other Problems

In many of the earlier earthquakes in

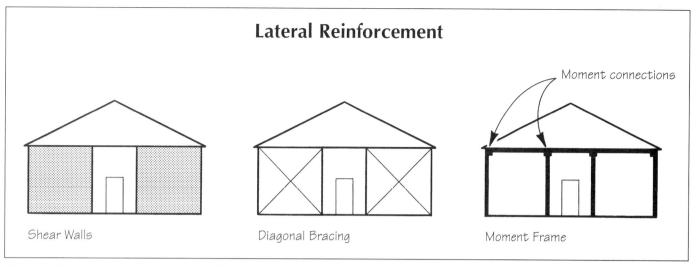

Figure 8. *Three common ways to provide lateral strength are shear walls (left), diagonal bracing (center), and moment frames (right).*

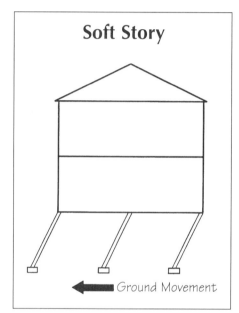

Figure 9. *A structurally weak story in a multistory building, such as a parking garage under an apartment building, can collapse in an earthquake, causing the upper stories to come down on top.*

this country, buildings simply slid off their foundations because they were not connected to them. This was especially true of houses. Today, buildings — even houses — are bolted to the foundation to keep them from sliding off.

Soft story. Some buildings have a floor that designers call a "soft story" (Figure 9). Usually, this means there is one story with fewer interior walls than the others, so that one story has reduced interior shear strength, resulting from the smaller number of shear walls. This condition can be caused by features such as an open story for a parking garage (sometimes made worse by too many garage doors on one side of the building) or a building with an open interior space for a lobby or stores on the first floor, with offices above. Such buildings sometimes act like an upside-down pendulum in an earthquake, with the result that the buildings tend to collapse into the soft story. They can also twist at the soft-story level, with essentially the same result: All floors above the soft story collapse into the soft story.

The solution lies in innovative designs for buildings that eliminate the soft story. For instance, garages can be built underground, instead of at ground level. Hotels can be designed with a tower for the rooms, but with the open lobby located in an adjacent yet still connected building beside the tower instead of under it. The guests think they went to the end of the building to enter the elevator, but they really left the lobby building and entered the tower, which contains the rooms and has no soft story.

Liquefaction. Buildings have collapsed in earthquakes because the ground under the foundation literally turned to liquid. This condition, known as liquefaction, results from too much ground water in the unconfined sandy soil at the time of the vibration from the earthquake. The soil suddenly consolidates and moves downward from the earthquake's vibration, and the liquid in the soil moves upward, resulting in liquefaction of what had been relatively firm soil under the foundation. The water has to be able to move for this to happen. This phenomenon can be prevented by confining the soil and the accompanying upward movement of the water. The water and sandy soil are usually confined by placing a layer of less porous soil over the sandy layer and then constructing the building on the more stable material.

Building configuration. When a building has a floor plan that is not symmetric, the shock waves from an earthquake will cause different stresses in different parts of the building. An L-shaped building might have one wing parallel to the shock waves and the other perpendicular to the shock waves. This kind of floor plan can create additional stresses and torsional forces that add to the potential damage to the building. The *Uniform Building Code* now recognizes the importance of building configuration and establishes design requirements that result from the configuration.

By Richard Mayo, P.E., a certified professional engineer, and director of Graduate Studies at the Del E. Webb School of Construction at Arizona State University. This article was adapted with permission from CFMA Building Profits *magazine.*

Shear Wall Construction Basics

Ever since the 1989 Loma Prieta Earthquake, which registered 7.1 on the Richter Scale, seismic codes for new construction in California have become stricter. It's now common to see engineered seismic bracing details even on plans for a simple addition. The most common detail is the shear wall, a framed wall with structural sheathing, designed to resist the lateral forces that an earthquake exerts on a building (Figure 10). Any wall properly sheathed with plywood has considerable shear strength, but the term "shear wall" is reserved for walls designed by engineers. A shear wall incorporates special construction details and materials specified by code to resist the forces that rack a building during an earthquake or high winds (Figure 11).

In a typical newly built two-story home in California, often all the exterior walls are shear walls and there may be a few interior shear walls as well (see "Typical Shear Wall Details," page 63). Exterior shear walls are usually sheathed with plywood; interior shear wall surfaces may be either plywood or drywall. Shear walls differ from conventional framed-and-sheathed walls in several ways:

- Both interior and exterior shear walls have rigorous nailing schedules. Nail spacing may be as close as 2 to 3 inches at panel edges.
- All edges of surface panels must be fastened to solid blocking. At panel joints where edge nail spacing is 2 or 3 inches, the code typically requires a minimum 3x framing member.
- All shear walls must be mechanically attached to the foundation. The plans usually specify metal hold-downs, metal straps, closely spaced anchor bolts, or some combination of the three.
- Shear walls usually tie in at the second floor and the roof with blocking and/or metal connectors.

Shear Wall Materials

Materials for shear walls are the same as for conventional framed walls with a few exceptions.

The shear wall panel. The panel is the core element of a shear wall. Usually it is structural-grade plywood from $3/8$ to $3/4$ inch thick. Often you see $3/8$-inch plywood on a single-story structure and $1/2$-inch or thicker plywood on two-story and three-story buildings.

The specs may call for either "Structural I" or "Structural II" grade plywood. This is an important distinction, because Structural I is stronger, and more expensive, than Structural II. Other structural grades are also allowable, but make sure the grade stamp matches the specs on the plans.

Oriented strand board (OSB) can also be used in shear walls, but since it has less shear strength than plywood, it isn't often used on the West Coast for seismic bracing.

Another option for interior shear walls is drywall. Because of the cost savings, many engineers specify drywall shear walls where they can. Drywall shear walls should have their sheets running horizontally, with blocking along all edges. A 4-inch-on-center nailing schedule is typical. Drywallers can use either nails or screws, but I recommend screws because they do less damage to the board. Where the hammer fractures the drywall around the nail head, it reduces that nail's shear value.

Plywood length. Buying the right length plywood will help speed production of shear walls. In many cases, if you order 4x8-foot sheets, you may be making extra work for yourself. In California, we typically run our shear panels vertically. By using 4x9 and 4x10 plywood, we can sheathe from the mudsill to the middle of the second-floor rim joist with a single sheet.

However, if you are framing a single-story home on a slab floor, 4x8 sheets may work fine. You can also use 4x8 sheets where the plans require you to run the plywood horizontally. This gives the wall additional strength against racking because the majority of the layers in the plywood have the grain running perpendicular to the studs. But it means that you'll have to install horizontal blocking at the joints between sheets.

Framing material. Shear walls are typically framed with 2x4s or 2x6s 16 inches on-center. When framing shear walls, the main thing to keep in mind is that all plywood edges must be blocked. This is because shear forces are transferred at panel edges. The blocking and close nail spacing required along the edges ensures that the plywood remains stiff under shear loads and can resist racking. Tests conducted by the American Plywood Association show that without edge blocking, shear panels under load tend to buckle and fail.

Shear walls with especially tight nailing schedules usually require a 3x stud where two sheets meet. The wider stud can stand up to the many nails required at panel edges without splitting. Since 3x studs are not widely available, I typically

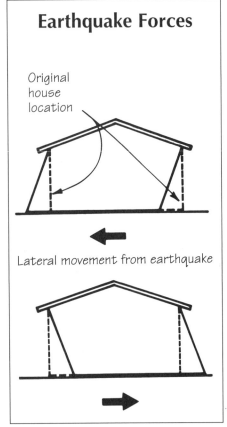

Figure 10. *An earthquake produces lateral movements in the ground that cause a building to snap back and forth. This stresses joints and connections and causes the building to shift and sometimes collapse.*

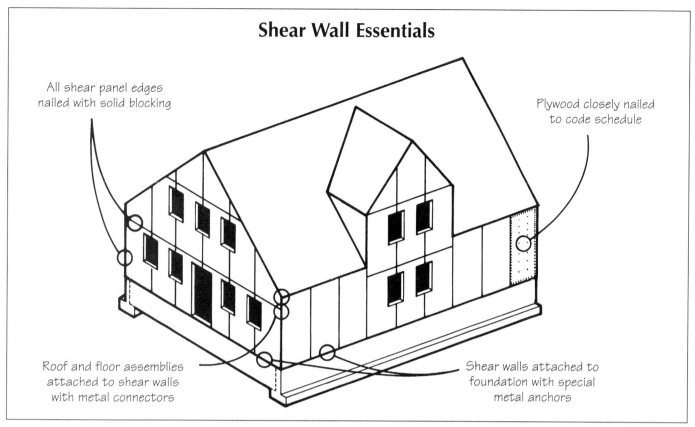

Figure 11. *Plywood shear walls resist an earthquake's lateral forces by preventing racking. Shear walls get their stiffness from close nailing patterns, solid blocking at all panel edges, and hold-downs and metal connectors at the foundation and roof.*

use 4x4s instead. If you run your sheets horizontally, the edge blocking at joints often must be 3x or 4x material as well.

Four-by-fours (or doubled 2x4s) are also usually required for attaching the hold-downs that connect to foundation anchor bolts. These are typically located at building corners, on the sides of large door or window openings, and at the ends of a shear wall.

Nails. Using the right nails is one of the most critical aspects of shear wall construction. Tests have shown that most shear wall failures occur because the nail head rips through the plywood. The size of the nail head, as well as the nail's diameter, is therefore important in determining a nail's shear value. Engineered plans should clearly specify the type and length of nail required.

The 1988 Uniform Building Code (UBC), which we follow in California, recognizes three types of nails in its shear wall calculations: common, galvanized box, and galvanized casing. Either common nails or galvanized box nails are good for installing plywood sheathing. Though box nails have a slightly smaller diameter shank, their heads are the same diameter as common nails and both types are assigned equal shear values.

For plywood siding, where either common or box nails would be too noticeable, the UBC allows the use of galvanized casing nails. These have a smaller head and consequently lower shear values than common or box nails. If you use casing nails you may find that you have to tighten the nailing pattern to make up the needed shear resistance.

Pneumatic nails. With the widespread use of pneumatic nailers has come a whole slew of different types of fasteners, which only adds to the confusion when it's time to nail off a shear wall. Fortunately, the code organizations have addressed this situation in National Evaluation Service Committee Report No. NER-272, available from the International Code Council (5360 Workman Mill Rd., Whittier, CA 90601; 800/284-4406; www.iccsafe.org).

In addition to the three types of nails mentioned above, NER-272 gives shear values for four kinds of pneumatic fasteners: round head nails, modified round head (or "clip head") nails, T-nails, and staples (Figure 12). The report also includes ring-shank and screw-shank nails. NER-272 is an important document for anyone having to decide what fasteners to use in a shear wall.

However, don't assume that because the code allows a nail type your inspector will approve it. For example, some inspectors will not allow clip-head nails, because of their reduced head size.

Metal framing connectors. Shear wall plans usually require a variety of metal connectors to tie the major framing members together. I usually buy these by the case at the beginning of the job. These clips, which install with several 1 1/2-inch joist hanger nails, strengthen the connections where the shear wall ties into floor and roof assemblies. Various hold-down straps and plates are also common for tying the shear wall to the foundation (Figure 13).

Estimating Shear Walls

When estimating materials and labor for shear walls, study the blueprints meticulously. As many framing contrac-

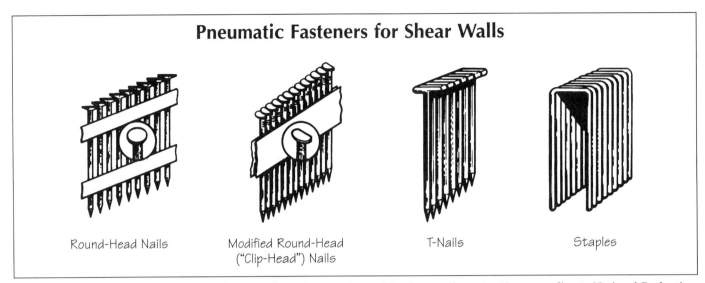

Figure 12. *The four kinds of pneumatic fasteners shown here can be used in shear wall construction, according to National Evaluation Service Committee Report No. NER-272.*

Figure 13. *Two types of metal hold-downs are common in shear wall construction. Right-angle plates (left) bolt onto threaded foundation anchors and attach to 4x4 posts in the wall. Hold-down straps (right) are embedded in the concrete foundation and nailed to 4x4s through the sheathing.*

tors have found out the hard way, you can't bid a shear wall the same way you bid sheathing. For one thing, you can trust a carpenter's helper with a sheathing job, but a shear wall needs the attention of an experienced carpenter.

But even with skilled carpenters, shear wall construction is labor intensive. Not only are there a lot of nails to drive, but the hardware for connecting shear walls to the foundation and roof takes time to install. Allow for these tasks when estimating.

When estimating interior shear walls, if the shear-wall panel is specified as plywood, and is shown as only part of the interior wall length, plan on sheathing the whole wall. This is so the interior wall treatment can be applied to a flush surface. Also remember that interior shear walls often require extra underfloor blocking that is better off completed before the subfloor goes on. Finally, it is to your advantage to nail on the plywood immediately after framing the wall so that the shear panels won't be interrupted by intersecting walls.

Construction Tips

While I prefer to nail on plywood while the wall is on the deck, this is often not possible when installing shear panels. Vertical panels have to extend from the mudsill over cripple walls, band joist, and wall framing. Also, shear panels have to tie intersecting walls and floor levels together. So, in most cases, the shear panels have to be installed after the framing is erected. This can complicate the sheathing process. Here are a few tips for an efficient job.

Measuring panels. To determine the panel length for a single-story gable wall, pull your tape from the mudsill to the center of the double top plate. For the first story of a two-story platform-framed house, pull your tape from the mudsill to the center of the rim joist or joist blocking.

Cutting. Once I have determined the panel length, I gang-cut several sheets at a time. Working on a full stack of plywood, I set the depth of my circular saw blade so that it penetrates several sheets and barely scribes a cut into the last one so I can use it for the next cut.

Layout. Before I begin sheathing, I mark the stud locations on the foundation with a pencil or crayon, being careful to mark hold-down post locations, which are often placed off of the standard 16-inch on-center spacing and can be missed when nailing the plywood.

Typical Shear Wall Details

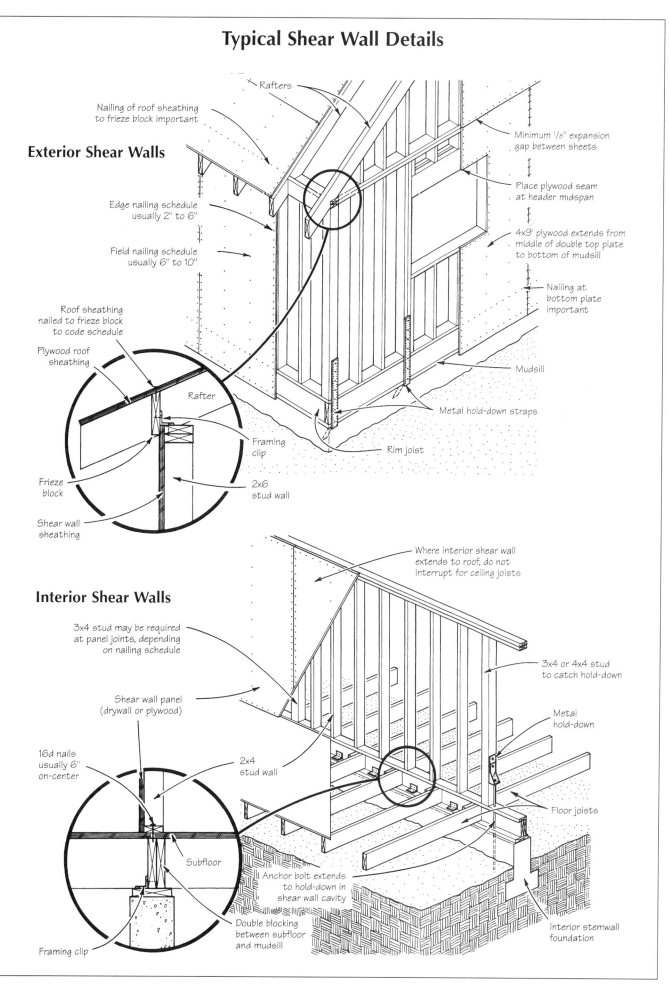

Shear Wall Construction Basics

A poorly placed seam can significantly reduce the strength of a shear wall, especially at openings, which are weak points. This is most critical at large openings, like garage doors or a set of windows. Therefore, I pay close attention to where the seams end up on the wall, making sure, for instance, that seams fall at the midspan of window and door headers, not at the edge of the opening.

Placing panels. Starting at a corner of the building, I partially drive a few 16d nails in between the mudsill and the concrete. This gives me a ledge to rest the sheet on. I also use two 16d nails tacked between sheets as spacers to get the required 1/8-inch expansion gap.

I tack the sheets in place with 8d nails, making sure I work the waves out of the plywood. If you fail to work the waves out, the plywood may not make full contact with the studs, which could result in a shear wall failure.

Before nailing off the panels, I walk around the building with a crayon marking the nailing schedules of the different shear walls. This not only helps the carpenter who ends up nailing off the shear wall, it also shows the inspector that we paid attention to the plans.

Nailing off the plywood. Even though nailing off a shear wall seems mindless, doing the job right demands careful attention. To pass inspection, a shear wall must have a consistent nailing pattern, according to the code schedule. Also, by code, nails must be installed with the heads flush with the surface of the plywood. This is easier said than done, especially with pneumatic nailers, which are set at the factory to countersink the nail head by as much as 1/8 inch. Most carpenters try to control nail head depth by adjusting the air pressure, which is a finicky process. I try to adjust air pressure to leave the nail heads 1/4 to 1/2 inch above the plywood surface, then finish driving them with a hammer.

Another solution is to adjust the driver in the nail gun. Some models have a depth adjustment right on the gun. For models that don't, however, you can return the gun to the manufacturer to have the driver shortened so it drives the nail flush. This is a permanent modification, but if you nail a lot of shear walls, it's probably worth dedicating a gun for this purpose.

When using staples, the crowns must be installed parallel to the long dimension of the framing member. That means when you're stapling plywood to studs, you have to hold the stapler horizontally, not vertically.

Finally, be careful nailing at plywood edges and keep nails 3/8 inch back from the edge. If you get too close to the edge, the nail will split the edge off the sheet and undermine the strength of the shear panel.

By Jim Hart, a foreman in Mountain View, Calif., and a former editor with The Journal of Light Construction.

Installing Seismic Framing Connectors

In the early '80s, new codes in California began calling for a variety of metal brackets and straps to help resist earthquake damage to wood-framed houses. These metal connectors work together with plywood shear walls to hold the framework of a house together at the weakest points of the structure (see "Shear Wall Construction Basics," page 60).

Though very much needed, metal connectors can be a nightmare for framers. Depending on the type of connector used, special nails, machine bolts, washers, and threaded rods may be needed. When nailing together and lifting walls, the last thing a framer wants to do is break the rhythm of the job to drill bolt holes and ratchet down a lot of nuts. Instead, many carpenters choose to ignore these connectors until the day before the framing inspection. I discourage this approach. Connectors that are poorly planned out require hours or even days to install properly.

I lead remodeling jobs for a general contractor in the San Francisco Bay area. On a typical 1,500-square-foot, two-story addition, we may be required to install as many as 24 foundation connectors, not to mention straps over the second-floor rim joist, ridge connectors at every rafter and scores of framing clips (Figure 14). Even though the total material cost for each foundation connector is less than $20, the head scratching and labor it takes to get them properly installed can add up to more than $100 each.

Over the past several years, I've familiarized myself with the different manufactured connectors available. I've also learned a few tricks, the hard way, that have helped me install these connectors without interrupting the flow of the framing. While there is a wide variety of metal clips and hangers, I will focus here on the connectors that are most difficult to install.

Reading The Plans

The architects and engineers we work with will usually indicate on the structural page of the blueprints the metal connector location and model number. Sometimes an installation detail is included, but I've found more often that it is up to the framer to figure out how to put them in.

Before pouring the foundation, I use a colored marker to highlight on the plans all the specified metal hardware. I sometimes color-code them — red for foundation connectors, yellow for wall connectors, and blue for roof connectors. Color-coding helps remind me of things like, "The red ones go in before the pour" or "The blue ones need to be in before the roof is covered with felt." I also make a list of all the hardware we need and get everything on site as soon as possible. Even a large building supply warehouse may not stock everything, so ordering hardware well in advance of the pour will prevent slowing down the job.

I also number each of the foundation connectors on the plans. Numbering the connectors helps to make sure they aren't forgotten from the estimate and the order list. Starting at one corner of the foundation plan, I mark the first connector as (1), then move clockwise along the perimeter of the foundation, counting off all the connectors. I finish up the count by numbering the interior metal connectors.

Foundation Connectors

On a typical two-story house, there are three weak points that often require metal connectors: where the foundation meets the first-story wall, where the first-story wall meets the second-story wall, and where the second-story wall meets the roof. By far the most time-consuming and difficult to install are the foundation connectors.

Hold-downs. Two types of metal anchors are commonly used to strengthen the connection between wall and foundation. The most common is a large L-shaped bracket that usually sits on the mudsill and attaches with two or more machine bolts to a wood stud or post (Figure 15). The bracket connects to an anchor bolt that is embedded in the concrete stemwall or slab.

On site we call this type of connector a hold-down, after Simpson Strong-Tie's brand name "holdown." Kant-Sag also markets a similar product under the brand name "anchor down." Depending on which brand an engineer is familiar with, hold-downs are referenced on plans as either HD or AD, followed by a number that indicates the size. The two brands are similar, but you should clear it with the engineer if you use the brand not referenced.

Engineers will also specify the type of anchor bolt to use with a hold-down, the rod diameter, how deep it is to be embedded, and how it terminates in the concrete (Figure 16). There are four types of anchor bolts I install regularly:

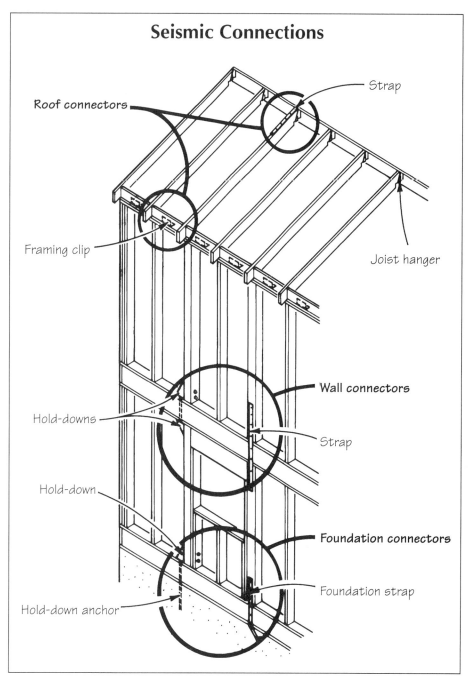

Figure 14. *On a new two-story house on the West Coast, a variety of metal connectors are used to tie the structure together and strengthen weak connections against earthquake uplift.*

Installing Seismic Framing Connectors

Tips for Installing Hold-Downs

About 11 years ago, I installed my first set of foundation connectors. These pieces of steel are much different than your garden-variety anchor bolts. They commonly use 1-inch-diameter or larger threaded rods and heavy-duty bolts. As a framing contractor, I wasn't used to dealing with such massive hardware. Without the right tools and experience, my early hold-down installations were less than pleasant, and rather costly. Hundreds of hold-down installations later, they are still far from fun, but I have refined my methods so at least installing this hardware doesn't cost me a fortune.

Tools. To deal efficiently with these beasts, I carry a few tools that help to make the job go more smoothly:

- A set of 6-inch-long high-speed metal bits $1/2$ inch to $1 1/4$ inch in diameter, in $1/8$-inch increments
- A set of self-feeding bits $1 1/2$ inches to $2 1/2$ inches long
- A $1/2$-inch capacity drill (a right-angle drill can be helpful)
- A ratchet with a good selection of sockets (deep sockets are helpful)
- A couple of good adjustable wrenches
- A copy of the hardware manufacturer's catalog

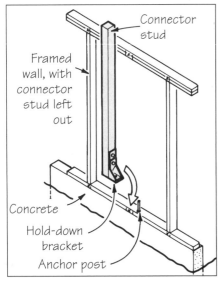

When framing walls, Dunkley recommends leaving out the studs that connect to hold-downs and installing them after the walls have been tipped up. This allows you to accurately attach the hold-down bracket to the stud, then place the stud and bracket over the embedded anchor bolt.

Working with the concrete sub. I rely on the concrete subcontractor to locate and install the anchor bolts for the hold-downs. Most of the subcontractors I deal with can read the plans well enough to place the anchor bolts in the right places. But for complex jobs, I always make a trip to the site before pour day and locate the hold-downs for them.

Unfortunately, one thing I can't control very well is how plumb the anchor bolts are and how far above the concrete they extend. There's nothing more aggravating than walking the perimeter of a newly-poured foundation viewing bolts that look more like crocodile teeth than fasteners.

If, when framing, I discover the anchor bolt is too short, I make use of a threaded-rod extender, which is a threaded sleeve that allows me to add a short piece of rod to the embedded piece without compromising the strength of the connection. If the anchor bolt ends up long, I either raise the hold-down bracket off the mudsill (which also has no impact on design load values) or cut the bolt with a Sawzall. Don't use bolts that are too long, as they prevent you from getting a socket on the nut, forcing you to use an adjustable wrench.

Installing the bracket. Depending on the size of the hold-down, between two and five bolts are needed to make the connection to the wood member. The length of the bolts depends on the dimension of the wood. As a general rule, add $1 1/2$ inches to the width of the wood member to find the length of bolts needed. This added length will cover the thickness of the washer and the hold-down bracket, and leave enough thread left over to get a nut on.

My system for hold-down installation is fairly simple and works for small or large anchors. It also accommodates, with little hassle, anchor bolts that are poorly placed.

The catalog, which I use like a bible, gives me the distance between the anchor bolt and the connector stud. When I lay out the plates, I mark the stud location and drill the hole through the bottom plate for the anchor bolt, but I leave out that stud. After the wall is lifted in place, I come back and temporarily place the stud in its location and dry-fit the hold-down on the anchor bolt. Depending on the type of hold-down, the bracket might rest flush on the bottom plate, or it might have a built-in gauge that lifts it off the bottom plate a specific distance.

I then trace the holes for the machine bolts on the stud. This eliminates the possibility of making a measuring error and drilling my holes in the wrong place. I remove the post, comfortably drill the holes, and bolt the hold-down bracket to it. With the hold-down fastened, all I have to do is place the stud back in the wall over the anchor bolt, toenail the stud ends, and cinch down the hold-down nuts (see illustration). This method of preinstalling the bracket eliminates the aggravation of tightening the stud bolts when the anchor bolt is in the way.

At king studs and outside corners. Hold-downs are often specified at king studs of window and door openings and on outside corners. In these cases, I am forced to nail these studs in place before I lift my wall and deal with the hold-downs later. After the wall is framed, I position the hold-down bracket on the stud and drill through the holes in the bracket. Sometimes I have to Sawzall through the nails holding adjacent studs to the plate and temporarily move them out of the way to give some room to drill the holes.

On an outside corner, I recess the bolt head about $3/4$ inch so it won't bulge the exterior siding or create a weak point in the stucco. But be careful: Don't count the $3/4$-inch recess as part of the required stud thickness. If you need to recess for the bolts, add extra thickness to your stud. Plan for this before the concrete guys set the anchor bolts.

Before insulation, I make a point of going around and tightening all the hold-down nuts. When working with green or wet lumber, the studs and plates will shrink and I'm always surprised whenever the nuts I cinched down tight with a ratchet a couple of weeks ago can be turned by hand.

By Don Dunkley, a framing contractor in Cool, Calif.

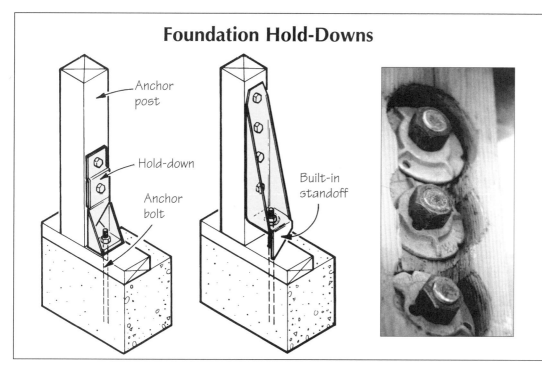

Figure 15. *Foundation hold-downs form a strong connection between a 2x or 4x wall stud and a concrete-embedded anchor bolt. One type of hold-down (left) rests directly on the mudsill or bottom plate; another type (center) has a built-in standoff that lifts it off the plate far enough so the bolt holes don't split out the end of the post. When installing a hold-down at an outside corner (right), be sure to countersink the nuts so they don't bulge the siding.*

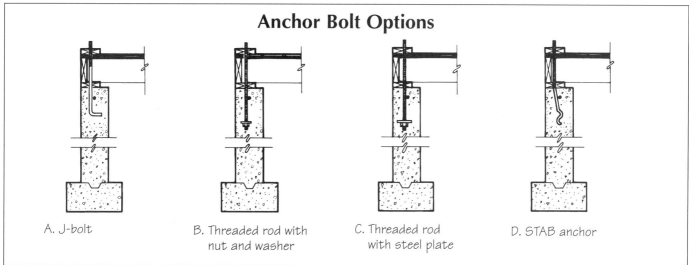

Figure 16. *There are four types of anchor bolts for use with foundation hold-downs: (A) A threaded rod with a 90-degree bend in the end, like a large J bolt; (B) a threaded rod or long bolt with a nut and washer on the end; (C) a threaded rod with a 1/4-inch-thick steel plate bolted to the end; and (D) a manufactured anchor bolt, such as Simpson Strong-Tie's STAB anchor.*

- A threaded rod with a 90-degree bend in the end, like a large J-bolt
- A threaded rod or long bolt with a nut and washer on the end
- A threaded rod with a 2- or 3-inch-square, 1/4-inch-thick steel plate bolted to the end
- A manufactured anchor bolt, such as Simpson Strong-Tie's STAB anchor

Threaded rod diameters range in size from 5/8 to 1 1/2 inches; the most common are the 5/8-, 7/8- and 1 1/8-inch sizes. STAB diameters come in either 5/8-inch or 7/8-inch diameters. The STAB anchor provides some advantages over the threaded-rod options. For one thing, it is plumb above the concrete but angles away from the outside of the foundation within the concrete, which adds strength. Also, engineers like STAB anchors because the load values have been tested, whereas threaded rod strength is based on calculations. I like them because the angle makes it easier to avoid hitting the embedded horizontal rebar.

The trick to installing a hold-down is getting the anchor bolt embedded in the concrete in the right place. This usually means coordinating with the concrete sub (see "Tips for Installing Hold-Downs"). But on additions we usually do our own concrete work, so we have to deal with placing the anchor bolts, too.

If a hold-down is located in a wall without any door or window openings, positioning the anchor bolt is usually not as critical — you just have to make sure you position it in the center of the wall. It's when a hold-down needs to be positioned to either side of a wall opening or at a corner that proper placement becomes crucial.

I increase the chances of proper placement on a building corner by making a

Installing Seismic Framing Connectors

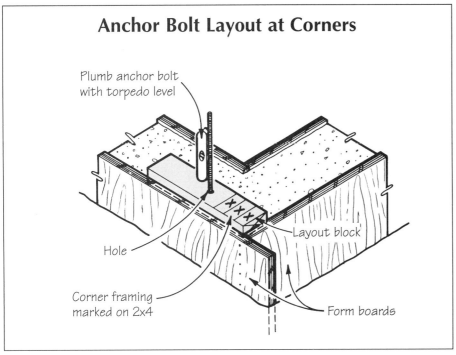

Figure 17. *To ensure the exact position of a corner anchor bolt, the author uses a layout block with a hole to hold the anchor bolt in the foundation form. The block is also useful for hold-downs at door and window openings.*

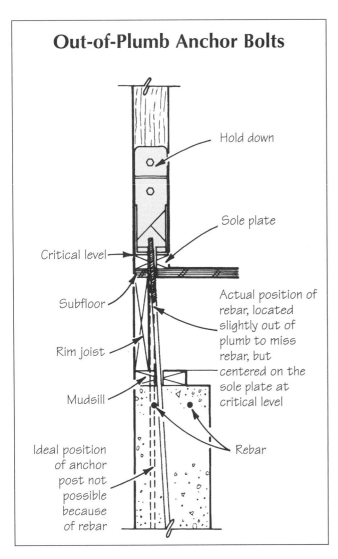

Figure 18. *You may install an anchor bolt slightly out of plumb if the rebar is in the way. This will not affect the strength of the hold-down, but you must make sure the top end of the bolt is still centered where it comes through the sole plate.*

layout jig — a simple block with a hole to position the anchor post (Figure 17). If possible, I wire the anchor post directly to the rebar in the right position before the concrete pour. But if the rebar isn't in a workable place, I lay the anchor posts near where they are to be placed so they won't be forgotten.

Sometimes I've found it impossible to install an anchor post plumb. If the rebar happens to be in just the wrong place, you have little choice but to place the anchor slightly out of plumb. But as long as you take into consideration the hold-down's final placement, you can get away with being out of plumb. This requires picturing the framing exactly, and juggling a tape measure and blocks if necessary to mock up the location of the rod (Figure 18). You have to be fairly accurate—engineers and inspectors don't like seeing a dramatically bent anchor bolt because it weakens the metal.

Foundation straps. The other common type of foundation connection is a concrete-embedded metal strap that nails or bolts to the face of a stud member (Figure 19A). We generally call these "foundation straps" on site. Simpson and Kant-Sag both refer to these as PA series anchors, and they are referenced on the plans as PAHD, MPAHD, and HPAHD.

Compared to hold-downs, the PA series anchors are easy to install. Because no drilling is usually needed, all studs can be nailed in when lifting walls.

The big advantage with PAHDs is that you don't have to deal with an anchor bolt sticking up through the subfloor. The PA series straps simply nail to the form before the pour. Also, a typical installation only requires a handful of nails. The manufacturers call for 16d common nails to get the full design loads, but you can use 16d sinkers as well — this reduces the design load values by one eighth. (A common nail has a larger-diameter shank and a larger head than a sinker.)

There are some disadvantages with foundation straps:

• Design loads for straps are significantly less than for hold-downs. The largest hold-down can handle up to 17,000 pounds of uplift, whereas the largest

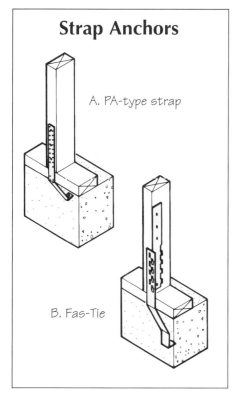

Figure 19. *Foundation straps, like the PA series (A), are nailed to the foundation forms during the pour, then nailed to the framing through the exterior sheathing once the walls are up. A new type of strap anchor, Simpson's Fas-Tie (B), has two pieces. The upper piece nails into the wide face of the stud, then interlocks with the part embedded in the concrete.*

foundation strap in the PA series can handle less than 5,000 pounds of uplift.
- If the concrete isn't vibrated, you may leave surface voids beneath the strap. I make a point to rap the form with a hammer at each of the strap locations to make sure the concrete settles beneath each connector.
- The studs that the straps attach to can sometimes split under the heavy nailing. This will seriously reduce the design load, and you may have to replace the stud with a doubled or 4x member.

Simpson has new strap anchors — the Fas-Tie series — that improve the design by allowing nails or bolts to be installed on the wide face of the stud (Figure 19B). Because you have more nailing, this type of connector generally forms a stronger connection. I haven't yet used this product, but I can see a number of advantages: First, the design load value is comparable to the smaller threaded-rod hold-downs, so in some situations it can replace the more labor-intensive installation of a threaded-rod connector. Second, there's more flexibility with this strap connection because it doesn't require as much space as a threaded-rod anchor.

A disadvantage is that the Fas-Tie can interfere with the plywood sheathing. Either the plywood needs to be lumped over the strap, which might create a visual problem with the exterior siding, or the bottom plate must be notched to position the strap within the wall surface, which might cause a problem with the inspector.

Wall Connectors

There are a variety of wall connectors that are used to reinforce the connection between first- and second-story walls. The most common are hold-downs used in pairs with a large threaded rod between them (Figure 20). When installing these, re-member to align the studs between floors so that the threaded rod misses a floor joist. I drill an oversized hole for the threaded rod to pass through the subfloor. Even if the studs don't perfectly align with each other, as long as the rod is not bent, the threaded rod can be installed a little out of plumb without affecting the design value.

Metal straps of different gauges, widths, and lengths are also common for strengthening the connection between stories. Both Simpson Strong-Tie and Kant-Sag offer a variety of configurations. We frequently use Simpson's MST or ST series. Strap ties require either 16d or 10d commons, but can be replaced by 16d sinkers, reducing design load by one eighth.

When laying out a wall that requires one of these larger straps, I usually use 4x4s as the connecting studs, even if the plans require only a single or double 2x4. With a $1^{1}/_{2}$-inch-wide strap, I have had problems with the nails splitting the 2x4 studs. And over a double 2x stud, a shift of only $1/4$ inch may mean that you nail into the crack rather than into the center of each member. Also, a 4x4 offers a little more flexibility if the studs don't line up perfectly. With 2x4s, if one stud

Figure 20. *Wall connectors come in two varieties. You can use hold-downs bolted together through the floor assembly (top) or nail strap ties into the studs (bottom).*

is bowed, the strap doesn't always center over the meat of the stud.

Many strap ties are nailed onto the exterior of the plywood. To help locate the strap position, I drive a few 8d nails on either side of the upper and lower studs from inside the house. This allows

Installing Seismic Framing Connectors

me to center the strap on the studs to get the strongest attachment.

Roof Connectors

In our area of California, we don't see nearly as much roof hardware as we do foundation anchors. Nevertheless, the engineers do spec some.

Framing clips. In most cases we use metal clips (Simpson A35 or Kant-Sag MP-A1) to connect the frieze block to the double top plate. With a framing clip on every frieze block and the roof sheathing nailed to the frieze block, a strong connection is made between the wall and the roof. It's not difficult to install these. Just be sure you do it before the roof is sheathed.

Hurricane clips. Seldom do we have to install rafter ties, though occasionally we're required to strap a hip to a wall. However, rafter ties are very common in coastal wind zones. Simpson and Kant-Sag make a variety of "hurricane clips" that reinforce the connection between rafters and plates.

There is a temptation when setting rafters or trusses to just wrap the tie over the top of the rafter or top chord of a truss, bang in one nail, and move on down the line. But you must follow the manufacturer's nailing schedule, which includes nailing along the side of the rafter or top chord.

Ridge connections. While we don't see many hurricane clips, we are frequently required to strengthen the rafter connection across the ridge with either joist hangers or straps across the top of the rafters. The joist hangers force the carpenter to make a horizontal notch at the top of the rafter so that the rafter seats firmly in the hanger. Usually the straps simply nail on over the roof sheathing before the roof is felted. But these can be a real pain for a couple of reasons. The number of nails required to hold two sheets of plywood sheathing together over one rafter, plus the nails for the strap, can make mincemeat of a 2x rafter. Also, these straps can easily be forgotten in the mad rush to felt the roof against the weather.

—*Jim Hart*

Hold Down Problems and Solutions

Building shear walls is an everyday affair out here in California, where I work. Shear walls are one of the main design elements engineers use to help wood-frame buildings resist the lateral loads imposed by earthquakes. Shear walls are also becoming increasingly common in coastal zones throughout the U.S., where new high-wind codes often require them. For the carpenter, shear walls require some changes to standard wall framing. The most obvious is that the plywood sheathing gets installed with a lot more nails, especially around the edges. That's the easy part. There are also heftier framing members that have to be included — typically 4x4 posts at the shear wall ends. And there are the hold-downs, those bothersome pieces of hardware that have to be embedded in the concrete foundation and connected to the posts, creating another level of layout and coordination for the framer to worry about.

Nailing off plywood shear panels is easy enough to do right, but hold-downs are a little trickier. Here, we'll look at some hold-down installations that were muffed for one reason or another. But before discussing what's wrong with these installations, let's review how a shear wall is supposed to work.

Shear Basics

When the wind pushes a house or an earthquake shakes it, the force is delivered to the top of the shear wall (see Figure 21). At the bottom, where the wall is attached, there is an equal resisting force in the opposite direction. The "overturning moment" (a moment is a force times a distance) equals the force at the top of the wall times the height of the wall.

So that the shear wall does not overturn, there has to be another counterbalancing pair of forces. This is provided by the hold-down and tension post on one side and the compression post on the other side. (When an earthquake shakes a building back and forth, there is a cycle — first one post is in tension, then in compression, then in tension, and so forth.) With the counterbalancing pair of forces applied, the shear wall remains stationary.

For the shear wall to work, the hold-down must be correctly sized to handle the tension force generated. Any additional force, and the wall will begin to overturn. The hold-down must also be installed so that it transfers the force into other parts of the building — such as a slab foundation, for instance.

The Good

With those basics in mind, let's look at some actual installations. Figure 22 shows a reasonably good installation. The hold-down is slightly off-center on the post because of the foundation anchor bolt's position, but not enough to be a problem. The bolt positions were carefully marked on the 4x4 post to ensure accurate drilling and thus equal bearing at each bolt.

Note the coupler nut used to extend the anchor bolt. Is this a sign of trouble? The coupler nut itself is not a problem, but its presence where a nut for a standard sill bolt would go makes me question whether this is an ordinary sill anchor that has been modified for use as a hold-down anchor. Hold-down bolts are typically longer than sill anchors — the deeper embedment allows the uplift forces to be distributed more widely in

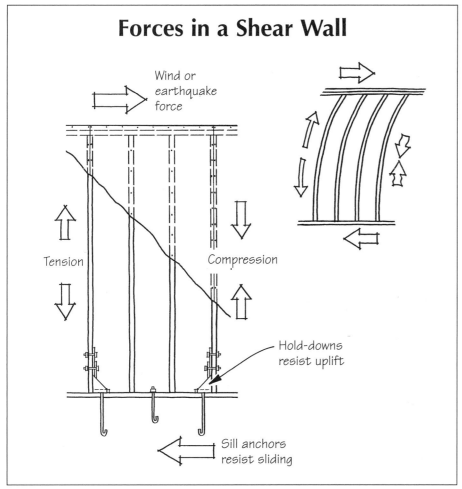

Figure 21. *When wind or an earthquake exerts a force, the load is delivered to the top of the shear wall, along the plates. As the shear wall resists the load, one edge is put in tension, the other in compression.*

Figure 22. *Assuming the anchor bolt is properly embedded, this hold-down is correctly installed.*

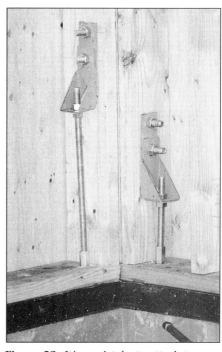

Figure 23. *It's a mistake to attach two hold-downs to a single post. The two shear walls meeting at this corner should have been properly connected with nails, and one correctly sized hold-down attached.*

the slab footing or stem wall.

It's a good idea to mark hold-down bolts with spray paint or colored cap nuts before concrete placement begins: This will help ensure the right bolts are used at shear wall locations.

The Bad

Figure 23 shows an outside corner, with two hold-downs bolted through the end studs into the same post beyond. The two shear walls are sharing the post, which in itself is not a problem. But a dangerous assumption has been made — that the two hold-downs can act together to resist uplift forces.

First, the positions of the hold-downs relative to the direction of overturning will affect their relative degree of stress: The one farthest away from the compression end of the wall will see the maximum force. Second, we must expect some variation in the workman's skill in drilling the two pairs of holes. In an earthquake, the bolts in the hold-down that feel the load first will receive the total force. Only after some yielding occurs will the other hold-down's bolts achieve maximum bearing. The solution, when sharing a post between shear walls, is to use only one larger connector, with adequate anchorage to the foundation.

A Collision of Purposes

The shear wall in Figure 24 has several strikes against it. The most obvious is the way the stem wall, sill, and sole plate have been butchered to make way for a radiant heat feed. Given the size of the notch, the bolts don't have enough end distance (relative to the cutout), so their lateral shear capacities must be reduced.

Beyond that, the length of this shear wall is a problem. Under the 1997 Uniform Building Code, a shear wall must have a maximum height-to-length ratio of 2:1. But even under the former rule of 4:1, it's doubtful this wall would

Figure 24. *This shear wall is not long enough (plate length) to meet code, and the plate has been so badly mangled by the hvac sub that its lateral strength is compromised.*

Figure 25. *The window opening at the bottom left of the photo was enlarged, pushing the first-story hold-down post out of line with the second-story post. In a quake, the uplift forces will likely rip the long threaded rod through the header or pull the header out of position.*

Figure 26. *A triple-2x4 post makes it difficult for the carpenter to know where to place the perimeter nailing. The author recommends nailing off all three studs with the edge nailing schedule.*

have passed. The contribution that such a short wall can make to resisting overturning forces is very small, but something about the design of the building forced the engineer to use every bit of wall to resist lateral forces. It gets even more interesting if you consider that this particular wall section is at the other end of the header shown in Figure 25.

The Ugly

Of greatest concern in Figure 25 is the hold-down on the left. Where is the tension force supposed to go? The threaded rod doesn't even have a properly sized bearing washer on the bottom face of the header. If tension does act to pull the rod upwards, the small washer and nut, acting as a wedge, could fracture the header. But even if it holds, it is doubtful that the few nails in the 16-inch cripple studs above will transfer force to the 4x4 post. Notice also the way the header is nailed to the post with four bent-over nails — not much good against uplift.

There's another issue here: The engineered rim board is not strong enough to handle the kind of compression forces that may be exerted by the posts in the event of a quake. Nor are there any floor joists coming in perpendicular on the other side to provide this support.

So Close

Besides the problem the hold-down in Figure 26 will have in mating with a hold-down above, this installation also illustrates some of the challenges of working with engineered lumber in shear wall construction. Just as in Figure 24, neither the engineered rim board nor the wood I-joists that meet it are strong enough in compression to handle the forces that will be transferred through the hold-down and post. The framer must either bring the post above all the way down onto the top plate (which would interrupt the second floor sole plate and require notching the subflooring), or use triple pony studs between the floors to give solid bearing between the posts. Using solid blocking with the grain oriented parallel with the plates would be a mistake, since the shrinkage relative to the engineered lumber floor system would leave a gap in the load path.

One final note: Using multiple studs instead of a post at hold-downs can make plywood nailing confusing for the carpenter, since there is no one stud taking the uplift and downward shear force at the wall boundary. I recommend following the perimeter nailing schedule at each stud in this case.

By Scott McVicker, P.E., a consulting structural engineer in Palo Alto, Calif.

Fixing Shear Wall Nailing Errors

Tests done at the American Plywood Association have shown that nailing is the controlling factor in shear wall performance. Shear walls fail in one of three ways: the nails bend, the plywood (or OSB) buckles and pulls through the nail head or pulls out the nail, or the framing lumber fails. The lumber failure — typically splitting — is also a result of nailing. The lumber splits either because the nails are placed too close to the edge, or because too many nails have been driven along a grain line, wedging the lumber apart (see Figure 27). This problem has prompted current codes to require wider framing members at panel edges where shear forces exceed 350 pounds per foot. This affects shear wall framing where nails are spaced closer than 6 inches on-center.

Because the controlling component in shear wall design is the nailing, it's no surprise that the correction for field problems generally involves installing additional fasteners. Often you can continue to use nails for the fix, but where the nails are closely spaced or where the framing and sheathing have dried out with age (making them more likely to split), wood screws are the right choice. In some cases, sistering on more lumber to provide a sound nailing surface is your best choice.

Nailing Too Close to the Edge

One of the most common field mistakes is nailing too close to an edge — either the edge of the plywood panel or the edge of a stud or plate. Code requires nails to be 3/8 inch back from the edge of the plywood. Where two panels break on a stud, you have to leave a 1/8-inch gap and make sure both panels have a strong 1/2 inch of bearing. In a perfect world, the sheathing would always fall dead center (minus a sixteenth) on the stud. In the real world, it doesn't always happen that way and perfect by-the-book shear nailing can be difficult to achieve. Fortunately, if you ever get red-flagged, there are some simple solutions (Figure 28).

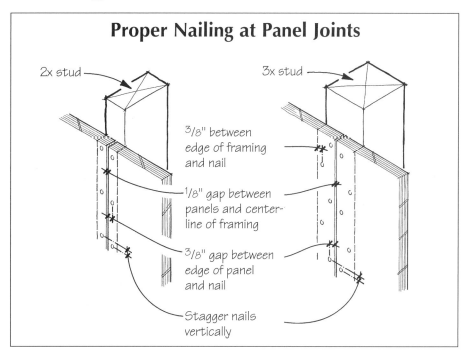

Figure 27. *Where shear panels break on a stud (left), leave a 1/8-inch gap and hold nails back 3/8 inch from the panel edges. Where nails are spaced closer than 6 inches on-center, code now requires 3-by framing; the reduced nail spacing would tend to split a 2x4.*

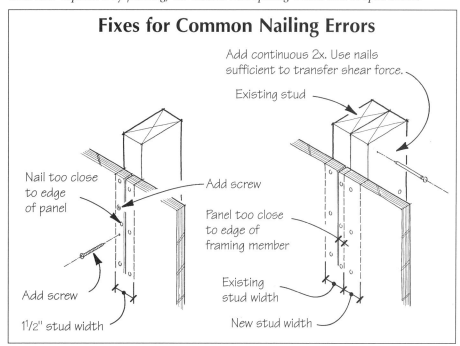

Figure 28. *If you nail too close to the edge of a panel (left), install two screws of the same diameter in predrilled holes on both sides of the stray nail. When the panel edge falls so close to the edge of the stud that it's impossible to nail correctly (right), nail on a second framing member for backing. The two studs must be securely attached to one another to transfer all the shear forces.*

Overdriven Nails

Nails that rupture the outer plies of the structural panel decrease the shear strength of the assembly because there is less of the nail shaft bearing against the plywood and resisting lateral forces (Figure 29). Overdriven nails are also more likely to allow the plywood to buckle and pull through. Pneumatic clip-head nails are especially prone to

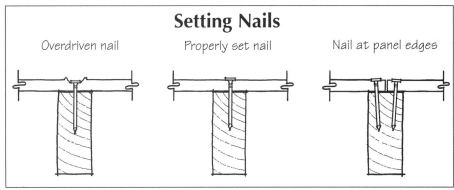

Figure 29. *A nail that ruptures the outer plies of the panel (left) has less shear strength and should be reinforced with screws, as in Figure 28. A properly set nail should either sit snugly on top of the plywood (middle) or slightly dimple the outer ply. When panels break on a 2x4 (right), set the nails at a slight angle to ensure proper embedment into the framing.*

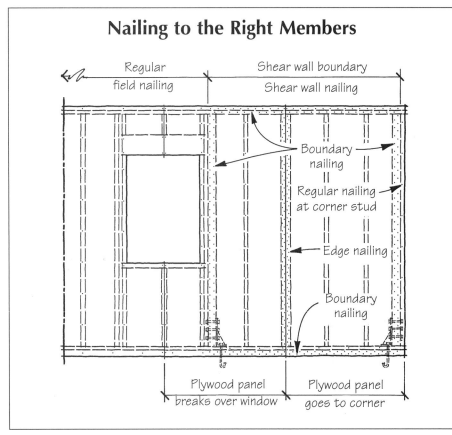

Figure 30. *This illustrates code nailing for a shear wall designed to handle forces above 350 pounds per foot. Note that where plywood edges fall within the shear wall boundary, 3-by framing is required to prevent splitting from the close nailing pattern. Where a plywood edge falls outside the shear wall boundary, however, as at the window opening, standard sheathing nailing is permitted.*

this problem. The fix is to predrill the holes and add a screw on both sides of the overdriven nail.

Nailing to the Wrong Member

Another common mistake happens when the plywood sheathing gets tacked up over the shear wall framing, and the guy coming along to shoot in the shear nailing forgets to carefully mark where the shear wall boundary falls (Figure 30). Putting the boundary nailing schedule into the wrong member weakens the shear wall. Fortunately, the fix is easy enough — more nails in the right place. Taking the time to snap some chalk lines in the first place is the cure.

Note in Figure 30 that one sheathing panel falls completely within the shear wall and one extends beyond it. Code requires boundary nailing at the boundaries and at all panel edges within the shear wall, but if a panel extends beyond the shear wall, you may be able to use field nailing at the panel edge outside the shear wall boundary. An exception would be if the designer intended to use the framing above and below the window opening in conjunction with the adjacent full-height walls to create one big shear wall with an opening in it. In that case, the design might call for additional blocking or strapping and closer nailing all the way to the plywood edge.

Nailing Into Too Narrow a Stud

The 1997 Uniform Building Code requires 3-by studs at panel edges where the design shear exceeds 350 pounds per foot. This is a change from the earlier code, which would have allowed you to use 2-by framing with 4-inch on-center nailing for this condition. Word of code changes is slow to get around, so if you get caught with a 2-by stud with 4-inch nailing, there are a couple of things you can do.

You can add studs to each side of the offending stud, with plenty of 16-penny nails to transfer the forces. Then provide boundary nailing to the new studs. Check with your designer for the specs. Or you can add sheathing to the inside face of studs, under the gypboard, with edge nailing to different members (to avoid splitting) at 6 inches on-center. The combination of interior and exterior sheathing will make up for the loss of strength caused by the narrow boundary stud.

Of course, it may be that the plans called for 4-inch boundary nailing because the design shear just exceeded the allowable shear for 6-inch on-center nailing. As long as the design shear is less than 350 pounds per foot, the assembly is okay.

Choosing Nails

Finally, use the right nails. The Uniform Building Code accepts galvanized box nails as equivalent alternates to common nails for shear wall nailing. The galvanization increases the diameter of the thinner box nail, making the plywood think it is bearing on a common nail. If you use untreated box nails, the shear wall will have less strength.

You can also use many of the available pneumatic nails and staples, but you have to be careful. These fasteners have varying strengths depending on shank diameter and gauge (code report NER-272 provides this information). Rather than trying to make nail substitutions on your own, the best plan is to let your designer know the brand and type of sheathing nail you like to use, and let him or her design accordingly. And as I mentioned earlier, I would avoid clip-head nails.

—*Scott McVicker*

Chapter Five
EXTERIOR DECKS

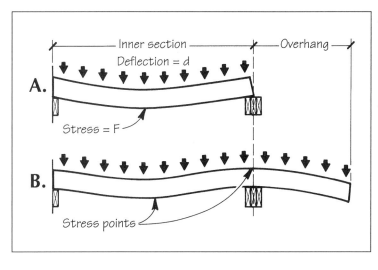

Deck Support: Making the
Crucial Connections 78

Overhanging Decks:
How Far Can You Go? 79

Deck Support: Making the Crucial Connections

"It looked like a battlefield," shouted the headline of the *Atlanta Journal and Constitution* for April 9, 1995. There was no field of combat involved, however. The scene of this eye-catching story was the backyard of a prominent Atlanta citizen in a well-to-do neighborhood: His deck had collapsed during a party, with 63 people on top.

Sixty-three people may seem like a lot, but some easy calculations show that the load on the deck was no more than the building codes say a deck should safely support. The collapsed structure was 25 feet long and extended 12 feet out from the house: 300 square feet of deck.

The codes typically require that decks be able to support the same live load as the rooms to which they connect. In the case of a residence, that's no more than 40 pounds per square foot (psf). For a 300-square-foot deck, that's a total of 12,000 pounds. Dividing by 63 people, that comes to an average of a whopping 190 pounds per person. I'm willing to bet that even throwing in the weight of the tables, chairs, and food that had been set out, the total live load (in addition to the weight of the deck itself) was less than 12,000 pounds.

So if the deck wasn't overloaded, what went wrong? I didn't inspect the failed deck, but I do know from the photo that the failure occurred at the connection between the deck and the house.

This type of failure is not surprising, because almost none of the decks that I do inspect have enough bolts connecting the deck band joist to the house. Let's take a look at what is really required.

How Many Bolts Are Required?

When a bolt (or nail, screw, staple, or other mechanical fastener) is being pushed down on one side by one piece of wood and is being held up on the other side by a second, it's what engineers call a single shear condition (Figure 1). To calculate capacities for a bolt in single shear, there are a few variables you need to know: What is the bolt diameter? What is the density (which you know if you know the species) and thickness of the wood? Is the load applied parallel or perpendicular to the grain of the wood? What is the duration of the load? Armed with this information, an engineer can look up capacities for single shear in the National Design Specification for Wood Construction (NDS) published by the American Forest and Paper Association (202/463-2700).

Take, for example, a 2-by Southern Pine side member, such as the deck band joist attached to the house framing. With a $1/2$-inch-diameter bolt and single shear applied perpendicular to the grain for "normal" duration (a factor that includes full live loading but not high wind or earthquake forces), the allowable shear load in the bolt is 330 pounds. Of course, the bolt itself could handle many times that load, but what will fail first is the wood: That's what limits the capacity to 330 pounds.

For a deck with joists spanning 14 feet with 2x10s 16 inches on-center and 2x6 decking, the weight of the deck itself is approximately 7.3 psf. Add this to our design live load of 40 psf, and the total load is 47.3 psf. One-half of this load must be supported at the band at the house. Simple multiplication (14 ft. x 47.3 psf x $1/2$) gives a load along the band of 331 pounds per foot. Therefore, to transfer this load from the deck band to the house, $1/2$-inch bolts spaced 12 inches on-center are required. That may seem like a lot of bolts. It's certainly more than I'm used to seeing on the decks I inspect. On the other hand, it's cheap insurance if 63 people ever happen to be eating barbecue on that deck.

Table 1 gives the code-required bolt spacing for the conditions described above: $1 1/2$-inch-thick Southern Pine bands loaded across the grain. It gives an acceptable bolt size and spacing for spans from 6 to 16 feet. If you have more load than the standard 40 psf, or if you frame with a less dense wood (Spruce-Pine-Fir, for example, is less than 80% as dense as Southern Pine), then you can have your engineer develop a similar table.

Lag screws and nails. The table also gives a schedule for 16-penny nails, for builders who so choose. I prefer bolts because they're easier to inspect and easier to take out if you ever have to remove the deck. Where you can't get access for a through-bolt, lag screws are my second

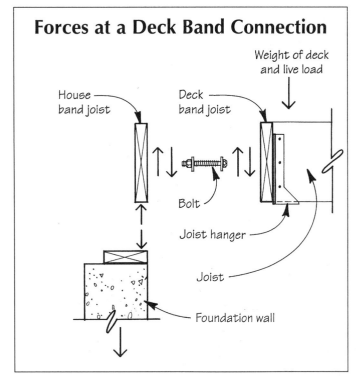

Figure 1. *The attachment of a deck band joist to the band joist at the house is a "single shear" connection—two members side by side transferring a vertical load. Bolts (with washers) are the best way to carry this load.*

choice; substitute the same diameter as the bolt required. Either lag screws or bolts are obviously much beefier than nails, so will resist long-term corrosion better than nails.

Nailed Ledgers Inadequate

There are two more important points to make. First, if $1/2$-inch bolts 12 inches on-center are required for a 14-foot deck span, how many nails do you suppose are required to properly transfer the load from joists to band through that 2x2 ledger strip you may have used in the past to support deck joists? Trust me, it's more than you want to put through that skinny, split-prone member. Please use joist hangers in this situation; they're far more reliable.

Also, the bolt capacities given by the NDS assume that you're dealing with good sound material. Rot and termites can reduce those numbers to zero in a hurry. Like the decking and joists, the deck band is always pressure-treated. Buy yourself a little more cheap insurance and install a pressure-treated band at the house wherever a deck (or screened porch or similar space) will be bolted to it. It won't do much good if the deck band is sound but the one at the house rots out.

Ounce of Prevention

Unfortunately, CABO and other residential codes aren't explicit enough about proper deck support. Even so, most decks with less bolting than is required by code are still standing. But that may only be because there are safety factors built into the design values and because, more significantly, not everyone hosts a 63-person picnic.

Fortunately, no one in Atlanta died from the collapse of that deck. But proper bolting in the first place might have avoided the trouble altogether.

By Christopher DeBlois, P.E., a structural engineer with Palmer Engineering, in Chamblee, Ga.

Table 1. Bolting Schedule for Deck Bands

Joist Span	6'	7'	8'	9'	10'	11'	12'	13'	14'	15'	16'
Bolt Size		1/2"	1/2"	1/2" 5/8"	1/2" 5/8"	1/2" 5/8""	1/2" 5/8"	1/2" 5/8"	1/2" 3/4"	1/2"	5/8"	5/8"
Bolt Spacing		24"	24"	18" 24"	18" 21"	16" 18"	12" 18"	12" 16"	12"	12" 16"	12"	12"
16d Nail Spacing		9"	8"	7"	6"	5"	5"	4"	4"	4"	3"	3"

Note: *This table assumes a deck design load of 50 psf (40 psf live load, 10 psf dead load), and Southern Pine 2-by dimension lumber. As an example, for a deck spanning 8 feet, you can use $1/2$-inch bolts on 18-inch centers or $5/8$-inch bolts on 24-inch centers.*

Overhanging Decks: How Far Can You Go?

When the outer edge of a deck is supported by a girder, the joists are ordinary beams (Figure 2A). They follow the expected laws for simple beams with uniform loading, so you can select the proper joist sizes by looking them up in the tables. A simply supported beam has a single point of maximum bending stress, right at the center of the span. This is also the point of maximum deflection.

To design the deck shown in Figure 2A, you would use a span table to look up the right joist size and species. Many codes require a 40-psf live load, 10-psf dead load, and a deflection limit of 1/360th the span, so you would choose the span table accordingly.

A more common deck construction is shown in Figure 2B. Here, the inside ends of the deck joists are attached to the house and the outer ends rest on and extend slightly beyond the girder. In fact, many builders take the edge of the deck way beyond the girder, to gain an extra 50 or 100 square feet of deck space. But how far should you go? Let's find out.

Out in Space

When the outer end of deck joists go beyond the girder, the loading situation changes. Instead of a simple sag, the joists take on an "S" curve, as in Figure 2B. There is a second point of bending stress directly over the girder. There is also a second potential point of high deflection at the outer edge of the deck. As we extend the deck, the load on the outer end lifts the middle of the span and lessens the bending stress there; but it also increases the bending stress at the girder.

When the overhang is about 40% as long as the inner section (Figure 2C), the stress at the girder and the stress at midspan are equal, and at only two-thirds the bending stress that would occur in the middle of the simple span.

How about that? You make the beam 40% longer and the stress goes down! This means we can extend the deck by at least 40% of the inner section without having to increase the strength of the system.

But what about deflection? As it happens, when the overhang is equal to 40% of the inner section, the deflection along the uniformly loaded deck joist becomes zero. When the overhang is shorter than 40%, the outer end is actually lifted a little by the load on the center span.

Practical limits. When we exceed this 40% overhang, however, things start to change. At 50% overhang (Figure 2D), the bending stress at the girder equals the stress that would occur at the middle of the simple span. When the over-

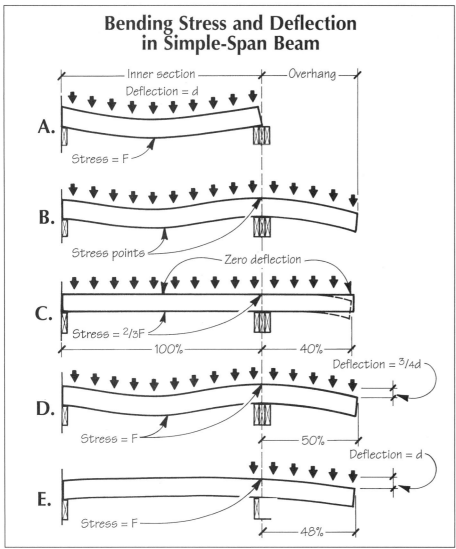

Figure 2. *Sketch A shows bending stresses and deflection in a uniformly loaded simple-span beam. Sketches B through E show the changes in bending stresses and deflection when the beam is cantilevered.*

hang is 70%, the bending stress is double the simple span stress. Stretch the overhang any further and things deteriorate in a hurry.

The situation is a little different for the deflection, but generally in the same order. With a 50% overhang, the deflection at the end is about three-fourths that of the simple span; with a 59% overhang, the deflection of the outer end equals the simple span deflection. With a 68% overhang, the deflection is double that of the simple span.

Moving Loads

Our analysis is not quite complete. What if the deck is loaded only outboard of the girder, like a deck I was on this summer? We all moved toward the outside railing to watch the Fourth of July fireworks. When a uniform load is applied only on the overhang (Figure 2E), the bending stresses are actually the same as they would be if the entire deck were uniformly loaded.

The deflections are different because there is no inner load to lift the outer end. With a 48% overhang, the deflection of the loaded overhang equals the deflection you would have if only the inner section were loaded. A 60% overhang doubles this deflection.

Back to the Span Tables

Having compared the overhang loads with the simple span loads, we don't really need new formulas; all we need are the tables for simple joists. Then, with data from the tables, we just scale things up. It looks like 50% is our magic number: A 50% overhang has bending stresses equal to the bending stresses of the simple span, and roughly the same deflection (rounding up the 48% number from Sketch E). This means a span table designed for the simple span in Sketch A would actually work for the overhang in Sketch D, as long as the overhang were no longer than 50% of the inner section.

As an example, let's say the girder is 8 feet out from the house but we want to go farther with the deck. Using the Span Tables for Joists and Rafters (American Forest and Paper Association; 202/463-2700), the table for 40-psf live load, 10-psf dead, and L/360 deflection says I can span 8 feet with 2x6s on 16-inch centers. The required Fb is 855 and the required E is .9 (x 106 psi). Here in the Northwest, I might use treated No. 2 Hem-Fir for deck joists. It has an Fb of 1,270 and an E value of 1.1, so we're okay.

Based on the discussion above, we can extend the deck beyond the girder 50% with no increase in stress or deflection. So our deck can actually go to 12 feet.

Judgment Call

Some engineering judgments: In keeping with my personal view that floors should be a little overdesigned, I typically use a combined load of 70 psf for residential decks. I would actually use 2x8s in the example above. If you think the deck you're building is going to get heavy use, you too might want to increase the design load. The AFPA span tables make this easy to do — just use the tables for 50- or 60-psf live loads (plus 10-psf dead load).

Finally, with large overhangs and light inside loads, there can be a lifting force at the house end of the joists — a "pry bar" effect. This force is real, but slight, and normal nailing of the ends of the joists to the hangers provides sufficient hold-down.

—*Harris Hyman*

Section Two
FIELD TECHNIQUES

Chapter Six
FLOOR FRAMING

Leveling the Deck 82

Floor Framing with Wood I-Joists 85

Floor Framing With Open-Web Trusses ... 87

A Fix for Bouncy Floors 93

Leveling the Deck

In a perfect world, every foundation would be dead level. In reality, though, you're eventually going to face a foundation that's so out of level that you'll have to take corrective measures. This article describes the methods that work for us in such situations.

When we "shoot" a foundation with a transit, we're looking for any variations in readings that exceed 1/4 inch. Since lumber dimensions often vary by that much, it doesn't make sense to demand closer tolerances.

If we discover any variations that exceed 1/4 inch, we carefully check the top of the foundation at 4-foot intervals. The person behind the transit jots down all the readings on the blueprint at the location they were taken, while the person with the rod marks the readings on top of the foundation. After we've taken all the readings, we put our heads together and decide on the quickest approach to correct the problem without compromising the quality of the floor framing.

Going Up or Down?

How we proceed is generally governed by how far out of level the foundation is. When faced with a foundation that is off by 3/4 inch or less, shimming up the low spots is usually the most sensible approach (unless you're matching an existing floor height, which we'll cover later).

All shims are not created equal. When using shims to level a floor system, it's important to consider any concentrated point loads that will be present after the house is completed. The cedar shims that most builders use are softer than typical framing lumber. For example, cedar will support only about 60% of the weight that SPF (Spruce-Pine-Fir) lumber is capable of supporting before the fibers start to crush. So in areas of concentrated loads, we use shims made of OSB or plywood of the appropriate thickness (see Figure 1).

Going down. When the variations in the foundation exceed 3/4 inch, we "scribe" the floor system to the out-of-level foundation. This is a labor-intensive process, but allows us to frame a level floor system without the use of shims.

To scribe the floor system, we determine the difference in height between the highest and lowest spots on the foundation, then cut a scrap block of equal thickness. Next, we cut and tack all the rim joists together, letting them sit loose on the sill plate.

Starting at the highest point of the foundation and working toward the lowest point, we block up the rim joists so they sit level (Figure 2). Using the scrap block as a gauge, we rest a pencil on top of the block, and slide it along the sill plate to mark the scribe line. We then disassemble the rim joists, make our cuts, and reinstall them, nailing them off as we go.

After the rim joists are in place, we "fit" each individual floor joist by cutting a notch at the end of the joist so it will sit flush with the top of the rim joist. By measuring from the sill plate to the top of the rim joist, we determine

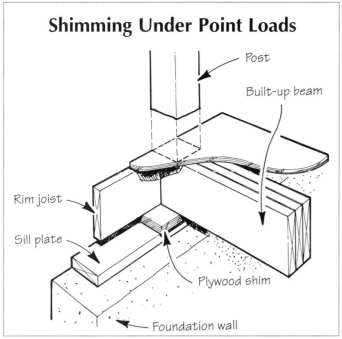

Figure 1. *Under point loads, plywood or OSB shims are better than cedar, which crushes easily.*

how much of the floor joist should be left "standing." Keep in mind that notches at the ends of floor joists should not exceed one-fourth the depth of the joist.

Hybrid approach. Occasionally it may make sense to use a hybrid approach, which combines shims and scribing. If, after shooting the top of a foundation, we discover that one 4-foot section is 1/2 inch high and another 4-foot section is 1/2 inch low, we typically scribe the floor system at the high spot, and shim the floor system at the low spot.

Matching Existing Floor Heights

With additions, the new floor system usually has to finish out at the exact height of the existing floor. The first step is to establish the height of the existing

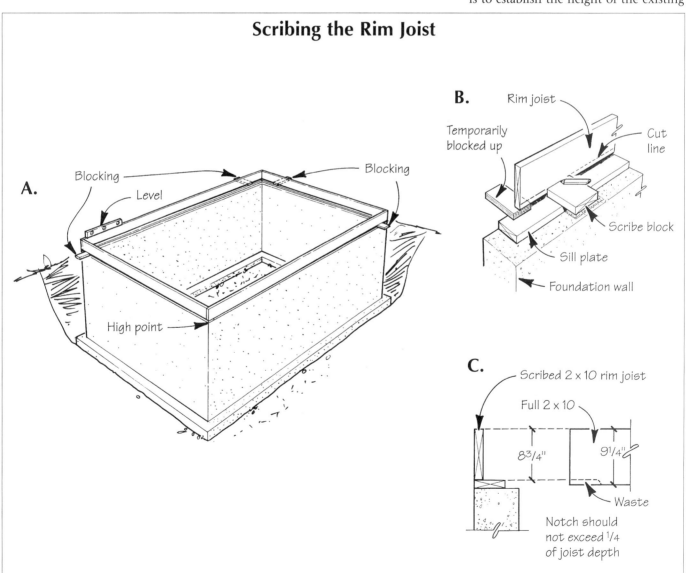

Figure 2. *To scribe the rim joists, first tack and level them in place (A). Then, using a scribe block, mark the cut line around the entire foundation (B). Floor joists are then notched flush to the rim (C).*

Recipe for a Good Foundation

The accuracy of the excavation and footings sets the tone for the entire foundation and for the rest of the building.

Avoid the Pick and Shovel

Foundation contractors who arrive at the job and are forced to perform a lot of pick and shovel work often develop a bad attitude. If you won't bother to get the hole dug accurately, why should they bother to build the foundation accurately?

As the GCs, we always make an effort to be on site and available to take readings as the hole approaches final depth. Variations in level of up to 2 inches can be accommodated fairly easily when forming footings. Even in stony soils, there is seldom a good reason why the excavator can't work to these tolerances. We make sure we like the way the hole looks before the excavator and his machine leave the site.

If you're not comfortable directing the excavator as the hole is dug, arrange to have the foundation sub on site to answer any questions the excavator may have. Budget a few extra dollars to cover the sub's time, and as the machines run, use any slack time to go over foundation details like beam pockets and knockouts for water and septic lines.

Forming Footings

Though it may not be the solution for everyone, we've concluded over the years that forming our own footings is the best way to guarantee the accuracy of the foundation. It's not a job we particularly enjoy doing, but it places the final responsibility in our hands. In the past we've left the job to foundation subs, only to find on occasion that they had "subcontracted" the footings out. As a result, the footings were inaccurate and we had to scramble to make last-minute adjustments.

We always work with the same mason when forming footings. The process goes quickly, and it gives us an opportunity to discuss details of the foundation as we work. We typically budget 16 man-hours to cover layout and form time. If it takes longer than expected, we don't panic. Spending a few extra hours forming the footing sure beats shimming or scribing the first-floor system. When it comes to foundations, an ounce of prevention is worth a yard of cure.

— R.W. & R.W.

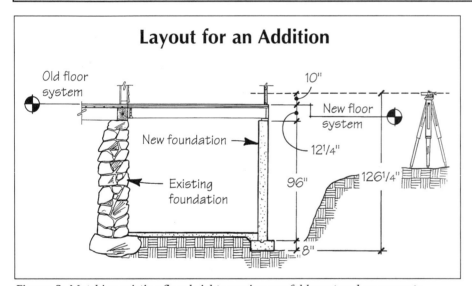

Figure 3. *Matching existing floor heights requires careful layout and an accurate foundation hole. Add the benchmark height (10 inches in this example), the thickness of the new floor system and sill plate, the height of the foundation wall, and the thickness of the footing to determine the excavation depth.*

finish floor in the area where the two floor systems will meet. Don't assume that the floor in an existing building is level, and if a doorway will be cut through to the existing building, take the transit reading at that point. Subtract the thickness of the floor system from this existing finished floor height (include the sill thickness, floor joist depth, subfloor, and finished floor thickness), and you'll have the desired height of the new foundation wall.

When New Meets Old

When building additions, we plan on spending an hour or two establishing the layout heights for excavations, footings, and finished wall heights. To avoid confusion, we drive concrete nails into existing walls that indicate the target height of the new wall.

To establish the target heights, we start by taking a transit reading of the existing finish floor at a point where the new will meet the old (Figure 3). At this stage, it pays to double-check your math; ten minutes with a pencil and calculator can save hours later with a pick and shovel. A good operator can work to final depth tolerances of plus or minus an inch — even in the stony soil we encounter in northeastern Pennsylvania.

When matching an existing floor height, make sure your foundation sub understands that low is preferable to high. It's much easier to raise a floor system by ripping a 1/2-inch strip of plywood and nailing it on top of the sill plate than it is to lower the floor system by scribing the rim and notching all the joists.

Dealing with an out-of-level foundation is never any fun. And the longer it goes uncorrected, the more trouble it will cause. A smart builder will straighten it out right away and get on with the business of building.

By Ron and Roger Whitaker, builders in Hartford, Pa.

Floor Framing with Wood I-Joists

Wood I-joists perform better than wood in most cases, due to their better dimensional stability and greater resistance to deflection. But they take some getting used to. Wood I-joists require special attention in handling, cutting, fastening, and bracing. Here are the lessons we've learned in five years of working with them.

Storage and Handling

You need to handle wood I-joists carefully, beginning with the way they're loaded at the yard. The availability of long lengths is one of the attractions of using wood I-joists, but not every supplier is equipped to handle them properly. This can result in damaged or weakened members.

Most lumberyards follow the manufacturers' recommendations to keep wood I-joists on edge while moving or storing them. But if they're handled while lying flat — both individually and in bundles — they can bow several feet at each end. This is especially true when long I-joists are lifted at the center with a standard forklift. Severe bowing, even for a short time, can cause splits in the flanges and webs that weaken the joist. Unfortunately, these defects are difficult to detect after the joists have been delivered to the site. Until you're sure the joists are being handled correctly at the lumberyard, plan to be on hand while they're loaded and unloaded. If your supplier doesn't have a wide-spread forklift or a boom truck with a two-point sling, ask to have the I-joists loaded by hand.

At the site, keep the stack of joists off the ground with a 2x4 every 8 feet or so. Stack the joists on edge (one next to the other or nested) and nail a 2x4 or a piece of strapping across the tops to keep the whole row from falling over). Finally, be sure to cover the stack to protect the joists from the weather: Wood I-joists (especially those with OSB webs) are more susceptible to water damage than sawn lumber.

Most of the time, we work with joists no longer than 24 feet. Individually, wood I-joists of this size don't bow enough to make much difference, so we lay them out on the walls just as we do with sawn joists. But long lengths can bow several feet if left unsupported at midspan. One solution is to use the interior partitions for support by framing them before laying the joists for the floor above. Another possibility is to build a simple header or erect pipe staging to temporarily support the I-joists in the middle.

Cutting

Crosscutting a wood I-joist is tricky because the surface isn't flat. We use scrap plywood to make a two-layer template that fits between the flanges (Figure 4). The bottom layer lies in the web flush with the flanges, and the top layer acts as a fence for the circular saw to ride against. The template also allows us to mark the cut along only one edge of the joist. A single template built for the narrowest I-joists should work for larger standard widths as well, but you should build a separate template for each circular saw used for crosscutting or else designate one saw for all crosscutting. (Rafter templates work the same way, but must be built to the angle of the plumb cut.)

Don't assume that one end of a wood I-joist is a factory end. Many suppliers stock wood I-joists in 60-foot or longer lengths, and fill orders for shorter lengths by cutting down the full-length pieces. Often this is done by gang cutting a bundle of joists with a chain saw. If your order is the first to be taken out of longer lengths, then one end is prob-

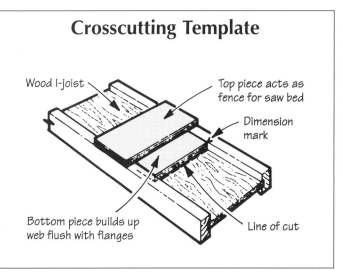

Figure 4. *A two-layer plywood template makes crosscutting wood I-joists a breeze. The bottom piece builds up the web flush with the flanges, and the top piece acts as a fence for the bed of the saw.*

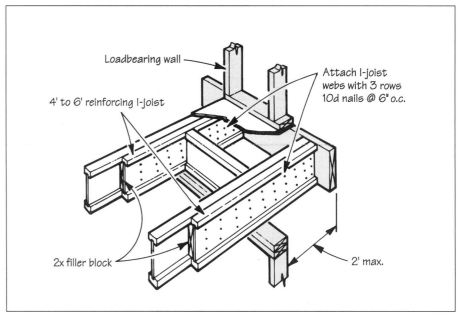

Figure 5. *Wood I-joists need "squash blocks" to support point loads or bearing walls (left). Extensive blocking is also necessary at cantilevers and where bearing walls are offset (right).*

ably square. But the second piece cut out of longer lengths is likely to have two wild ends. We make it a practice to check both ends for square before pulling dimensions for crosscutting.

Blocking

The installation instructions that come with wood I-joists show which joist intersections require extra blocking. Since blocking is time-consuming, however, we try to use details that don't require it. We use wood I-joists for rim joists, for example, even though thinner, less expensive material will do, because, according to the manufacturer, this detail usually needs no blocking. This also avoids any dimensional mismatches between the wood I-joists and sawn or composite rim joist material, like Parallam, Micro-Lam, or Timberstrand.

At stair openings, we use Parallam or Micro-Lam beams for the headers. The extra cost of the material makes up for not having to do all of the blocking required when engineered joists are doubled up or intersect. Parallam and Micro-Lam sizes, however, do not always match the joist sizes, so we sometimes have to rip the header material down to size. It's still easier and cheaper than blocking out the joists.

There are certain situations, however, where blocking is necessary. We install "squash" blocks, for example, on either side of wood I-joists at point loads and under bearing walls (Figure 5). A load-bearing wall that doesn't align perfectly with a wall below must be treated as a cantilever, which requires extensive blocking. And at joist hangers that don't support the top flange, you usually have to pad out the I-joist web (Figure 6).

Sometimes we screw blocks and filler strips to the joists before installation. But, for most applications, it's just as easy to fasten them after the joists are in place. We've used nail guns for this, but we switched to screws because we didn't like the way the nail points poked through the web.

Nailing

When fastening to beams or plates, wood I-joists will split if you nail closer than 1 1/2 inches from the ends, just like sawn lumber. But even at midspan, a nail that is started too close or angled too sharply toward where the web meets the flange will split off the corner of the dado. Nailing at a slight angle out near the corner of the flange seems to work best. We use 8d or 10d box nails, and we can usually get only two nails in without splitting the flange.

The joist hangers made for use with wood I-joists have fluted holes that start the nail at the ideal angle for nailing through the flanges. But if you're not careful, the action of nailing can lift the joist up slightly so it isn't resting on the stirrup of the hanger. This creates squeaks later. Make sure you put some weight on the joist while nailing the hanger.

When nailing deck plywood into wood I-joists with a pneumatic nailer, if you happen to shoot the nail straight through the flange into the web, the head won't always set and will need to be finished off by hand. This is not a

Figure 6. *The web of the wood I-joist must be padded out when joist hangers don't support the top flange.*

Sources of Supply

Boise Cascade
P.O. Box 50
Boise, ID 83728
208/384-6161
www.bc.com

Georgia Pacific Corp.
133 Peachtree St. NE
Atlanta, GA 30303
404/652-4000
www.gp.com

Louisiana-Pacific
805 SW Broadway, Suite 1200
Portland, OR 97205
800/648-6893
www.lpcorp.com

Snavely Forest Products Corp.
600 Delwar Rd.
Pittsburgh, PA 15236
412/885-4000
www.snavelyforest.com

Tecton Laminates Corp.
P.O. Box 587
Hines, OR 97738
541/573-2312

Trus Joist
2910 E. Amity Rd.
Boise, ID 83716
208/395-2400
www.trusjoist.com

Weyerhaeuser Company
P.O. Box 9777
Federal Way, WA 98063
253/924-2345
www.weyerhaeuser.com

serious time waster, but it's a nuisance.

But there are much more troublesome aspects to wood I-joists when laying deck plywood. First, wood I-joists are more flexible laterally than sawn joists. This makes them wobbly, so you can't walk on them at all until you brace them. In fact, you shouldn't even stack materials on unbraced joists because there's a chance the joists will roll over under the load.

Second, the flexibility of wood I-joists can give you trouble when aligning the plywood decking. With a sawn joist, you can usually straighten a bow with the first piece of plywood you lay, and the joist is stiff enough to hold position until all of the plywood courses are nailed off. Wood I-joists, however, are not stiff enough to hold position until almost the entire length is nailed off. If you pull one over to align it with the first course of plywood, you may still have to pull it over for the next course. In fact, if you're careless with your layout marks, it's fairly easy to pull first in one direction and then the other, creating an S-shape in the joist (Figure 7).

Our solution is to mark the joist centers on the plywood before laying it in place, using a drywall square to extend the lines across the whole sheet. Then we're careful not to force the joist too much.

Shrinkage

If they are kept dry to begin with, wood I-joists don't change much in their height dimension. I have had I-joists expand about $1/8$ inch in height from moisture, and then return to their original dimension when they dried. (This kind of movement is more noticeable in

Figure 7. *Wood I-joists tend to bow — if you aren't careful, you can push and pull an I-joist into an S-shape when nailing decking.*

joists with OSB webs than in joists with plywood webs.) I have also had wood I-joists shrink in length as much as $3/8$ inch over 60 feet. It may not seem like much, but in one case it was enough to bow the flush-framed Parallam beam that supported the joists at one end.

I've also had some difficulty matching wood I-joist sizes to Micro-Lams and Parallams. The joists are usually accurate and uniform, but the other engineered products sometimes vary in size and have to be ripped down to get a flush fit. Still, it's better to use engineered beams than to mix in dimensional lumber, which will eventually shrink at a different rate.

By Ned Murphy, owner of E.J. Murphy Builders in Framingham, Mass.

Floor Framing With Open-Web Trusses

Parallel-chord floor trusses offer builders many advantages. A floor truss can span 38 feet with no intermediate support. It does this not with massive timbers, but by efficiently using smaller dimension lumber.

A truss makes better use of labor, too. There's no cutting, so the framing goes faster; duct runs can be built into the webs, speeding up hvac installation; and pipes and wires can be quickly routed through any part of the open webs.

A $3^{1}/2$-inch-thick truss also has more lateral stability than a $1^{1}/2$-inch-thick joist and presents a wider nailing surface for decking. As engineered components, trusses are more predictable than dimensional lumber. They rarely warp, and they're generally stiff enough that if you glue and nail your subflooring, you'll never have to worry about squeaks.

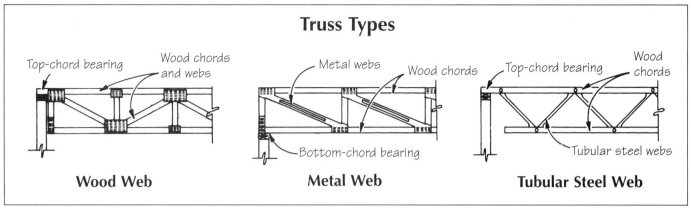

Figure 8. *Trusses are classified by their web material and whether they bear on their top or bottom chords.*

Of course trusses aren't right for every project. Like any material, they have their drawbacks and limitations. Knowing when to use them and how to use them correctly demands some knowledge of how they carry a load.

Truss Types

Structurally, a floor truss resembles an I-beam in that it puts most of its material along its top and bottom edges, where the stresses are greatest. But instead of having a solid web, a truss uses a series of rigid triangles, usually wood but sometimes metal. Because triangles are an inherently stable shape, the truss effectively transfers all loads to the bearing points with little deflection. Additional vertical members are installed wherever a truss encounters a shear force, such as where the truss rests on a wall.

Trusses are broadly classified by what their webs are made of and whether they bear on their top or bottom chords (see Figure 8). A bottom-chord bearing truss sits directly atop a wall plate or foundation sill, just like a standard joist. A top-chord bearing truss hangs from the bottom of its top chord and can rest either directly on a bearing surface or on a raised setting block (the setting block can be built into the truss). This flexibility gives the designer control over floor and ceiling heights.

Truss web members can be made from lengths of 2x4 held in place with metal connector plates, from tubular steel that's bolted through the chords, or from 20-gauge steel that's formed with the connector plates integral to the webs.

Metal-web trusses have the advantage of being lighter and easier to carry than their wood-web counterparts. And though most metal-web trusses are sized to match conventional 2x10s and 2x12s, 18-inch-deep trusses are available for very long spans. Some fabricators even make 18-inch-deep metal-web "stock" trusses that range in length from 24 to 28 feet. They're bottom-chord bearing, can be cantilevered up to 2 feet at one end, and are designed to carry live loads of 40 or 50 pounds per square foot. They're marketed to builders of standard ranch and Cape-style homes and are competitive in price with stick-framed floors.

Wood-web trusses. Unlike metal-web trusses, which come in just a few sizes and standard designs, almost every wood-web truss is a custom product. Not only can it be made in almost any configuration, but its performance depends on variables such as the strength of its members, and the location, size, and strength of each connector plate. Because of this, fabricators claim that wood trusses are over-engineered to give them a greater margin of structural safety than their metal-web counterparts.

In fact, the strongest residential trusses — girder trusses — are usually made completely of wood. A pair of girder trusses is often used to flank an opening such as a stairwell and to support the weight of a header that in turn carries several standard floor trusses.

Design and Planning

Most trusses are custom-made for a particular job which means that both the fabricator and the builder must spend extra time with the plans. A trussed floor requires more forethought than one framed with I-joists or solid lumber. Before ordering a set of trusses, for instance, you should know the location of all ductwork and plumbing. Space for these is figured into the truss design and, once in place, the trusses can't be altered without being reengineered.

These constraints may be unacceptable on some custom building jobs, where midstream design changes are the rule. Even small changes like moving a tub or toilet can cause problems on a trussed floor — especially if the drain lands over a top chord.

Remember also that deep trusses change the normal height calculations between successive floors. If you're used to framing with 2x8s and switch to 16-inch trusses for instance, you'll have to either lower your ceilings or add an extra tread and riser to any stairways. And if you make the stairway longer, you'll have to do the same to the stairwell, for headroom. Include the extra space in the floor plan.

Also, be aware that most floor trusses are designed to support typical residential live loads; the standard truss won't carry structural dead loads at midspan, which are part of many conventional framing systems.

In general, structural loads should land on the exterior walls, not on the truss. Potential problems include attic kneewalls that help support a set of rafters, or a load-bearing partition that

Figure 9. *The job foreman on this multi-family building estimated that the floor framing took half as long with trusses as it would have with standard dimensional lumber.*

Figure 10. *Floor trusses are usually lifted into place by hand (left), though fork lifts and cranes can speed the process (right).*

supports ceiling joists. While these loads are unlikely to break a truss, they could make it deflect enough to crack the interior wall finish. Fabricators *can* engineer a truss that will carry structural loads, but it will cost more than for a standard truss.

Ordering Trusses

The lead time for ordering floor trusses can run up to two weeks. You can order them through a lumberyard or directly from the fabricator. But regardless of where you place the order, choose your fabricator carefully.

One key thing to look for is good technical support, especially if you've never used trusses before. The fabricator should have a technical representative available to meet on site with the contractor, architect, or engineer. A good tech rep will search for the most economical way to truss a building. For most residential jobs, the next step is for the fabricator's design staff to review your plans and alert you to any possible problems. For complex designs, the fabricator may want you to hire your own engineer.

For complicated orders you can also have the fabricator send you a set of shop drawings for approval before the trusses are made. Although this happens in less than a quarter of all residential work, it's a worthwhile step if you want to make sure that the trusses are the right length or that hvac runs and special structural elements like bearing blocks (the extra vertical web members needed over support points) are in their proper places.

As soon as you receive a shipment of trusses, inspect them for handling damage. Look for tight joints that are fully covered by connector plates, and make sure that the plates themselves are centered over the joint and firmly embedded in the wood. (By the way, never try to refasten a loose plate: You can't restore the bond once it's been broken).

Reject any trusses with excessive splits, with loose knots next to the metal plates, or with loose or deformed plates. Also reject any that show evidence of having been damaged and repaired. And beware of warped or wet lumber, which can cause dangerous stresses as it shrinks and dries.

Working With Trusses

Most builders find floor trusses simple to install. One reason is that they're usually spaced 24 inches on-center, instead of 16 inches, meaning fewer to install. Another reason is that trusses don't have to be cut or measured; they fit together like a puzzle. In most cases, they're simply hauled up to the plates and toe-nailed down.

The four-unit condominium pictured in Figure 9 is a good example. The building was framed with lightweight metal web trusses 20 feet long and 13 inches deep. The 15 or so trusses for each unit were hand-lifted to the plates, rolled

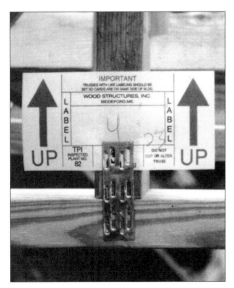

Figure 11. *In order to work properly, a floor truss must be installed right-side-up, as labeled. An upside-down truss is a recipe for failure.*

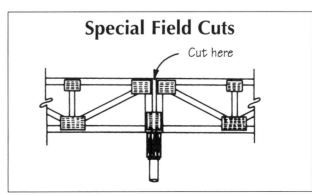

Figure 12. *In some cases, a bottom-chord bearing truss that crosses a girder or bearing wall may be engineered as two simple beams. The tag will instruct the builder to cut through the top chord after installation so that the truss does not see-saw across the girder.*

Floor Framing With Open-Web Trusses

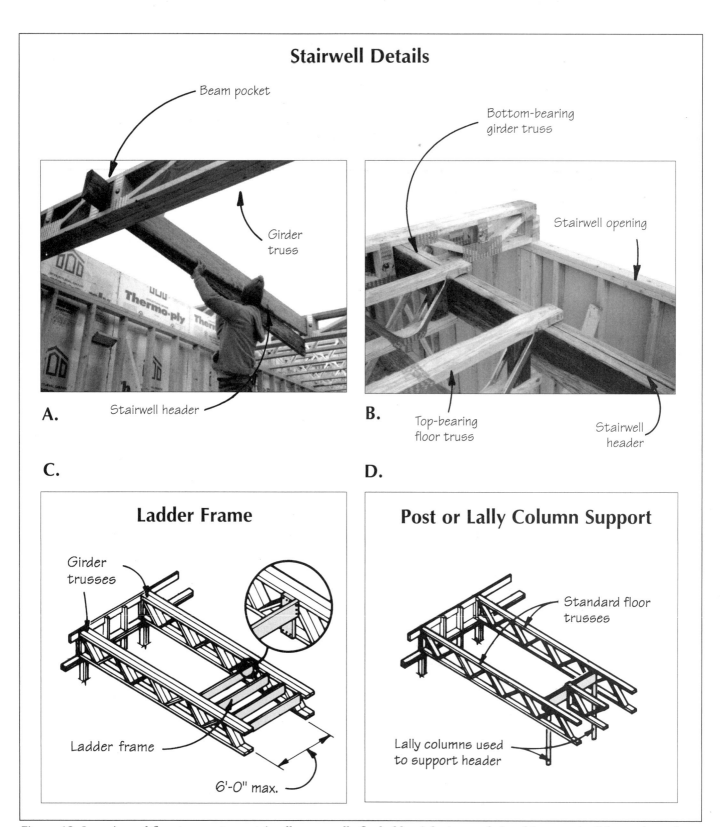

Figure 13. *In engineered floor truss systems, stairwells are usually flanked by girder trusses designed to support solid or engineered lumber headers. The header slips into a beam pocket (A) in the girder truss, and is shimmed tight to the underside of the girder's top chord. The header, in turn, carries the top chords of several floor trusses (B). Combining top-chord bearing trusses with a girder that bears on its bottom chord keeps the framing flush.*

Where design flexibility is needed, you can ladder-frame the floor area between the two girder trusses, which allows you to adjust the width or length of the stair opening (C). Or you can forego girder trusses altogether and support the staircase header with lally columns (D).

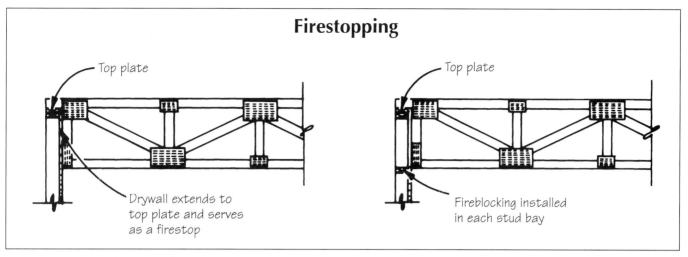

Figure 14. *Many codes require firestopping at the intersection of enclosed vertical and horizontal cavities. With bottom-chord bearing trusses, this means extending the drywall past the truss to the top plate (left) or installing a 2x4 firestop inside each stud bay just below the truss's bottom chord (right).*

into place, and fastened in about an hour. The job foreman estimated that the floor would have taken twice as long to frame with standard joists.

This job was typical. It's rare to see trusses craned into place on a residential job (though a long, wood-web girder truss can be quite heavy). With a little sweat, most can be set by hand (see Figure 10). If you're lucky enough to have a crane however, be sure not to lift the truss sideways: The excess bouncing and flexing can loosen the connector plates, setting up an eventual failure. (For more on using a crane to lift floor trusses, see "Setting Floor Trusses", page 92).

Watch for tags. Most trusses come with tags or labels that tell the builder how to install them. For instance, every truss should have a tag that shows which way is up (see Figure 11). Never ignore it. Each web member is designed to be in compression or tension, but not both. Installing a truss upside down defeats the design and will lead to failure.

When necessary, trusses should also include labels that say where to fasten lateral braces or that alert the builder to any special structural conditions. These include the locations of concentrated loads, cantilevers, and interior bearing points. One example is a bottom-chord bearing truss that crosses a girder or bearing wall and that's designed to function as two simple beams. A note attached to the truss will tell the builder to cut the top chord after installation (see Figure 12). If this cut isn't made, then a load applied to one end of the truss will cause uplift on the other, making the truss act like a seesaw.

Stairwells. Stairwells are usually headed off with solid lumber or with LVL or parallel-strand headers. Figure 6 shows three methods of doing this. One way is simply to place a column below either end of the header. If columns are unacceptable, an alternative is to flank the opening with a pair of doubled girder trusses. If the length of the stairway must remain flexible, then the space between the girder trusses can be filled in with a 2x6 or 2x8 "ladder frame." If you know the size of your stairwell opening, however, a better method is to order girder trusses with built-in beam pockets to catch the ends of the header.

Firestopping and draftstopping. Firestopping slows the spread of fire by preventing a structure from acting as a duct system for smoke, flame, and hot gases. On platform-framed buildings, the top and bottom wall plates double as fire stops. But a top-chord bearing truss short-circuits the top plate. Common solutions are to extend the drywall past the truss to the top plate or

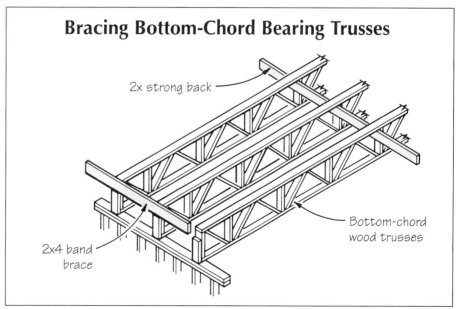

Figure 15. *The ends of bottom chord bearing trusses are tied together with a 2x4 band brace. Horizontal strongbacks at 10-foot intervals distribute loads and prevent twisting, which can loosen truss plates.*

Floor Framing With Open-Web Trusses

Setting Floor Trusses: Three Men and a Crane

I've set second-floor trusses both by hand and with a crane, and I'm convinced that the crane is the only way to go. On the last job where I put them up by hand, we installed about twenty 2x4 wood-web trusses. They were 28 feet long and 22 inches deep, and weighed about 150 pounds each. Someone had to stand underneath one end and hand it up to the carpenter on the wall. This is stressful work, and if the person on the top plate pulls too hard, the truss can slip off the opposite plate — possibly while someone is standing beneath it. This job, which took three men most of a day, would have taken only an hour with a crane. Plus I wouldn't have had an aching back for the next three days.

By comparison, while building a 5,000-square-foot custom home a few years ago, it took three of us two hours to set 45 second-floor trusses — with the help of a crane. These were all bottom-chord-bearing 2x4 trusses, 22 inches deep, and ranging in length from 24 to 32 feet. The job site had a large flat area with nothing in the way of positioning the crane or swinging the trusses once they were picked up. We stored them about 40 feet from the house, so the crane would have a clear swing.

We put one man, Bill, by the trusses, and John and myself at the front and rear walls of the house. Bill separated out three trusses at a time, and slipped cable loops around them at two points (see illustration.) He then signaled the operator to take up the slack and lift the bundle enough to see if it was in balance. If it rode level, the operator hoisted it up above the wall and swung it into place, taking care to keep the bundle from rotating.

John and I paid close attention while the bundle came down to the plate, catching the ends and steering them to the middle layout mark, making sure we had good bearing at each end. One of us would then release the loop and send it back. While Bill was getting the next bundle ready, John and I moved the trusses to their layout marks, flushed them up to the outside of the walls, and nailed them off. It was easy to set three trusses in about five minutes, just in time to receive the next bundle.

Before the crane operator left, we braced across the top chords and had him lift the 3/4-inch plywood for the subfloor in two stacks. What could have been an hour of heavy, miserable work dissolved into ten minutes of focus and handwaving.

Next, we plumbed, squared, and braced the first truss in the layout. We then attached the band brace, which we had marked with the same layout as the top plate. We aligned the ends of the trusses with the layout on the brace and nailed them off. Before the day was over we had finished laying the double plywood subfloor.

Like most things in building, using a crane requires good planning. Give careful thought to how and where you'll store your materials before the crane gets there. Some sites, espe-

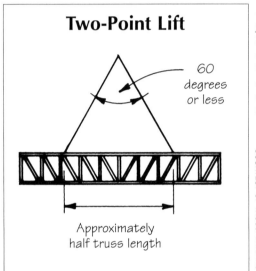

Use a two-point lift for trusses up to 30 feet long. One-point lifting can damage the truss.

cially in remodeling, may present difficult or insurmountable obstacles, like utility wires, trees, or other buildings, and will have to be handled in some other fashion (one option is a fork lift). Make sure the crane operator, his safety engineer, or an independent safety engineer looks at a difficult site beforehand — it'll save you a lot of time, money, and anxiety.

By Richard Dempster, a builder and remodeler in Asheville, N.C.

---

to insert a separate 2x4 firestop inside each stud bay just below the truss's bottom chord (see Figure 13). Check with your local code official for firestopping requirements in your area.

Draftstopping slows down the spread of fire through a large concealed open space like a truss-framed floor by dividing it into two or more smaller spaces. The BOCA National Building Code limits horizontal areas to 1,000 square feet; anything larger must be partitioned off. The code requires the use of 1/2-inch thick drywall or 3/8-inch thick plywood for draftstopping. The Uniform Building Code has the same requirements and also requires that any draftstopping divide the space in question into approximately equal halves. Installing draftstopping is just a matter of fastening the drywall or plywood to one side of a truss, although pipes and ducts in the floor cavity will slow the process.

Bracing and spacing. A bracing system must keep a set of trusses from bending, twisting, or otherwise deforming. On bottom-chord bearing trusses, the ends are tied together with a 2x4 band brace that doubles as a nailing base for the perimeter of the plywood deck (see Figure 14). Underneath, a horizontal strongback — a 2x6 or 2x8 laid on edge — should run continuously through all the truss webs at 10-foot intervals. The strongback also serves the same purpose as bridging in a standard floor, distributing concentrated loads over a wider area.

The 24-inch spacing of most floor trusses can also lead to problems in the finished floor if you don't compensate for it. When placed over that span, a 3/4-inch-thick plywood subfloor can deflect enough to cause waves in vinyl flooring and cracks in ceramic tile. Some builders tackle the problem by installing two layers of subflooring or using 1 5/8-inch-thick Sturdi-Floor panels (be sure to use a nailing pattern that's designed for 24-inch on-center framing). You'll also be better off with flexible grouts and smaller tiles, whose joints more readily absorb slight deflections.

Finally, you may have to approach the insulation a bit differently when using trusses. It can be tough to properly insulate the 3 1/2-inch-thick open webs with fiberglass batts. If you need a tight insulation job, consider using a blown-in product.

By Charles Wardell, a former editor at The Journal of Light Construction.

A Fix For Bouncy Floors

A remodeler recently described to me a common problem: a springy great room floor, framed with 2x10s 16 inches on-center spanning 15 feet on both sides of a center girder. The existing condition poses no real safety issue; it's just plain too bouncy. What's the best way to stiffen this floor, he asked, with minimal disturbance to existing finishes? Upstairs there is tile in the great room kitchen area and carpet in the living and dining area; downstairs, in the basement, there is a dropped ceiling.

The usual methods would be to either double the joists or add a second girder in the middle of the 15-foot span. Adding joists becomes less attractive because of the presence of the kitchen plumbing pipes, while posts for a second girder would intrude upon the living space below.

There is another option worth considering: to "skin" the bottom of the joists with a layer of plywood, effectively turning the floor system into a plywood stressed-skin panel. I have used this technique successfully in a similar remodeling situation to protect a tile floor upstairs where existing plumbing, electrical, and structural conditions downstairs made sistering the joists difficult. The resulting floor system has strength and stiffness comparable to a floor framed with the next larger size of 2-by joists.

How Stressed-Skin Panels Work

The added strength of the stressed-skin panel comes from the composite action of the joists and plywood, which causes them to perform as if each joist were a T-beam or, in the case of two-sided panels, an I-beam. The plywood acts like the flange of the T (see Figure 16) while the lumber member acts as the web.

Two conditions must be met for this to work: The plywood must be laid with the strong direction (surface grain) parallel to the joist and continuous for the length of the joist, and the connection between the plywood and the joist must be strong enough to resist any slippage or shear between the joist and the plywood due to bending forces. This is accomplished using a combination of screws and epoxy or resorcinol glue. (Typical floor sheathing, which runs perpendicular to the joists and is nailed down, does not meet these criteria.)

In a case like the one described above, I chose 5/8-inch five-ply Group 1-Exposure 1 or Group 1/Exterior plywood.

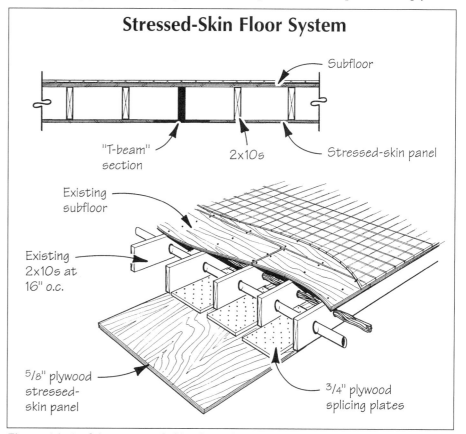

Figure 16. *Applying a stressed-skin plywood panel to the bottom of the floor joists is a good way to stiffen a bouncy floor when plumbing and mechanicals prevent you from sistering the joists. The technique works because each lumber member acts as a T-beam (top). The plywood must be glued to the lumber with a waterproof structural adhesive to achieve this composite strength.*

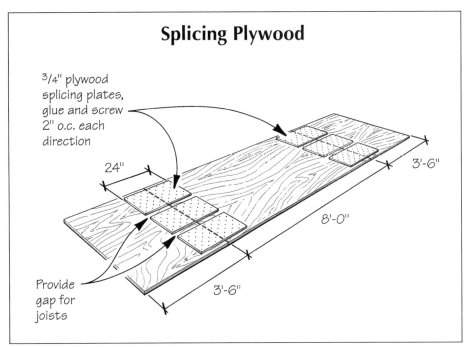

Figure 17. *To make long plywood panels for retrofit stressed-skin applications, butt the shorter lengths of plywood and join them with glued-and-screwed splice plates. Avoid placing a splice at midspan, where bending forces are greatest.*

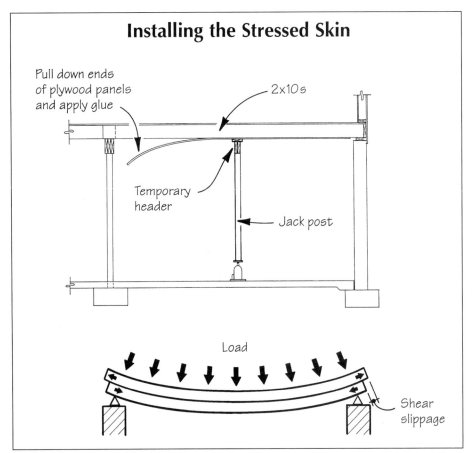

Figure 18. *To install the plywood panels, tack them temporarily, then pull down the ends one at a time and apply glue to the joists and the plywood. Then reposition the plywood and screw it to the joists with 1 5/8-inch deck screws, working from the center out. If the existing floor has sagged, jack it to a little above level first, using a temporary header. Although it is impossible to spread glue under the header, this is of no consequence, because there is no shear slippage between the plywood and joists at midspan (bottom sketch).*

The APA rating ensures that the plywood veneers are of the highest-strength group and that the glues used are waterproof enough to resist delaminating in the event of a spill or plumbing leak.

Fabricating the Panels

Since it's usually hard to find a ready source for the extra-long plywood needed (15 feet, in this case) to extend continuously the length of the span, you can splice it on site (Figure 17). Make the butt joints carefully, with the joints placed as shown. Though it might seem like less work to use one splice right in the middle of the span, this is the point of maximum bending. Since the plywood in a stressed-skin assembly carries bending forces, I've placed the splices at the quarterpoints where the bending moment is less.

Cut 24-inch-long splice plates from 3/4-inch plywood. For a flat ceiling that will be drywalled, arrange the splice plates so the floor joists can pass by, as shown in Figures 16 and 17. In the remodeler's case where there is a dropped ceiling to cover the plywood, one 4-foot-wide splice plate on the bottom may work. Make sure the plywood is clean, dry, and free of foreign material, then sand all mating surfaces at the splices parallel to the grain with medium grit sandpaper. (Because the plywood will later be glued to the underside of the joists, this is also a good time to sand 2-inch strips on the plywood where the joists will fall.)

Glue the splice blocks with epoxy or resorcinol, using screws 2 inches on-center in both directions to maintain clamping pressure while the glue cures. Follow the manufacturer's instructions for mixing and curing time and temperature.

Installation

Sand the underside of the joists with a belt sander and a medium-grit belt, keeping parallel to the grain. Remove all pitch, paint, and the dark oxidized surface to expose clean, fresh wood.

When the spliced panels have cured properly, lift them into place, position them, and tack them up with a few temporary nails. If the existing floor has sagged, install a temporary header with jack posts beneath the plywood, positioned to raise the center of the span a bit past level (Figure 18). I suggest overjacking about $1/1{,}000$ of the span (in the case of a 15-foot span, this would be 180 inches/1,000 = .180, or about $3/16$ inch).

With the panels thus positioned, you can then lower the ends of the sheets one at a time to spread the glue on both the plywood and the bottom of the joists. Again, use either epoxy or resorcinol glue. Because of the header, it may be difficult to get glue all the way to the center of the sheet. Don't worry about this: Shear slippage is practically zero at midspan (Figure 18).

Now raise the plywood back into place, and fasten it with $1^{5}/8$-inch deck screws, working from the center outward to the ends. Place the screws on 6-inch centers in the middle but reduce the spacing to $2^{1}/2$ inches on-center for the last 3 feet at each end, where the maximum shear forces occur. Leave the jack posts in place until the glue has completely cured. When the jacks are removed, the floor will drop slightly, ending up pretty close to level. However, the floor system will now be almost twice as stiff as before, or a bit stiffer than if the original floor had been framed with 2x12s.

Using 1x6s

As an alternative to the plywood stressed-skin, you can use No. 1 or 2 nominal 1x6 southern pine boards to get a comparable strengthening and stiffening effect. If 15-footers are locally available, their installation might be easier than the plywood method, because you could avoid splicing. Follow the same installation guidelines as for the plywood, installing one 1x6 per floor joist.

Limits

This technique can be expected to increase both the strength and the stiffness of 2x6, 2x8, or 2x10 floor framing to approximately the equivalent of the next larger size joist. However, it shouldn't be relied on where the existing floor framing is undersized for the usual design loads — at least, not without the guidance of an engineer.

For any structural designers interested in reading further, the Plywood Design Specification Supplement 3, Design and Fabrication of Plywood Stressed-Skin Panels, is available from APA — The Engineered Wood Association (206/565-6600). This is an excellent design guide for specific applications, but is written for engineers.

—*Robert Randall*

Chapter Seven

WALL FRAMING

Layout Tricks for Rough Openings 98

Framing Rake Walls 102

Site-Built Panelized Walls 105

Building Stiff Two-Story Window Walls . .109

Bracing Foam-Sheathed Walls 113

Framing for Corner Windows 116

Layout Tricks for Rough Openings

An experienced framer goes beyond the blueprints and knows how to detail the frame to make things easier for the subs who follow. Here are a number of tips for framing R.O.'s that can save your company time and money.

Stairwell Openings

Explaining how to size rough openings for every stair design is beyond the scope of this section. But rough openings for straight-run stairs are fairly simple. If the stairwell opening is placed between an upper and lower flight of stairs (as when the main stairs are stacked over the basement stairs), make it 2 to 3 inches shorter than the total run of the stairs (Figure 1). Then make a seat cut at the bottom of the upper stringers to lock them into place.

To find the width of the stairwell opening for a closed stringer stair (one with a wall on each side), add the width of the finished stair assembly, including the skirtboard, if any, to the thickness of the finished wall surfaces on both sides of the stairs (Figure 2). Then add another 1/4 inch for play. This will leave enough room to drop a set of pre-assembled stairs through the hole, then to slip the drywall and skirtboards behind the stair stringers.

A word of caution: If the plans give a dimension for the stairwell opening,

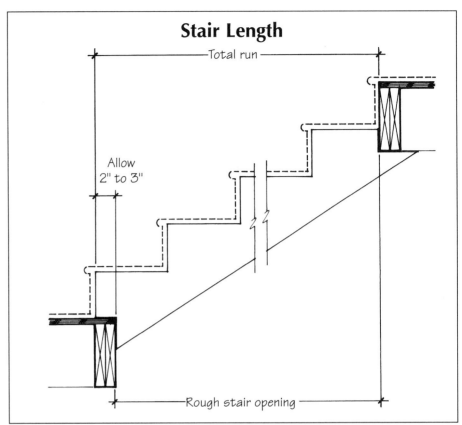

Figure 1. *When a flight of stairs lands on a stairwell opening, make the opening 2 or 3 inches shorter than the total stair run. Then notch the bottom of the stringers to lock them into place.*

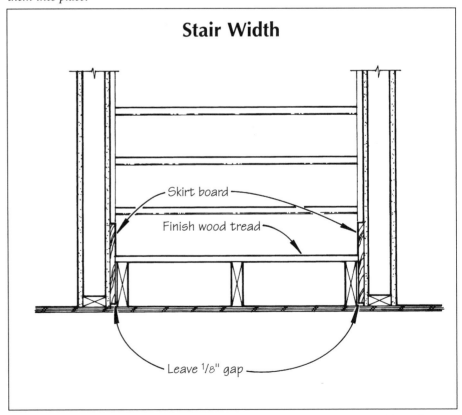

Figure 2. *To size the opening width for a closed stringer stair, take the width of the finished stair assembly plus the drywall, then add 1/4 inch working room.*

Chapter Seven: Wall Framing

ignore it. Chances are, it's wrong. In the case of complicated stairs, the only rough opening dimensions I would rely on are those of the stair builders. If your only responsibility is as a rough framing sub, diplomatically ask the general contractor to verify the stairwell opening with the stair builder.

Interior Hinged Doors

Finding the rough opening width for a hinged door is pretty straightforward: Add 2 inches to the door size. This means that the rough opening width for a 2-foot 6-inch door will be 2 feet 8 inches. But finding the rough opening height can get tricky. Not only must you know the thickness of the finished floor (including any additional underlayment), but you have to decide how much space you want between the finished floor and the door bottom (Figure 3).

If the floor is to be finished with tile installed over a second layer of plywood and a layer of cement board, you'll have to increase the opening height accordingly. If the finish floor thickness varies throughout the house, you have to consider how different rough opening heights will affect the trim detail. If you want all of the head casings to align, I recommend figuring R.O. heights based on the lowest finish flooring height. Where the floor thickness is greater, you will have to trim the bottoms of the doors.

Pocket Doors

Though the rough opening height for a pocket door depends on the track hardware, it's always taller than the opening for a standard hinged door. Rough opening heights vary from manufacturer to manufacturer. The pocket door hardware I use requires a rough opening height of 84 1/2 inches, and will accommodate a finished floor thickness of 3/4 inch. If you don't have the manufacturer's rough opening height, you need to know how much space is taken up by the track and wheel assemblies. Ask the door supplier for the required rough opening for the finish floor material you will be using. Better yet, get the track hardware before you start framing

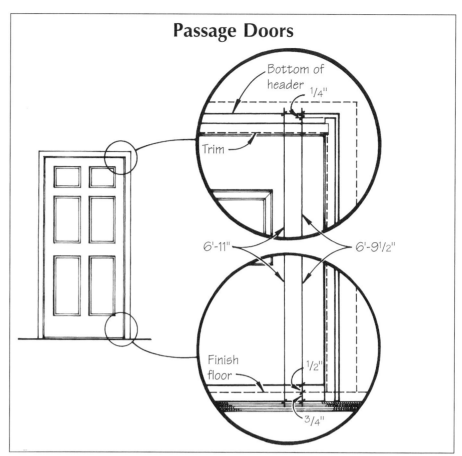

Figure 3. *To find the rough opening width for a hinged door, add 2 inches to the door size. To find the rough opening height, you must know the thickness of the finished floor, as well as the gap below the door. Start with 6 feet 9 1/2 inches — the height of the door plus the thickness of the head jamb and the "ears" of the side jambs that stick beyond. Then add the thickness of the finish floor (3/4 inch in this example), the desired gap below the door (here, 1/2 inch), and a 1/4-inch adjustment space for leveling the door head. The rough opening height in this case is 6 feet 11 inches.*

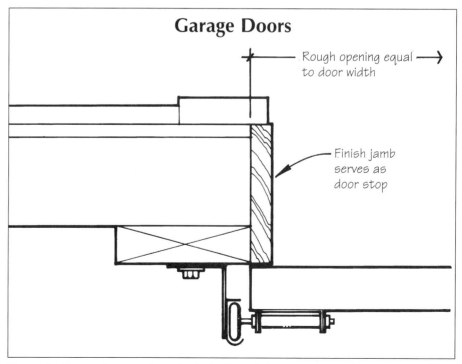

Figure 4. *Garage door openings should be framed the same size as the garage door. The finish jamb will then cover the edges of the door.*

Layout Tricks for Rough Openings

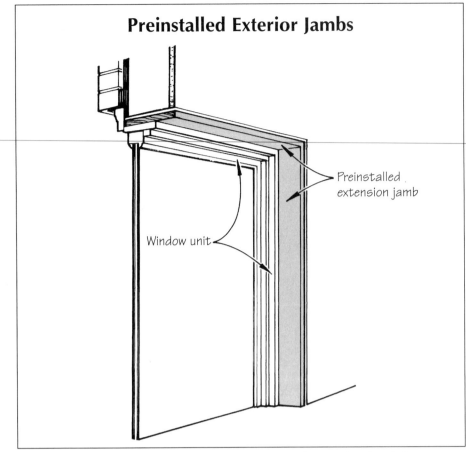

Figure 5. *To ease the process of installing extension jambs, oversize the window openings by 1 1/2 inches and wrap the window frame with a nailer.*

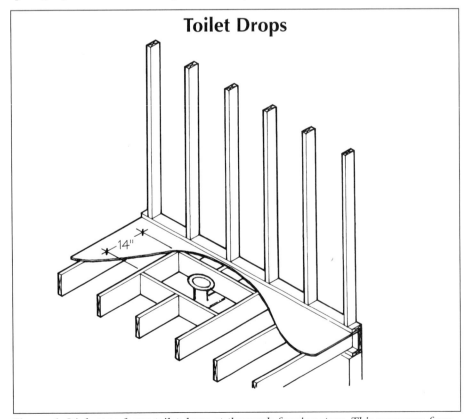

Figure 6. *It's best to frame toilet drops at the rough framing stage. This saves you from having to come back later and head off joists that the plumber has cut. Making the drop about 14 inches wide leaves the plumber enough room to adjust the pipe a few inches in either direction.*

the opening. Then take your R.O. dimensions directly from the instruction sheet.

Garage Doors

These are about as easy as it gets. Ready? Frame the rough opening the same size as the door (Figure 4). The door size should be on the blueprints or in the door schedule. The finish jambs will then cover the edges of the door. If you use a steel I-beam as a framing header, you'll need to increase the rough opening height by 1 1/2 inches. When you fasten a 2-by nailer to the underside of the beam, the opening will be the correct height.

Window Walls

If you're faced with a long run of individual windows of equal height, be alert. Even if you accurately plumb and level each window, if you don't level them in relationship to one another you may run into problems. It may be impossible to get the sills and head casings to line up, and to get the siding on the outside to line up easily. One way around this is to make the rough openings 3/8 inch taller, to allow a little more play. Also make sure that windows are installed carefully.

Extended Extension Jambs

You can usually find rough opening dimensions for windows in the manufacturer's catalog. But there are some instances in which it may be better to deviate from these. One example is where you'll need wide extension jambs. You can save time by wrapping the outside of the window frame with a nailer that will later provide something to fasten the extension jambs to. A second alternative is to install the extension jambs themselves on the window before you install it (Figure 5). Both techniques eliminate the labor-intensive blocking that's usually needed (an especially tough task in older masonry openings).

If you wrap nailers around the window, you'll need to increase the size of the rough opening by the combined thickness of the nailers. This presents a problem with windows that use narrow nailing flanges: Because the nails have to be placed at the outermost edge of the flange, the window won't sit solidly in the opening. Windows with brick mold work better because the brick mold is stiffer than an aluminum or vinyl nailing flange.

Tubs and Showers

When laying out wall studs on the fixture side of a tub or shower, be sure to center a stud bay on the centerline of the faucets so the plumber can rough them in. Otherwise, you'll eventually have to move a stud.

And don't forget to leave a large enough path into the bathroom for the tub/shower unit. While this is usually a simple matter of leaving out a couple of studs, it occasionally requires some real planning. For example, one job we did called for a 3x6-foot cast-iron soaking tub. Rather than wrestle this through the house, we framed and temporarily sheathed an opening in the exterior wall that was big enough to slip the tub through. This made it easier to unload the tub, and allowed it to be delivered much later in the construction schedule.

Toilet Drops

Want to make a plumber happy? Sometimes all it takes is to frame the toilet drop (Figure 6). It can be tough to determine the precise location, so take your best guess based on the floor plan and the particular model of toilet and head off an opening about 14 inches wide. This gives the plumber plenty of room to work, even if the rough opening misses the mark by a few inches. The alternative is to let the plumber remove the offending portion of the joist. Unfortunately, the result tends to look like it was cut with a dull chain saw, and someone has to come back and install a header anyway.

Fireplaces

Fireplace and chimney openings are governed by code. The National Fire Protection Association calls for a minimum 2-inch free air space between combustibles and a masonry fireplace or chimney, and a 4-inch clearance at the back of the fireplace. Framing headers

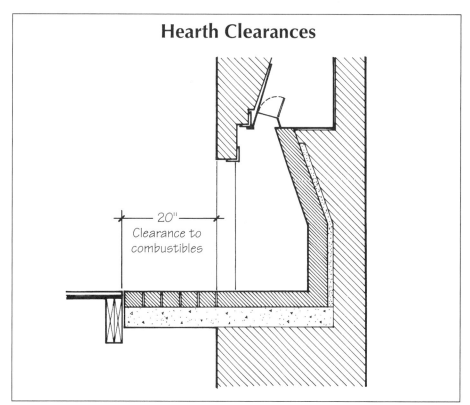

Figure 7. *Headers that support hearth extensions must usually be 20 inches from the firebox. Because fireplace openings are strictly regulated by code, the best policy is to run your plans by the local inspector so you don't have to frame the opening twice.*

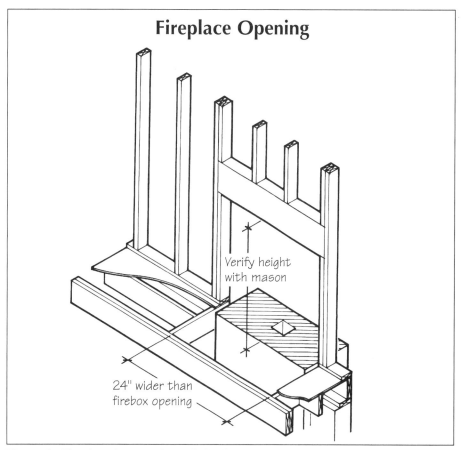

Figure 8. *The size of an exterior wall fireplace opening will depend on the dimensions and design of the finished face of the fireplace, as well as on the mason's methods of building fireboxes, throats, and smoke chambers. If you can't get this information, then oversize the rough opening and fill it in later.*

Layout Tricks for Rough Openings

that support masonry trimmer arches or concrete hearth extensions should be at least 20 inches from the face of the finished firebox (Figure 7). However, it's best not to frame any fireplace or chimney openings until you find out what clearances your building inspector wants. Local codes vary, and inspectors may interpret portions of the code differently than you do. Unless you enjoy appealing decisions, follow your inspector's recommendations.

When a fireplace is on an exterior wall, the rough opening size is seldom called out on the plans. You'll need to get it from the mason (or from the general contractor if you're the framing sub). More often than not, no one really knows what size the opening should be, so call early, to give them plenty of time to scratch their heads. The mason will need to know the size and design of the finished face of the fireplace before calculating a rough opening size (Figure 8). In fact, without the finished fireplace design, any rough opening calculations will be, at best, a guess. If the design hasn't been settled on, get it done before you start framing.

Every mason I've ever known had his own technique when it came to building fireboxes, throats, and smoke chambers. If you don't have a mason lined up, then your only recourse is to oversize the rough opening. Later on, it will be easier to add material to the rough opening than to remove it after the masonry has been laid.

By Carl Hagstrom, a builder from Montrose, Pa., and a contributing editor with The Journal of Light Construction.

Framing Rake Walls

Rake walls are common in house design, but building them is far from routine. Unless you have developed a standard method for framing these walls, the first time you encounter a complex design combining vaulted spaces with intersecting dormers, you'll spend a lot of time scratching your head. Here are the techniques I've developed after years of experience framing rake walls.

From the Top Down

When I'm getting ready to frame a complex house, I do my figuring from the roof down instead of from the sill plates up. Most framers look first at the foundation plan, then the floor plans, and so on; I look first at the roof. In a complicated house, it's easy for the designer to get lost in the elevations. Because I've seen my share of roof plans with eaves that pass through windows and hip beams that cut through doorways, I make sure all of the roof planes come together properly before I start calculating wall heights. Once I can visualize how the roof creates the spaces below, I can figure out how to frame the walls that support the roof. I rarely use the elevations provided on the plans without double-checking. The roof design determines the wall elevations; unless the roof changes, the wall heights can't change.

It's also important when framing rake and other tall walls to work in the proper sequence. Avoid leaving a tall wall standing alone on the subfloor with braces all over. Whenever possible, frame any adjacent walls first and stand the balloon wall between them.

To ensure that I don't miss any rake walls, I typically figure the elevations before I plate my subfloor. The more complicated the roof, the more you need to pay attention to how to plate your walls and the heights of those walls. I write down as much information as I can directly on the plates or on the subfloor nearby so that I don't need to keep referring to the plans when I'm building the wall (see Figure 9). Write down the lengths of the longest and shortest studs and the length of the top plate from short to long point, plus the length of trimmers and cripples, window rough openings, and header lengths. With an inexperienced crew, I may even cut the top plate, king studs, and headers, and position them on the subfloor near the layout marks.

In some cases, I also write down the kind of lumber that should be used. It's difficult to find 2-by material these days that is both long and straight, so I often use laminated veneer lumber (LVL). Also, because LVL studs are stronger and stiffer than sawn lumber, they are often specced by engineers for tall walls. With sawn studs, engineers sometimes spec doubled studs to add strength to the wall.

I usually block balloon walls every 8 feet to strengthen the diaphragm. I install the blocking on edge so that it also serves as fire blocking.

When it comes time to lay out a rake wall, first snap out the perimeter of the wall at full scale on the subfloor (Figure 10). Also snap out the king and common studs, as well as the headers and sills of all window and door openings. If there are any beam pockets in the wall, lay out their elevations on the subfloor as well. For instance, if I have a large ridge beam that needs to sit in a pocket in the rake wall, I draw a full-scale mockup of how the rafters connect to the ridge beam so that I know the exact elevations of the beam and post in the rake wall.

Rake Wall Math

I use a calculator to figure the elevations, and I double-check my layout as I go. Because calculators are easily damaged, I don't use an expensive feet-and-inch model; instead, I have a Texas Instruments scientific calculator. It has all of the trigonometry functions I need, plus three memories. Once you learn how to convert decimals to feet and inches on a regular calculator, it's easy to do all the job-site math.

Rake wall height. To figure the high point of a rake wall where it intersects the ridge, I first measure the deck and convert feet and inches to decimals. (The building footprint has a way of growing or shrinking a little, so I prefer to use actual measurements rather than reading the dimensions off the plans.)

For example, if the house is 26 feet

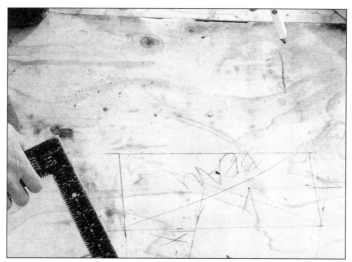

Figure 9. *While laying out rake walls, the author draws details at full scale on the subfloor. To avoid having to continually refer to the plans while framing, he also writes dimensions for plates, studs, jacks, trimmers, and headers directly on the deck.*

Figure 10. *The perimeter of a rake wall is snapped out on the subfloor, along with lines for all studs, and headers and sills for window and door openings.*

6 3/4 inches wide, the span (the measurement to the center of the ridge) is 13 feet 3 3/8 inches. Convert the inches to decimals (see "Feet-Inch Conversions," next page), and plug the results into the formula. In this example, the center of the ridge is at 13.28 feet which, when multiplied by the roof pitch, gives you the rise in feet of the highest point of the rake wall. For a 6/12 pitch, the formula is:

6/12 = .5
.5 x 13.28 = 6.64

Add this to the height of the wall at the eaves and you've got the elevation of the rake at its highest point. If the eaves are at 8 feet, the total height is 14.64 feet (8 + 6.64).

Whether or not you add the length of the rafter plumb cut to this dimension depends on how you treat the rake wall at the roof line. I usually cantilever lookouts over the rake wall, tying them into the rafters one or two layouts back. For a shallow overhang (less than 18 inches), I use 2x6s on the flat, so I typically frame the rake wall 1 1/2 inches shorter (measured square to the rafter) than the elevation of the rafter tops. For lookouts on deeper overhangs, I use 2x6s on edge, so I frame the rake 5 1/2 inches lower than the top of the rafters.

With some truss roofs, such as scissor trusses, I have seen framers build rake walls to the underside of the truss. But this creates a place where the wall can buckle, and you also have to remember to order a shallower gable truss so you can shoot the lookouts over the top without notching. With trusses, I prefer to omit the gable truss and frame the rake wall to the underside of the lookouts.

Length of top plate. To figure the length of the top plate from short to long point, I typically use the following keystrokes on the calculator, in this order: pitch ÷ 12, inverse tangent, cosine, 1/x. For instance, with a 6/12 roof

Feet-Inch Conversions

Feet-inch calculators make it easy to work with building dimensions, but you can convert between decimals and feet-inches using an ordinary calculator. The easiest way to explain the steps involved is to work through the sample dimensions I mention in the main article.

Converting feet-inches to decimals. The span in my example is 13 feet $3^3/_8$ inches. To convert this to a decimal, first solve the fraction; next, add the number of full inches, then divide by 12; finally, add the full number of feet. Enter the numbers or operations into the calculator in sequence, one after the other:

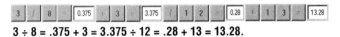

$3 \div 8 = .375 + 3 = 3.375 \div 12 = .28 + 13 = 13.28.$

In decimals, then, the center of the ridge is at 13.28 feet.

Converting decimals to feet-inches. The key to working in the other direction — decimals to feet-inches — is to remember that you're always working with either 12ths of a foot or 16ths of an inch. Let's work the problem in the article, which is to convert 14.87 feet to a feet-inch measurement. What we'll do is subtract out the feet and multiply the fractional part of the decimal by 12, which gives us full inches plus a remainder; subtract out the full inches and multiply the remainder by 16 to get sixteenths of an inch (again, key in the following numbers or operations in order):

$14.87 - 14 = .87 \times 12 = 10.44 - 10 = .44 \times 16 = 7.04$

Now add the whole numbers together and you get 14 feet 10 and just over 7 sixteenths inches. Simple.

Pulling diagonals. This method is also handy when you have to pull long diagonals to square up a foundation or deck. Following the Pythagorean Theorem ($a^2 + b^2 = c^2$), take feet-inch measurements of the two sides, then use the calculator to convert them to decimals and find the square. Add the squares together, then take the square root. Convert this decimal number back into feet-inches and check it against your measurement of the diagonal.

For example, say a foundation is 21 feet $8^7/_8$ inches on one leg, and 15 feet $9^1/_4$ inches on the other. The calculations to find the feet-inch dimension of the diagonal should go like this:

Long leg (a^2):
work the fraction $7 \div 8 = .875$
add full inches $8 + .875 = 8.875$
divide by 12 $8.875 \div 12 = .7395$
round up and add feet $21 + .74 = 21.74$
find the square $21.74 \times 21.74 = 472.6276$

Short leg (b^2):
$1 \div 4 = .25 + 9 = 9.25 \div 12 = .771 + 15 = 15.771$
$15.771 \times 15.771 = 248.7244$

Diagonal ($a^2 + b^2 = c^2$):
$472.6276 + 248.7244 = 721.352$
$\sqrt{721.352} = 26.858$

Convert to feet-inches:
$26.858 - 26 = .858 \times 12 = 10.296 - 10 = .296 \times 16 = 4.736$

The diagonal should be 26 feet 10 and just under 5 sixteenths inches long. Close enough.

— E.D.

---

pitch, the unit length (the length of the sloping plate per foot of horizontal run) of the top plate is:

$6/12 = 0.5$
Inv. Tan = 26.565 (degrees of pitch)
Cosine = 0.894
$1/x = 1.118$

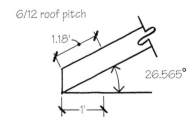

6/12 roof pitch

To find the plate length in decimals, first round off 1.118 to 1.12, then multiply by the span: $1.12 \times 13.28 = 14.87$ feet. Now convert to feet and inches. (See "Stacking Supported Valleys" in Chapter 8 for figuring rafter lengths, and for rake wall top plates as well.)

Framing the Wall

The first step in actually framing the wall is to cut the sole plate and toe-nail it on edge to the subfloor (Figure 11). This allows you to frame the wall right over your snapped layout lines. Since you've already done all the math while snapping the layout, physically lay studs on the deck and scribe the bevel cut where the stud crosses the top plate. Start with the longest and shortest studs; if there are a lot of windows, frame all the king studs next.

Before standing a tall wall, use a metal lumber strap to anchor the sole plate to the deck in a few places. These straps keep a tall wall from kicking out at the bottom as you lift it. Facing the edge of the deck, slide a strap under the sole plate, then bend it up and nail through the strap into the edge of the plate. I use a 16d nail fully set and nail the rest of the strap through the subfloor into solid floor framing in three or four places.

On most walls, I install all of the sheathing while the wall is still on the deck. If the wall isn't too large and heavy, I raise the wall up high enough to slide some sawhorses underneath. Then I nail a couple of long 2x4s or 2x6s to

Figure 11. *Framing begins by toe-nailing the sole plate to the deck (left). Before sheathing the wall (right), fasten steel binding straps around the plate and nail them to the joists. This will keep the wall from kicking out when you stand it up.*

the side of a post or king stud within the wall. (I avoid the studs at either end, because the push-sticks will get in the way later and have to be removed.) I nail the 2-bys with a couple of 16d nails, placed close together so that as we raise the wall, the push-sticks will rotate. The push-sticks add overhead leverage and balance, and serve as braces after the wall is standing.

Never use this method, however, on an extra-tall or very heavy balloon wall. It's just too difficult and dangerous to raise a wall when most of the weight is way above your head as you raise it. For some large walls, I use Proctor wall jacks (Proctor Products, P.O. Box 697, Kirkland, WA 98083; 425/822-9296), which are slow but safe, and give you complete control over the lift. If I have a lot of tall, heavy walls, I generally use a crane. In this case, I try to schedule the crane for a time when it can also be used to lift beams and trusses into place.

Once the wall is standing, plumb and brace the center of the wall and throw a few more braces on king studs or posts. Also nail diagonal braces to adjacent walls from top plate to top plate to add some strength to the wall until it's completely plumbed and lined. In Colorado, where I work, it gets really windy, and it sometimes seems that I use more braces than studs. But those tall rake walls are like sails and I wouldn't want to lose one.

By Eric Dickerson, a long-time framing sub, and owner of a general contracting company in Ridgway, Colo.

Site-Built Panelized Walls

In the mountains of southern California, where I work as a framing subcontractor, we build most of our houses on stepped footings that follow the grade of the hillside. We start by framing a "buildup" from the footings to a point that provides a level base for the first-floor deck. On many sites, this buildup to the deck can be 18 or 20 feet high on the downhill side.

When you're working at the very top of a house like this, it's a long way down. Installing windows and siding at these heights — whether from ladders or scaffolding — is slow and hazardous. So during 13 years of framing in these hills, my crew and I have developed some tricks for completing most of the exterior work while the walls are still lying flat on the deck. We've gotten to the point where when we stand a wall, it's almost finished — siding, windows, window trim, even chimneys and fascia are all done (Figure 12). About the only work we have to do from ladders is apply the corner trim on the walls.

The routine we've worked out is based on the materials we use. Most of our houses get T1-11 or hardboard lap siding. If you're using another kind of siding, you might have to adapt these techniques for that material.

Square Plus Level Equals Plumb

It's important to start with a level floor deck. If your deck is level and your wall is square, when you stand the wall the corners are sure to be plumb. We start by nailing the bottom wall plate on edge to a chalk line that marks the inside surface of the wall. We shoot 16d air nails at an angle through the plate every 3 feet or so to keep the plate from slipping around while we frame and square the wall.

As soon as our wall is framed up on the deck, we check the diagonal measurements from corner to corner to make sure the wall is square (Figure 13). Then we tack the top plates to the deck in a few places, to hold it square until the siding is on. This procedure is the same on a balloon-framed gable end. If we're using trusses, the last step before sheathing and siding is to nail the gable

Figure 12. *To avoid the dangers of working from scaffolds and ladders, the author finishes walls while they are lying flat on the deck. Even heavy gable walls — complete with siding, windows, rake trim, and wood chimneys — can be lifted with wall jacks or a crane.*

Figure 13. *Since the finished wall can't be racked after siding is applied, it's crucial to check diagonals to square the wall. When the wall is raised on a level deck, it will stand plumb.*

truss to the wall plate (Figure 14).

Strap anchors. The toe-nails in the bottom plate might not hold a heavy wall from slipping over the edge when we go to raise it. Some people nail a vertical 2x4 block to the outside of the building to stop the wall from kicking out over the edge, but I don't have much confidence in that method. Instead, we use the metal banding used to bundle lumber. We cut a foot-long length of banding, slip it under the plate, and bend it in an L-shape. We nail one leg of the metal strip into the bottom of the wall plate, then nail the other into the deck. We use two or three straps for a small wall, and four or more for a big heavy wall. When the wall's going up, the straps bend, allowing the bottom plate to rotate, but they won't let the wall slip. Just remember to get the straps in place before you cover up the wall with plywood — after that it's too late.

Windows

Once the wall is framed and squared, we move on to the windows. On walls that get hardboard lap siding, we first

Figure 14. *The author nails a gable truss to the top plate. The plan is to raise an entire second-floor gable wall, complete with windows, siding, and rake trim.*

Figure 15. *Applying shear-panel sheathing moves quickly when the wall is lying on the deck. Instead of pulling dimensions or figuring angles, the plywood is simply nailed in place and the excess cut off.*

nail plywood shear-panel sheathing to the studs for seismic strength (Figure 15), then staple on housewrap and install the windows. If we're using T1-11 siding, we first staple housewrap to the studs, then install the windows; the T1-11 will hold the wall square and provide the racking resistance required by the seismic code.

Sometimes we have to make adjustments. For example, the windows for the house shown in these photos were shipped late: We couldn't wait, so we raised the wall without windows. Usually, however, we put the windows in, too, because it's so much easier than installing them from a scaffold.

With the materials we generally use, the routine is pretty simple. The windows come with a nailing flange that nails to the rough opening, either directly to the wall framing or through the shear panel into the framing. We hold the bottom of each window down on the rough sill and nail the two bottom corners of the window flange to the framing. Then we square the window the same way we squared the wall — by checking the cross-diagonals and making sure they're the same. The high-quality windows we're using are almost always square, but occasionally one needs to be moved a sixteenth or so. When we're satisfied that the window is square, we nail off the rest of the flange.

Figure 16. *Aligning, fastening, and caulking the siding is a breeze when the wall is lying flat (top). As with sheathing, the author lets the siding run wild at the ends, then snaps a line along the edge and cuts off the excess (above).*

Siding

Next, we lay out the siding by snapping lines on the housewrap to mark the stud locations. If the siding is T1-11, we make a layout mark for the first piece by squaring up from a seam in the T1-11 on the wall on the story below.

If we're using lap siding, the starter course needs to be offset so that when we stand the wall, the first piece will hang down just the right distance to lap over the last piece on the wall below. For example, if the last piece on the wall below stops 2 inches below the floor, we set the first piece on the upper wall so that it will hang down about $1^{7}/_{8}$ inches. That extra $^{1}/_{8}$ inch allows for a little bit of play so that if something's off just a hair when we stand the wall, the siding

won't be too tight.

The lap siding we use has a shallow tongue-and-groove edge that makes it easy to lay up (Figure 16). When the starter course is nailed on just right, it guides the placement of succeeding courses. The siding is prefinished, so when it's nailed on, it's done.

We keep plenty of high-quality, all-purpose silicone caulk on hand, and caulk any joint where water might get in. The siding gets caulked at the butt joints, and we leave a $^{1}/_{8}$-inch gap around the windows that gets caulked carefully, too. Also, as we trim out the windows (usually with square-edge 1x6), we apply a bead of caulk over the top of the head trim.

It's easy to do a nice neat job when everything is lying flat and within easy reach. Lately, we've even been framing up the gable-end roof overhang while the wall is lying down. The ladder-shaped barge rafter gets nailed right to the gable-end truss (Figure 17).

Lifting the Wall

With all of this material nailed in place, the wall is heavy to lift. We sometimes lift small walls by hand, but usually we use Proctor wall jacks (Proctor Products, P.O. Box 697, Kirkland, WA 98083; 206/822-9296). Two wall jacks are enough to lift most walls.

There are a few things to keep in mind when you use wall jacks. First, make cer-

Figure 17. *Before lifting the wall, the barge rafter for the rake is fastened and the rafter tail is laid out and cut in place.*

Figure 18. *The author and another carpenter use wall jacks to slowly raise the heavy wall (top). Jacks must be placed to bear over a floor joist so they don't punch through the plywood. As the wall moves toward the vertical (above), permanent braces are nailed to the studs. The wall jacks can be removed after the wall plate is nailed off and the wall is securely braced to the floor.*

tain each jack is securely attached to the wall framing at a point that is structurally strong (Figure 18). Those attach points have to be strong enough to support the whole wall. Also, make sure that the base of each jack is bearing directly over a joist; otherwise, when you start cranking, the foot of the jack could break right through the plywood.

As you raise the wall, make sure all the jacks lift slowly and continuously at the same time. Don't move in jerks, and don't let one jack get ahead of any other. And don't let the cable build up on one side of the spool as you crank — when it slips off, there will be a sudden jerk that you'd just as soon avoid.

When the wall moves past 45 degrees and heads toward vertical, you can feel the weight coming off the jacks. At that point, someone should step under the wall and quickly nail a brace onto a couple of studs. Use a single nail so the brace can rotate as the wall rises. Don't spend long doing this — standing under the wall isn't safe.

When the wall is vertical, get the level on it and make sure it's plumb. Then get some more nails into the top of the braces, and nail the brace bottoms to a block that's nailed to the deck. (Make sure the block is nailed into a joist, not just to the plywood.) When the braces are securely fastened and the wall plate is nailed to the deck, you can get on a ladder and remove the wall jacks.

If a wall is really huge, we get a crane on site. When you're lifting with a crane, pay attention to attach points just as you do when using wall jacks. Often, we leave off a piece of siding near the top of the wall so that the lifting straps for the crane can go through the wall and wrap around a gable truss member. Each wall is different, but each has some way to make a strong lifting connection. Pay attention — if a strap lets go, it could be a disaster.

By Bill Nebeker, a framing subcontractor in Crestline, Calif. Photographs by Peter Rintye.

Building Stiff Two-Story Window Walls

Great rooms with high ceilings and walls full of windows have become popular in contemporary house design. But for the unprepared builder, a 16-foot-high wall can spell big headaches. When a strong wind hits an underdesigned wall, excessive flexing and racking may cause problems ranging from cracked plaster and stucco to cracked or rattling glass.

To prevent these headaches you need to understand how structures react to wind loads (see Figure 19). All walls have to resist three types of forces: flexing, which acts perpendicular to the wall, pushing or pulling it in or out; shear, which acts in the plane of the wall to cause racking; and normal gravity loads.

Most builders are familiar with gravity loads. Here, we'll look at ways to resist flexing and shear in two-story walls. Throughout, I'll assume a typical design wind load. In much of the country and in the Atlanta area, where I work, that's 75 mph, although the Great Plains and coastal areas face even stronger winds. Building in hurricane-prone areas is a different subject and won't be covered here.

Stiffening Tall Walls

Wind blowing perpendicular to a wall causes it to flex. To resist flexing, a wall should be stiff enough to limit the mid-height deflection under the design wind load to $1/240$ of the overall wall height (about $3/4$ inch for a 16-foot-high wall). In planning such a wall, there are four basic variables to consider:

- **Design wind speed:** The higher this is, the stiffer the wall must be.
- **Height of the wall:** With all things equal except height, a 12-foot-high wall is five times more flexible than an 8-foot-high wall; a 16-foot-high wall is 16 times more flexible than the 8-foot-high wall.
- **Thickness of the framing:** 2x6s are stiffer than 2x4s.
- **The number of studs** you have room for: This depends a lot on how the windows are laid out.

Window layout. Figure 20 shows

three window layouts, ranging from worst design to best. A regular pattern of windows leaves more room for more structural studs and beams than does an irregular pattern or one with enormous windows. If your mullions are narrow, you can fit more full-height studs into them by supporting any lower headers (which are nonstructural) on steel joist hangers rather than jack studs, as in Figure 21.

Balloon framing. In general, two-story walls should be balloon-framed. In a platform-framed wall, the hinge created by the horizontal plates adds too much flexibility. A common exception to the balloon-framing rule is a narrow two-story wall that's almost completely filled with glass (I see this most often in entry foyers), leaving insufficient room for full-height studs. Since the wall can't be balloon-framed, you may have to stiffen it with a horizontal beam placed at mid-height.

Steel channels. There will be times when none of these variables are work-

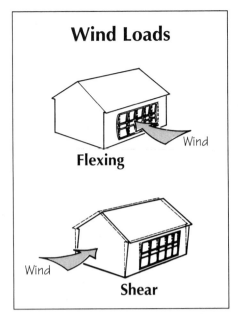

Figure 19. *Tall window walls must be able to resist wind loads in two ways: flexing, caused by wind blowing perpendicular to the window wall, and shear, from wind blowing parallel to the wall.*

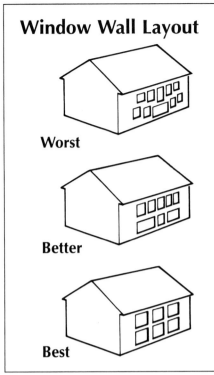

Figure 20. *A regular pattern of windows makes for stronger walls because it leaves room for more full-height structural studs and connecting beams than does an irregular pattern.*

Figure 21. *The more studs in a wall, the stiffer it will be. You can get two extra studs into a mullion by using joist hangers rather than jack studs to support the lower headers. In the example shown, this raises the number of studs from four to six, increasing stiffness by 50%.*

ing in your favor. If that's the case you may need something even stiffer than 2x6s. One option is to use steel channels (Table 1). These are structural steel members with U-shaped cross-sections that are bolted to the wide face of wood studs between the top and bottom wall plates. They're available in 3-, 4-, and 5-inch widths.

Steel channels are good for retrofitting flexible walls. Because they cover the wide face of the stud, however, they require you to remove the existing wiring, plumbing, and insulation. An alternative technique is to nail new studs flatwise to the 1 1/2-inch face of the existing studs. The resulting T-shaped members will increase both stiffness and strength without any internal changes to the existing wall. The design for either of these approaches should be handled by your structural engineer.

Using a Design Table

One way to make sure your wall will be stiff enough is with a design table (see Table 2). I've developed this table for use in Atlanta, in a wall framed with Southern Pine.

The table only covers walls that support fairly light loads. You can't support more than 1,000 pounds with a Southern Pine 2x6 stud and also expect it to have enough strength left to resist flexing. The issue is even more critical for 2x4s. Building codes don't let you use 2x4 studs longer than 14 feet 7 inches to hold up gravity loads. Beyond that height, the stud's capacity falls too rapidly. Finally, to keep them from buckling, codes require that tall studs be properly braced in the plane of the wall. You can satisfy this requirement with horizontal fireblocking or let-in steel straps spaced 6 feet on-center.

Resisting Shear

An even trickier problem presents itself when a two-story wall must resist shear, or racking forces. Unfortunately, the options for resisting shear are limited. The best solution, when possible, is a diaphragm of OSB or plywood shear panels. You can use shear panels in a

Table 1. Stiffness of Wood Studs vs. Steel Channels

Channel	Specifications	Compared to No. 2 Southern Pine 2x6	Compared to Stud Grade Southern Pine 2x4
C3x6	3 in. deep 6 lbs./lin. ft.	1.8 times as stiff	8 times as stiff
C5x9	5 in. deep 9 lbs./lin. ft.	7.8 times as stiff	35 times as stiff

One retrofit option for flexible walls is steel channels bolted to the wide faces of the studs. As this table shows, adding a 3-inch-deep channel to a 2x4 stud makes it as stiff as nailing eight additional studs to it. A disadvantage of channels is that they may require you to reroute the existing wiring and plumbing.

Chapter Seven: Wall Framing

Table 2. Design Table for Tall Window Walls

		Wall Height								
		14'	15'	16'	17'	18'	19'	20'	21'	22'
STUD TYPE	Stud So. Pine 2x4	5.7	4.6	3.7	3.1	2.6	2.2	1.9	1.6	—
	Stud So. Pine 2x6	21.9	17.7	14.5	12.1	10.1	8.6	7.3	6.3	5.5
	#2 So. Pine 2x6	25.1	20.3	16.6	13.8	11.6	9.8	8.4	7.2	6.3

Note: *This design table gives the stud spacing for a solid wall, which you can then use to determine how many studs are needed between windows in a tall window wall. This table was developed by the author for the Atlanta area, with a design wind speed of 75 mph and southern pine framing lumber. A structural engineer can develop an appropriate table for your area.*

To use the table:

(1) Find the window wall height along the horizontal axis and the stud type along the vertical axis. The intersection gives you the wall length in inches (measured parallel to the plate) that each stud can support. In Example A, a 15-foot-high wall framed with stud-grade 2x6s needs one stud for every 17.7 inches of wall.

(2) Now divide the window wall into sections that vertically bisect each bank of windows. Each section should enclose one window mullion and half of any adjoining windows. In Example A, that gives 3-, 6-, and 4-foot sections (36, 72, and 48 inches).

(3) Divide the width of each section in inches by the number from the table to get the approximate number of studs required in each mullion. Round the answers to the nearest whole number. In Example A, that's two studs for the 36-inch section on the left (36 ÷ 17.7 = 2.03), four studs in the middle (72 ÷ 17.7 = 4.07), and three studs on the right (48 ÷ 17.7 = 2.71).

If upper and lower windows don't line up, but overlap in such a way that you can't fit any studs between them, then you must treat them as a single bank of windows. In Example B, the overlapping windows on the right are treated as one 6-foot-wide window. The walls in Examples A and B are identical from a structural perspective.

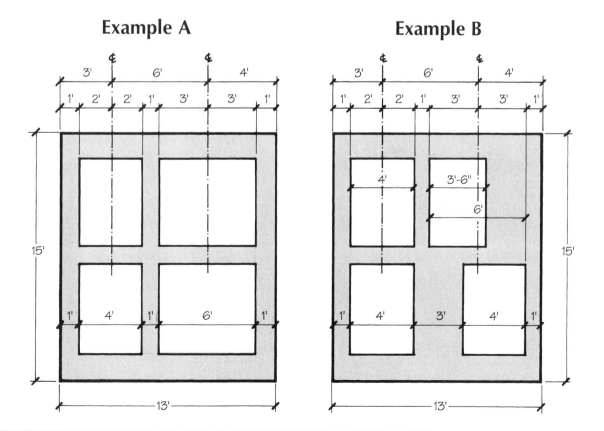

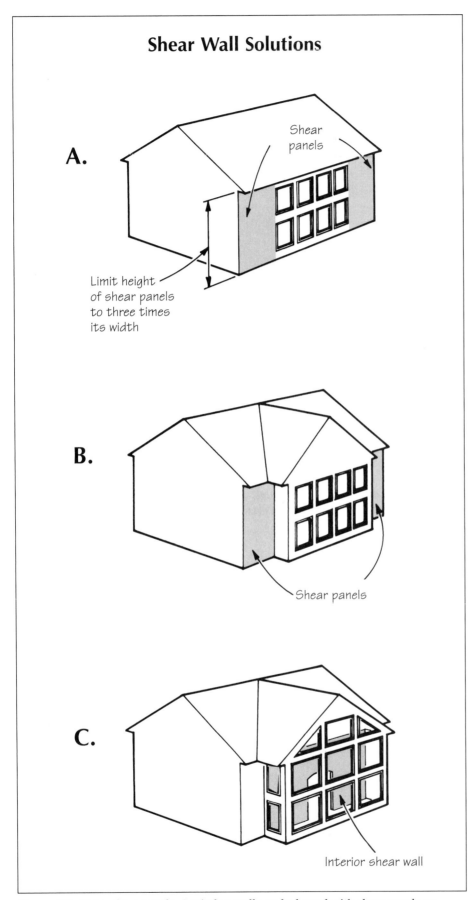

Figure 22. *Using shear panels: A window wall can be braced with shear panels on each side (A). With an addition that is not too long (B), shear panels on the main wall at each side of the addition may work. In some cases, an interior shear wall (C) is the best way to prevent racking.*

two-story wall where the windows don't take up the full width of the house. In this case, you flank the windows with shear panels. If the two-story wall is at the end of an addition, there may be room for panels at the intersection of the addition and the main house (Figure 22).

One crucial dimension is the shear diaphragm's height-to-width ratio. An 8-foot or 9-foot wall can get by with one shear panel at each corner, but the same isn't true for a very tall wall. A shear panel is just not effective against racking — and the building codes won't accept it — if its height exceeds three and a half times its width. A 4-foot-wide strip of sheathing won't meet the code requirements for shear for any wall over 14 feet tall.

The concept behind this becomes clear if you look at the plywood as a simple way to provide X-bracing. The part of the panel doing most of the work is an X from corner to corner; the taller and more slender the X becomes, the less effectively it will resist racking. I like to add a margin of safety by limiting the height of the shear panel to three times its width: If there isn't room for a 6-foot-wide strip of plywood at each corner of an 18-foot wall, it's time to rethink the house design. One obvious solution is not to let windows take up the entire width of one wall.

Engineered solutions. If you can't change the window layout, another alternative is to frame an interior wall parallel to the window wall and to sheathe it with OSB or plywood shear panels. Unfortunately, you can't use rules of thumb to find the required distance from the end wall to the interior shear wall; you will need a structural engineer.

If the design prohibits any type of shear wall, you may want to consider another engineered solution, a structural-steel frame known as a moment frame. "Moment" is bending in a structural member created by an applied load or force, and a moment frame is a structural frame that resists bending moments by using rigid joints. The difference between the joints in a moment frame and those in a standard wood frame is the difference between a framing square and a door hinge — one is rigid, the other isn't.

A moment frame is usually made from welded steel, whether I-beam, channel, or rectangular tube (for a comparison of steel types and sizes see "Steel Beam Options," Chapter 13). The material you choose will depend on the amount of racking the wall must resist, as well as on the interior and exterior finishes. For instance, the flat face of a rectangular tube is an easy surface to shoot strapping into. Whatever your choice, a steel fabricator can turn an engineer's design into the moment frame you need.

I've been asked from time to time if a moment frame can be made from wood with steel connectors. The short answer is no. It's theoretically possible by bolting heavy-gauge T plates to the face of each joint. But the size of the plates and framing members, as well as the number of bolts needed, would make the extra effort and expense greater than that required for a steel frame.

How Much Is Too Much?

One final piece of advice is to make two-story walls stiffer than you think necessary. This may raise costs, but it will also keep you out of trouble. One builder I've worked with in Atlanta learned that lesson the hard way. About five years ago, he framed a two-story window wall with 2x4 studs. It seemed stiff enough to him, especially after the interior finishes went on.

Unfortunately, the owners didn't agree. Because of visible flexing, the old wall came down and a 2x6 wall went up — at the builder's expense. Now he won't build tall walls with anything less than 2x6s. If he doesn't have room for a comfortable number of studs, he asks me to check out the design in advance. He's learned that tall walls are like that old oil filter ad: You can pay now or you can pay later.

—Christopher DeBlois

Bracing Foam-Sheathed Walls

All building codes require that walls be braced to resist racking caused by wind loads (Figure 23). Most prescriptive codes recommend one of three options: 1x4 diagonal bracing, 5/8-inch diagonal board sheathing, or 3/8-inch plywood sheathing. Other more progressive performance-based codes specify minimum design values for racking resistance, but let you achieve them with almost any materials you choose. For example, you can count the combined contributions of metal strapping, drywall, interior partitions, and ceiling diaphragms.

Some foam sheathing manufacturers point to these performance-based measures as proof that a foam-sheathed wall meets code. It's true that with the right floor plan, wall dimensions, and number of wall openings, an unsheathed or foam-sheathed wall might comply with code. But bear in mind that most code officials will require the builder to prove that such an alternative wall bracing system meets the requirements. Usually this means that an engineer must stamp the plans.

For conventional buildings (not those in high-wind or seismic zones), prescriptive bracing codes are based, in large part, on a Federal Housing Administration (FHA) interim standard

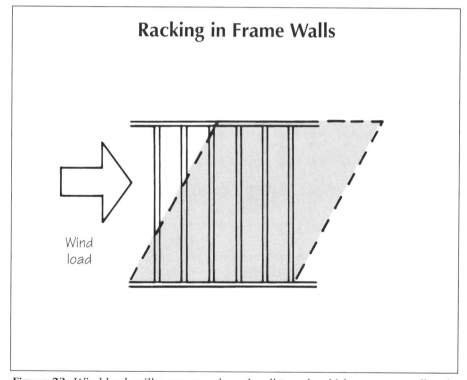

Figure 23. *Wind loads will cause an unbraced wall to rack, which can cause wall and ceiling finishes to crack and doors and windows to bind.*

that dates back to 1949 (a permanent standard was never introduced). This standard established 5,200 pounds as an acceptable level of racking resistance for wood-framed walls. This force does not easily translate to a typical wind speed. Rather, the 5,200-pound base load equals the racking resistance provided by wood-framed walls, sheathed with horizontal boards and braced by 1x4 let-in bracing — a common construction practice in 1949.

Certainly, 1x4 let-in braces have worked well in the past. Older homes

Figure 24. *Let-in braces gained code approval at a time when board sheathing was standard on all homes. With modern plywood sheathing, such let-in bracing is not necessary. With no sheathing or foam sheathing, research indicates that let-in bracing alone does not meet current design values. Where 1x4 braces are a part of your bracing system, use top-grade material, nail each brace with a pair of 8d nails in each stud and plate, and keep the stud spacing to a minimum of 16 inches on-center.*

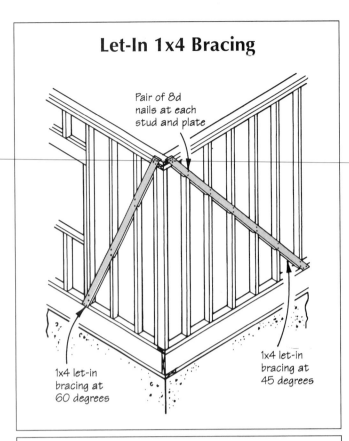

Figure 25. *Metal braces are intended only for temporary bracing during construction, not as a replacement for structural sheathing. Flat metal bracing only resists in tension. To resist wind loads in both directions, install the straps in an X-configuration, and wrap the ends over the top plates to keep the nails from pulling loose. Metal T-bracing is installed in a saw kerf in the stud. This type of bracing provides some resistance in compression, but not enough to replace a structural skin.*

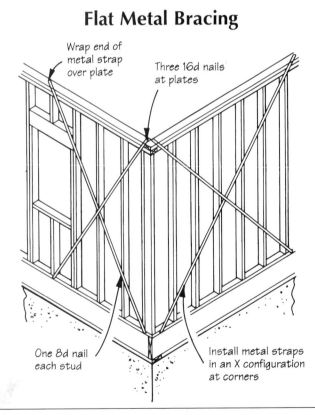

built with such bracing have a good track record. But usually these braces are used in conjunction with plywood or board sheathing. Questions arise when you build without structural sheathing.

In this section I will examine four common bracing systems for walls with nonstructural sheathing — 1x4 let-in bracing, metal bracing, diagonal stud bracing, and plywood corners. While builders often turn to let-in or metal braces as substitutes for structural sheathing, neither research nor mathematical calculations support the use of either in unsheathed walls. Using these bracing systems probably won't result in catastrophic failure, but it may result in a steady stream of callbacks for such things as drywall pops, cracked stucco, and windows that bind.

Let-In Doubts

In 1977, engineers with the U.S. Forest Products Laboratory (FPL), in Madison, Wis., studied let-in bracing. They found that much of a braced wall's racking strength is due to the interaction of board sheathing and let-in bracing. Even the clear, straight-grained 1x4 let-in braces used to support their unsheathed test walls provided less than two-thirds of the 5,200-pound value specified by the FHA standard (Figure 24). The #2 Common 1x4s typically found on the job site can hardly resist comparable loads.

Common-grade 1x4 stock is not structurally graded; it is graded only for appearance. Garden variety braces made from #2 Common 1x4s can have 2-inch-wide red knots. That means almost 60% of their cross-sectional dimension is nonstructural. Add to that the structurally weak cross-grain region surrounding the knots and you have a recipe for failure at low-level loading.

Additional FPL research in 1983 found that "off-the-shelf" #2 boards provided only 600 pounds of resistance to horizontal loads, such as those from wind, before failing. According to this test, it would take nine 1x4 let-in braces (without sheathing) to provide the FHA minimum 5,200 pounds of lateral resistance in a wall.

It's worth noting that typical design values for structural members carry a 2.5 factor of safety, reducing the value of 1x4 braces to 240 pounds of design strength. Also, the braces were installed at a 45-degree angle in the test. Braces installed at a 60-degree angle or greater provide far less resistance.

Metal Bracing

Several manufacturers of rigid foam

insulation recommend the use of metal bracing when building foam-sheathed walls. However, the product literature depicting this practice may be misleading. Tests conducted by both FPL and Simpson Strong-Tie indicate that metal bracing alone does not meet the FHA minimum standard of 5,200 pounds.

According to FPL researchers, a metal strap brace delivers 1,500 pounds of lateral resistance. Simpson Strong-Tie assigns an even lower value to metal braces, even to their more substantial T-shaped bracing. Simpson Strong-Tie's product manual clearly states that metal wall braces prevent walls from racking during construction and are not designed to replace load-carrying shear wall components.

In testing, metal braces usually fail due to the nails slipping. So it's the size and number of nails in each brace that determine the design values. In the FPL tests, the metal straps were wrapped over the top and bottom of the wall frame to minimize nail withdrawal.

Also, flat metal braces offer no resistance to compressive loads. So they must be installed in an X configuration at each building corner to resist lateral wind loads from all directions (Figure 25). This guarantees that one leg of the bracing will always be in tension.

Diagonal Stud Bracing

Diagonal stud bracing is a seldom-used but effective wall bracing system that deserves consideration when building foam-sheathed homes. With this technique, studs run diagonally at the corners within the wall cavity. The depth of the brace is perpendicular to the face of the wall.

This technique was once commonly used to brace stuccoed and plastered walls against cracking, but it has been largely abandoned by modern builders since it takes more time. For maximum strength, braces should be installed at a 45-degree angle and extend in one unbroken length from top plate to bottom shoe (Figure 26). Vertical studs are cut to fit around the continuous diagonal brace in the same way studs are installed under a sloping gable end.

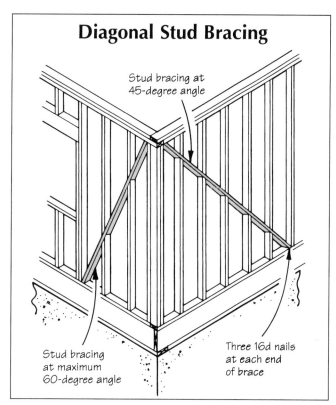

Figure 26. *Diagonal stud bracing is time-consuming to install but offers adequate racking resistance. For maximum strength, braces should be installed between a 45- and 60-degree angle and extend in one unbroken length from top plate to bottom shoe. Cut the studs to fit around the brace in the same way studs are installed under a sloping gable end.*

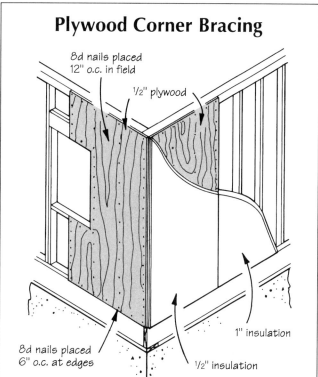

Figure 27. *Plywood corners offer adequate racking resistance. For best results, brace at least two opposing corners in each wall and fasten the plywood with 8d nails, spaced 6 inches at the edges and 12 inches in the field. If the wall has several windows or doors, also brace with plywood around the openings.*

This type of stud brace only works in compression, so braces should run in opposing directions from the top plate to the sole plate at each corner. In a 2x6 stud wall spaced 24 inches on-center, a "construction grade" 2x6 brace installed at a 45-degree angle will resist a 5,400-pound design load — above the FHA minimum. This load is computed using design stress values, which include a built-in safety factor. Failure in compression would most likely occur at a much higher level, especially if you use stress-graded lumber.

Plywood Corners

Many builders brace foam-sheathed homes with plywood corners. Typically they install 1/2-inch-thick sheets vertically at the corners and overlay 1/2-inch-thick rigid foam. One-inch-thick foam is then used to sheathe the remainder of the house, leaving the exterior wall surface flush.

Each corner panel will resist an ultimate load of 3,120 pounds when nailed with 8d nails, spaced 6 inches at the edges and 12 inches in the field of the panel. With two opposing corners braced, a wall can resist 6,240 pounds, so it seems to work nicely (Figure 27). Drywall on the inside will add a measure of safety.

Drywall Contribution

Gypsum wallboard is without doubt the most popular interior wall sheathing used in light-frame construction. But rarely is wallboard given the credit it deserves when evaluating the structural integrity of a wall system. Its success depends on orientation. In the FPl studies, 1/2-inch wallboard provided 150 pounds of resistance per lineal foot of wall length when applied vertically to an 8-foot-high wall, and 250 pounds per foot when applied horizontally. Studies sponsored by gypsum board manufacturers have yielded values as high as 660 pounds per linear foot. The Uniform Building Code recognizes a conservative 100 pounds per linear foot as a structural contribution.

Depending on the floor plan, some codes will also allow for bracing contributions from interior walls, provided the gypsum is fastened according to strict nailing schedules. However, this could be undermined by future renovations. No interior walls can be removed without disturbing the balance, much in the same way a truss can't be cut without interrupting the distribution of forces. While you may get some inspectors to approve such a design, it is much more prudent to rely on the strength of the exterior shell.

Window and Door Openings

Windows, doors, and garage openings will compromise any bracing scheme. Diagonal bracing in particular — whether it is let-in 1x4, metal, or stud bracing — works best at a 45-degree angle, and will add some resistance when installed at angles up to 60 degrees. But anything steeper is nearly worthless. This means that in any wall that relies on diagonal bracing, windows and doors must be kept 6 to 8 feet away from the corners for optimum bracing value. Any cutout for windows and doors also reduces the contributions you get from plywood and drywall. So regardless of the sheathing material you use, if you end up building a wall full of sliders and picture windows, or a second story over a two-car garage, have an engineer check your design. Otherwise you might have to repair more than a few nail pops and stuck doors.

—*Paul Fisette*

Framing for Corner Windows

Corner windows have been around for years, but until recently they were expensive novelty items rarely used in residential construction. Although they're still expensive, at least three major wood window manufacturers now offer corner windows in their catalogs of standard products (see manufacturers' list at end of article). It may be only a matter of time before you're asked to install one.

These windows use either bent or mitered insulated glass units in a typical wood frame. The catalogs give no structural details beyond pointing out the need for a cantilevered structural header designed by an engineer to prevent the window from carrying any load. This section provides such a header design suitable for several common framing conditions.

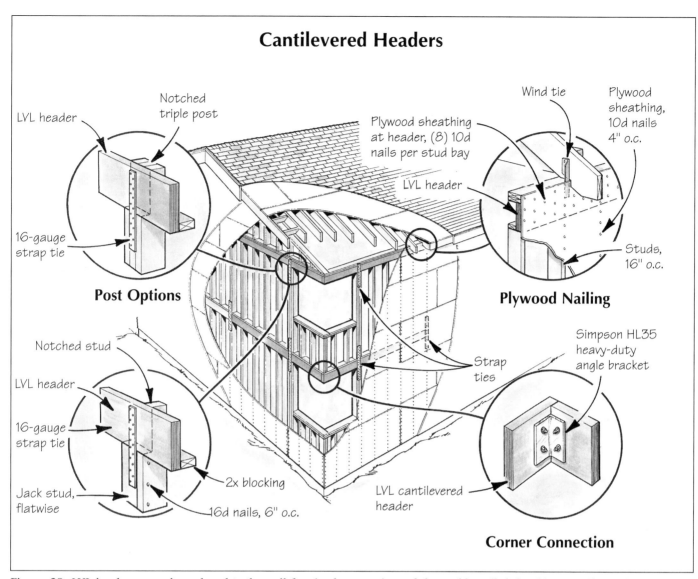

Figure 28. *LVL headers, securely anchored to the wall framing by strap ties and thoroughly nailed sheathing, cantilever above the corner window opening. For a two-story structure, the lower header may be incorporated as the band joist.*

Design Assumptions

Design loads vary greatly from region to region, so selecting design assumptions turned out to be a challenge. For example, for a typical one-story gable roof in the Sunbelt, with 2x6 rafters of #3 western woods, the roof load carried by the header might be only 230 pounds per linear foot. But up in 100-psf snow country, a roof framed with Doug fir 2x14s could exert a load of 2,510 pounds per linear foot — more than ten times as great. Similarly, the presence or absence of a floor above the header and the span and loading on that floor could result in a wide range of load conditions.

So while I couldn't include every extreme, I chose a set of design loads that would safely cover most conditions in the continental U.S. (see "Design Assumptions," next page).

Note that in using this design example (or any other), it is very important to review the design assumptions to make sure that they are appropriate to your situation. In this case, either a wider roof or heavier design loads would mean that this design might be unsuitable. If you are in the north woods or high in the Rockies, you might need to double up the LVL header.

Simple Framing

My basic design is quite simple (see Figure 28). The wall framing on each side of the corner window incorporates a 10-foot LVL header, with one end cantilevered over the approximately 40-inch rough window opening (the size of the largest corner unit offered by the manufacturers; check the manufacturer's literature for exact R.O. size).

The LVL header sizes for typical gable and hipped roofs are shown in Figure 29. For a two-story structure with corner windows on both floors, the lower header can actually be incorporated as the band joist, and trimmed to match the size of the floor joists if necessary.

At the corner, the headers meet and are bolted together with a heavy-duty angle bracket. This serves to keep them from twisting and also transfers load from one header to the other.

Header Support Posts

The cantilevered headers are sup-

Framing for Corner Windows

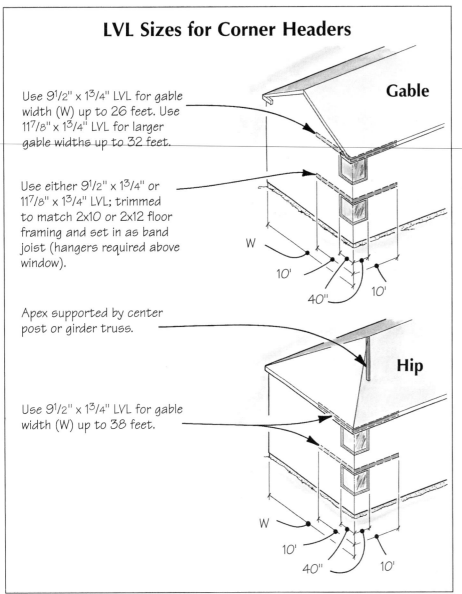

Figure 29. *Note that for the hipped roof, the acceptable span is greater while the headers are smaller. This is because the apex, or top end, of the hip rafter is assumed to be fully supported down to the foundation — either by a post or a rafter truss that carries the load to the exterior walls. The supported apex carries two-thirds of the roof loads, which would otherwise rest on the corner window headers. In the case of an unsupported apex (as you might have in a cathedral ceiling design), this design does not apply. For more on hip rafter support, see "Straight Talk About Hips and Valleys" (Chapter 3).*

ported by built-up posts at each side of the window opening. The posts must be notched to run past the header and up to the top plates. Because the LVL header bears on a small (1 1/2x1 3/4-inch) cross-section of the individual studs, we have to look carefully at compression perpendicular to the grain to be sure the load on the LVL header won't compress the wood fibers of the header where it rests on the posts. (This would allow the LVL header to drop.) The solution is to spread the load over a large enough area. There are two options: Either use a triple 2-by post, with all three studs notched for the LVL header, or notch a single stud and turn a second stud flatwise against the sheathing underneath the header.

The second method uses wood more efficiently and also somewhat reduces conductive heat loss, since you can insulate behind the flat stud. To assemble this L-shaped post, use 16d nails 6 inches on-center. It's not a bad idea to predrill the nail holes to prevent splitting, but don't drill a hole larger than 75% of the nail's diameter. For 16d common nail's, which are .135 inch in diameter, limit your hole to .1, or 3/32 inch.

However you build the posts, you must also notch the studs beyond the post to accommodate the header and provide nailing for drywall on the interior. Make sure you cut the notches on the post with care, because you want to avoid excessive sagging of the header assembly.

Allowing for Deflection

Make no mistake, though: The headers will sag under load. Under the full design load, calculated header deflection at the outside corner is slightly over 1/8 inch. But remember, too, that all wood is subject to a progressive sagging called creep, which means that in the long term, the total sag could be almost twice the calculated deflection.

The last thing you want is for the top of your $2,000 corner window to feel any load, resulting in fogged glazing or even cracked glass. The window head detail in Figure 30 (next page) is designed to prevent just that. The LVL header is placed 3/8 to 1/2 inch above the top of the window frame. The metal installation straps, available on special request from the window manufacturers, make for a secure but somewhat

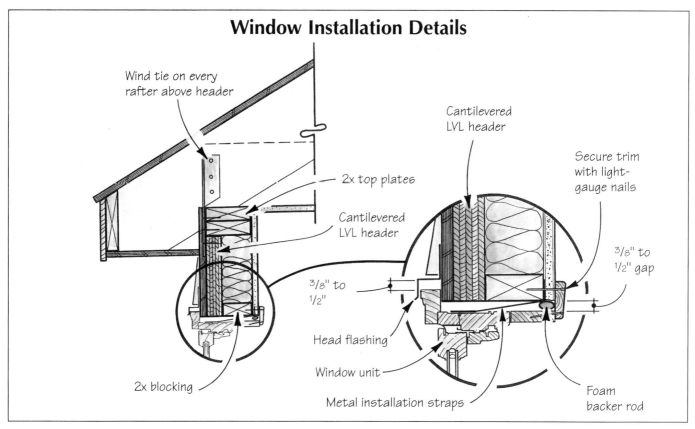

Figure 30. *To ensure that the header load never transfers to the top of the corner window frame, use foam backer rod as an infiltration gasket, and avoid nailing directly through the window frame into the header. Metal installation clips are available upon request from window makers.*

flexible connection. Note the use of foam backer rod, which makes an ideal compressible air infiltration gasket at the window head. Don't use spray foam in this area; it's too rigid.

Anchoring the Header

Recent articles have discussed the different roles that wall sheathing plays in the structural design of a house. In this case, the sheathing anchors the butt of the cantilever header where it is embedded in the wall framing.

Attach the sheathing to each stud with at least eight 10d nails (nail 4 inches on-center for horizontal sheathing), and continue this nailing pattern all the way down to the band joist at the foundation below. Also fasten the plywood to the header with at least eight 10d nails per stud (in other words, eight nails for every 16 inches running). Yes, that's a lot of nails, but it's cheaper than having to replace a fogged corner glass unit.

Preventing Wind Lift

The projecting corner of a roof overhang is subject to severe wind loading, far greater than other parts of the structure. And because a corner window breaks the structural continuity at the building corner, there is a redirection of the load path that concentrates a lot of force at the header posts. Strap ties are the simplest and most economical solution. Note that in the window head detail, hurricane ties are shown at each rafter over the window. In addition, I suggest installing 16-gauge galvanized steel strap ties such as Simpson ST6224s, which are 2 1/16 inches wide and 23 inches long. Use these at the locations shown in Figure 28.

—*Robert Randall*

Makers of Corner Windows

Marvin Windows & Doors
P.O. Box 100
Warroad, MN 56763
800/346-3363
www.marvin.com

Pella Corporation
102 Main St.
Pella, IA 50219
888/847-3552
www.pella.com

Weather Shield
1 Weather Shield Plaza
Medford, WI 54451
800/222-2995
www.weathershield.com

Chapter Eight
ROOF FRAMING

Layout Basics for Common,
Hip, and Jack Rafters 122

Laying Out Unequally-Sloped Gables 127

Joining Unequally Pitched
Hips and Valleys 131

Stacking Supported Valleys 136

Building Doghouse Dormers 142

Roof Framing with Wood I-Joists 145

Flat Roof Framing Options 151

Layout Basics for Common, Hip, and Jack Rafters

In my grandfather's day, you had to know the framing square to get anywhere as a carpenter. Today, sad to say, too many carpenters have no idea why a framing square has numbers on it.

But it's well worth learning about the framing square. Once you get comfortable with it, the square is the fastest and most reliable tool for figuring roofs. Knowing how to use it can save your day if your calculator batteries die, or if you lose your little blue rafter book. If nothing else, learn to use the square to check your results from the rafter table or calculator.

In this section, I'll focus on how to use the framing square to lay out hip and jack rafters, but will also cover the simpler common rafters.

With hips and jacks, you have to modify the layout method used for common rafters. As an example, I'll use a 24-foot-wide building with a roof that has a 6/12 pitch — 6 inches of rise per foot of run. We'll rely on the basic ratio of 12 to 17 every step of the way. (To understand why, see "The Magic Numbers" on the opposite page.)

Plumb Cut

To mark the plumb cut for a common rafter, you hold the 12-inch mark on the blade (or body) of the square and the 6-inch mark on the tongue (or arm), then draw your line along the tongue.

To lay out the plumb cut for a hip rafter, however, hold the same number on the tongue of the square, but on the blade, hold 17 instead of 12. Again, draw your line along the tongue (see Figure 1).

Side cut. The plumb cut on the common rafter is a square cut. But the plumb cut on the hip rafter must be a double bevel so that the rafter can fit into the corner where two common rafters meet the ridge. The line we've drawn represents the long point of this bevel. To find the short points, measure back square from the plumb line, half the thickness of the piece (for dimensional stock, that's 3/4 inch). Make a cut line parallel to the first line, and transfer that same line to the other side of the board. Set your saw on a 45-degree angle, and cut along both lines.

Rafter Length

Now you're ready to find the rafter lengths. For long rafters, I generally use the tables etched on the side of the square (Figure 2, page 124).

Common. Reading across the table labeled "Length of common rafters per foot run," and looking under the 6-inch mark, we see the numbers 13 42 — 13.42 inches. (Numbers that come out to even fractions are marked as fractions — the rest are decimals.) The run of our 24-foot-wide building — half the span — is 12 feet. So for the length of a common rafter, we multiply 12 by 13.42 and get 161.04. The rafter is 161 inches and a shy 1/16 inch long — call it 161 1/8 inches for good measure. (Rounding up to the nearest eighth is safer than rounding down: You can always shorten the rafter, but you can't lengthen it once it's cut.)

This is the distance to the exact center

of the building — a point that actually falls right in the center of the ridge board. We have to allow for the ridge board, so before we actually cut the rafter, we'll shorten, or "back," the rafter ³/₄ inch (for a 2-by ridge). Just measure back ³/₄ inch square to the first plumb line, draw a new plumb line there, and cut.

Hip length. You can find the hip rafter length the same way, by reading the table labeled "Length hip or valley per foot run." Under the 6, the table says 18 — an even number for once. Multiply 12 by 18 and you get 216 inches. To mark the rafter length, make the plumb cut, then hook your tape on the tip of the rafter and measure along the top edge of the stick. The mark you make at 216 inches shows you where the back of your birdsmouth will be.

As in the case of the common rafter, the length we calculate or step off would make the rafter long enough to reach the exact center of the building — a point that actually lies inside the ridge board. So the hip rafter also has to be backed — this time, by the diagonal thickness of the ridge board, or about ¹¹/₁₆ inches. For practical reasons, I make this correction after test-fitting the hip rafter (see "A Note on Accuracy," page 125).

Stepping Off

For a small roof, like a dormer, or for an unusual roof with a nonstandard pitch (a pitch that falls between those listed on the tables), I usually step off the rafter lengths.

Common. To do this for a common rafter, you start at the plumb cut line, holding 12 on the blade of the square, and the rise (6 in our example) on the tongue. As we step off the horizontal distance, we'll automatically come down the vertical distance of 6 inches per foot. Holding the tongue right on the plumb cut, you make a mark at the 12 on the blade. Then slide the square down the rafter until the edge is on the mark you made — you've stepped over a foot. Repeat the process until you have stepped off the whole 12 feet of our sample span. You should end up at the same point as when you measured: 161 inches down the length of the rafter.

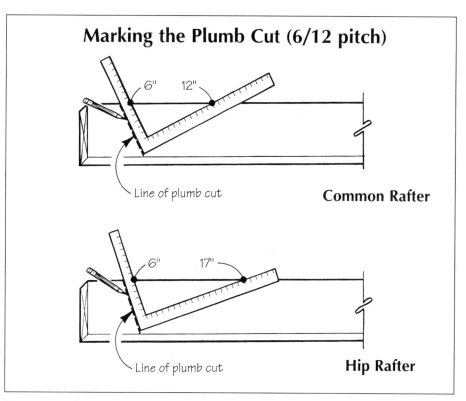

Figure 1. *To lay out plumb cuts for common rafters, align the number corresponding to the rise (in this case, 6 inches) on the tongue and the 12-inch mark on the blade with the edge of the rafter (top). For hips, use 17 inches, instead of 12, on the blade (bottom).*

The Magic Numbers

Long ago, a wise man named Pythagoras proved a basic mathematical fact: For every right triangle, the sum of the squares of the lengths of the two short sides of the triangle equals the square of the length of the long side.

Since then, a simple version of the Pythagorean Theorem has been used to square the walls of buildings: the 3-4-5 triangle ($3^2+4^2=5^2$, or $9+16=25$).

Another consequence of the Pythagorean Theorem is that the diagonal of a square 12 inches on a side measures 17 inches: 12 squared plus 12 squared equals 17 squared ($144+144=288$). The square root of 288 is 16.97 — close enough to 17 for framing. Take out your tape and measure between the 12-inch marks on the blade and tongue of your framing square. Yep — 17 inches.

That's why 17 is the magic number for framing hip roofs. The ratio between the length of a common rafter and the length of a corresponding hip rafter varies depending on the roof's pitch (which is why rafter tables are so complicated). But the ratio between the horizontal distances the rafters span — the run — is always the same: 12 to 17. For every foot your common rafters span, your hip rafter will span 17 inches (minus a whisker). You use the same units of run to figure every roof — all that varies is the rise.

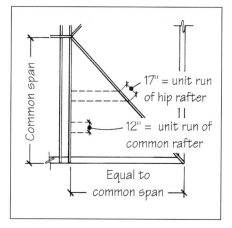

— R.D.G.

Layout Basics for Common, Hip and Jack Rafters

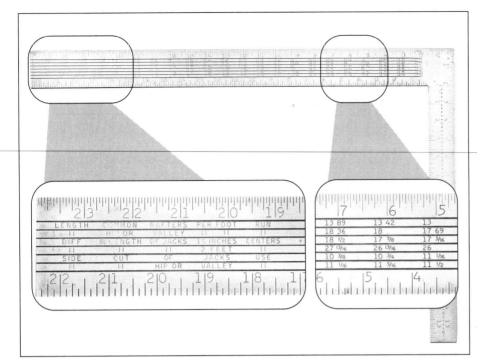

Figure 2. *To use the tables on the framing square to find the length of a common or hip rafter, read across the edge of the square to find the number that corresponds to the rise of the roof — 6, in this case. Multiply the number on the appropriate line — 13.42 for commons, 18 for hips — by the rafter's total run in feet.*

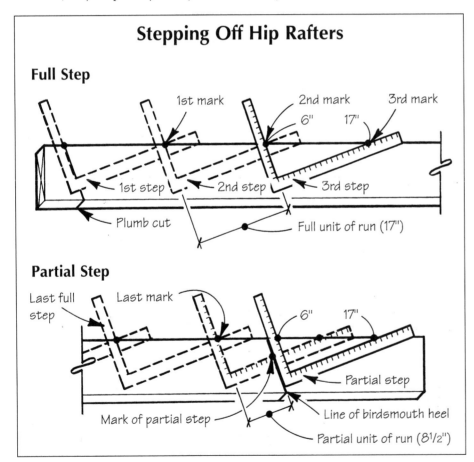

Figure 3. *Instead of calculating a rafter's length from the tables on the square, you can step off the length. For common rafters, each step accounts for 12 inches of run; for hip rafters, as shown, each step accounts for 17 inches of run.*

When the span falls on an uneven dimension (such as 12 feet 6 inches), use a partial step. For hips, maintain the ratio of 17/12. For example, a partial step of 6 inches for a common rafter would become a partial step of 8 1/2 inches on the hip, as shown.

(You can also lay out the birdsmouth first instead of the plumb cut, and step up the rafter instead of down — it makes no difference.)

Hip. To step off the hip (Figure 3), you take the same number of steps you took with the common rafter. But because the hip rafter spans the space diagonally, each horizontal step is 17 inches instead of 12, even though the vertical steps are the same. So you hold 6 on the tongue, 17 on the blade, and step off 12 times. Your last mark will be in the same place as if you had multiplied and measured: in our example, 216 inches.

Partial steps. If the run of your rafters isn't an even number of feet, you'll have to make one last partial step. For instance, if the building were 25 feet wide, and each common rafter had to span 12 feet 6 inches, your last step would be a 6-inch step. Holding the tongue and blade at 6 and 12 respectively, you would make one last mark at the 6 on the blade, then slide the square down to that point for your heel cut mark.

For hip rafters, that last partial step-off gets complicated — you have to maintain the proportions of 17 and 12. So you multiply the partial step by 17/12. Instead of 6 inches, your last step would be 17/12 x 6, or 8 1/2 inches (still holding 17 on the blade, and measuring along the blade, as shown in Figure 3).

The Birdsmouth

At the eaves end of the rafter, draw a plumb line exactly like the line you drew for the plumb cut, holding the number for the rise (in this case 6) on the tongue of the square, and either a 12 or a 17 on the blade (12 for a common rafter, 17 for a hip). The plumb line represents the point where the back, or "heel," of the birdsmouth snugs up against the outside wall.

Common seat. Next comes the seat cut, where the rafter sits on top of the wall. For the common rafter, you slide the square down so the blade intersects the plumb line, still holding 12 on the blade and the rise on the tongue. Your seat cut should be just as long as the width of the wall — for a 2x6 wall, you need 6 inches including the sheathing. Set the square so that the 18-inch mark (12 plus 6) crosses your plumb line, and draw along the blade from 18 to 12 — 6 inches.

Hip seat. The seat cut for the hip is a little more complicated. When the rafters are set in place, the top of the hip rafter must be even with the top of the common rafters in order for the roof sheathing to lie smoothly across them. So in laying out the cuts, the distance up the plumb line from the corner of the birdsmouth to the top edge of the rafter — the height above the plate — must be the same for both. Measure the height above the plate on the common rafter, and measure down the plumb line you made for the heel cut on the hip rafter. Mark the point, then make a line square off the heel cut for the seat cut (Figure 4).

Rafter Tails

You can make all your tail cuts before you set the rafters. But if you do, any little curves or bumps in your wall plates will telegraph out to the fascia and soffit. I generally cut all my rafter tails in place, after the roof is framed. That way, I can snap a line along the whole set of rafters and the cuts line up perfectly. But for anyone who wants to make those cuts in advance, here's how to lay them out.

Common. First, determine your soffit overhang (let's use 12 inches as an easy example). For the common rafter, hold 12 on the blade and the rise on the tongue, and slide the square until the 12 intersects with the mark for the heel cut. Then simply mark your fascia cut along the tongue. (Note: For this step, you'll have to flip the square around so that the tongue is on the downhill side of the rafter. The fascia cut is just like the ridge cut, only upside down.) Make sure you mark on the same side you measure on.

Hip. For the hip rafter, hold 17 instead of 12, then slide as before. To determine the distance from the heel cut to the fascia cut, take the number you used on the common rafter and multiply by 17/22. In this case, since we're using a 12-inch overhang, it comes out to 17 inches. Let the 17 intersect with the heel cut line, and mark the fascia cut along the tongue

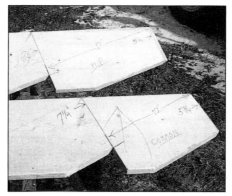

Figure 4. *When laying out seat cuts, it is critical to maintain the same height above the plate on both hip and common rafters.*

of the square (Figure 5, next page).

This is the short point of the cut. Mark the cut on both sides of the board by squaring across the edges and connecting the points. Set your saw on a 45-degree angle and cut on both sides of the board to form the pointed end of the hip rafter.

Soffit. To lay out the soffit cut, you'll need to know how wide your fascia boards will be and how far below the soffit your fascia will extend. Let's assume we're using 1x6 fascia board and we've decided that a 5 1/4-inch fascia cut will work well.

Once you know this measurement, you measure down the fascia cut line 5 1/4 inches and make a mark where the soffit cut will intersect. Then, holding 12 on the blade and the rise on the tongue, mark along the blade for the soffit cut.

For the hip rafter, you must make sure the vertical cut is the same length as the cut on the common rafter — just as you did earlier for the height above the plate. Measure down the fascia line 5 1/4 inches, the same distance as on the common rafter fascia, and make a mark. Holding 17 on the blade and the rise on the tongue, let the blade intersect the mark, and mark your cut line along the blade.

Jack Rafters

Jack rafters are just like common rafters, but instead of meeting the ridge they meet the hip rafter. So they have to be cut to a succession of different lengths, and their plumb cuts must be beveled to fit against the hip.

To find the lengths of the jacks, look on the square for the table labeled

A Note on Accuracy

Using the framing square is as accurate a method of laying out cuts as any other. It's certainly as accurate as using a speed square.

On the other hand, using the square to find rafter lengths isn't as accurate as a calculator. The numbers on the tables for length per foot of run are approximations (though a few are exact). When you multiply, you compound the error. By the same token, as you step off a length, you may accumulate an inaccuracy. (That's okay, though — just make sure you come out long rather than short. If it's long you can always shorten it, but if it's short you have to start over.)

For common rafters, if your test pair is a shade long, the joint where they meet will be open at the top. But don't recut the angle — fix the length instead by backing the plumb cuts a little (keep the same angle, but trim off 1/8 inch or so). When the pair fits right, go on and use them as a pattern to cut the rest of your commons.

The same is true of the hip rafter — you want one that's too long, not too short. The magic number we use for stepping off, 17, is an approximation that's 1/32 inch bigger than the actual number. Multiplied out, it can give you a hip rafter that's too long. That's okay — in fact, I never even back my hip rafters to allow for the ridge board.

Instead, I let them run about an inch too long. Then I hold them in place to see how they fit. Usually the cuts fit snugly into place, but because the rafter is a shade long, the heel of the birdsmouth is gapped away from the plate a bit, and the whole piece sits a shade high. Holding the rafter in place, I measure the gap at the plate. Then I take the rafter down and back the plumb cut just enough to close the gap and make a perfect fit.

— *R.D.G.*

"Difference in lengths of jacks, 16 inches on-center," or, if appropriate, "24 inches on-center." Read across to the number that falls under the pitch you're using. The number you find ($17^{7/8}$ for a 6-pitch roof with jacks 16 inches on-center) is the difference in the length of each jack. We'll subtract this number from the length of the longest jack to find the second longest, and so on until we have all the jacks.

First, we need to know the length of the longest jack. Instead of going through the geometry of figuring that out theoretically, I just take the simplest route: I use the length of the last common rafter from the heel of the birdsmouth to the point where it meets the face of the hip rafter. Starting with that number and subtracting over and over gives me the length of each succeeding jack rafter.

Stepping off jacks. For odd pitches that aren't included in the tables — say, an unusual pitch like a $7^{3/4}$ in 12 — you can step off the jacks. Hold the 12 and the rise, just like stepping off a common rafter. But instead of stepping horizontally one foot at a time, step 16 inches (or 2 feet if your layouts are on 2-foot centers). For partial layout steps, such as you might need to calculate jacks that frame in a skylight opening, you can make partial steps as you would for a common rafter (Figure 6). If the jack you need will be set 10 inches over from the one next to it, start with the length for the neighboring jack and step forward or back the 10 inches. The lengths will work out correctly.

You cut jack rafters in pairs: One pair at each length for each hip rafter. Set your saw on a 45-degree angle and cut along the line. Save the offcut — it already has the bevel on it for the opposite side of the hip, and needs only the birdsmouth cut to become a jack rafter, too. (Check the crown before you use it that way. If the crown's noticeable, you'll have to make the plumb cut the other way.)

I usually cut my middle pair of jacks first and test-fit them. They may need to be adjusted by $1/4$ inch to fit, though that's rare. Nailing the center pair on first straightens the hip rafter and locks it in place. The rest of the jacks should drop into place with no argument.

There are methods for making layout marks on the hip rafters to show where the jacks will fall, but they aren't worth the effort required. If your jacks are the right length, they'll fall in the right place. To check, just pull a measurement off one of the common rafters on the main roof (or off the center common if you're on that end). If you're within $1/4$ inch of 16-inch centers, the roof will be fine.

When you install the jacks, make sure each pair lines up across the hip, both at the top and bottom. If you nail them up so they match at the top but not at the bottom, you'll build a twist into the hip rafter.

When the hips have been cut and fitted into place, and the center jacks are nailed up, I like to pull a measurement from a common rafter and make sure my 16-inch centers are falling okay. If everything looks right, I feel comfortable handing the rest of the jack rafters up to the guys on the plate and walking away.

By Rob Dale Gilbert, a custom builder in Lyme, N.H.

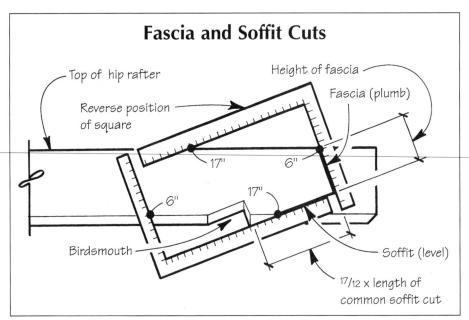

Figure 5. *The height of the fascia cut will be the same for both common and hip rafters (although the hip cut is double-beveled). The length of the hips soffit cut, however, will be 17/12 times that of the common.*

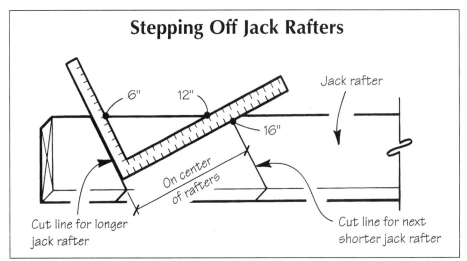

Figure 6. *To step off successive jack rafters, hold the rise and run on the tongue and blade of the square (in this case, 6 inches and 12 inches), but mark the rafter to correspond to the on-center spacing of the jacks (in this case, 16 inches). Cut along the line with your saw set at 45 degrees.*

Laying Out Unequally-Sloped Gables

Unequally-sloped gable roofs can be difficult to lay out, but the job is a lot easier if you have a proven method. Complicated mathematical calculations that use "inverse cotangents," "secants," and "adjacent angles" may be accurate, but they're not easy to teach to most tradespeople.

Over the years, I've developed a layout method that is fast and accurate and doesn't require any heavy math. To use this method, you only need to know how to add, subtract, multiply, and divide.

I'll assume that you can lay out a common rafter. You also need to know how to convert feet, inches, and fractions into decimal equivalents — or you can use the simple conversion ruler provided below in "Quick Decimal Conversion."

I treat layout for an unequally-sloped roof as the combination of two common

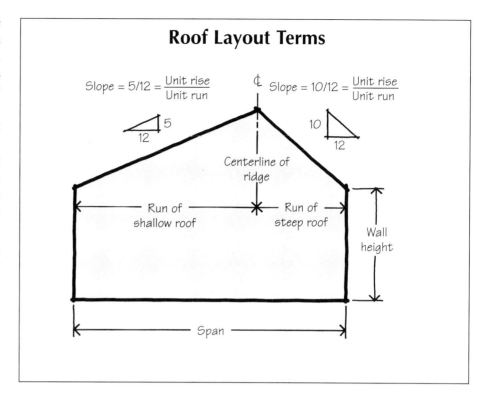

Quick Decimal Conversion

Most of the ways to convert feet, inches, and fractions to their decimal equivalents have one thing in common: too much math. To eliminate the need for tedious calculations, here's how to make a conversion scale that can be glued to a framing square.

The conversion scale is simply a standard 12-inch scale juxtaposed with a 12-inch-long 100ths scale. To use it, just locate the inches and fraction of the measurement you're working with, then read the decimal equivalent directly above. If you're converting a decimal figure back to inches and fractions, find the decimal on the upper scale and the inch-fraction equivalent is directly below. No batteries, no math — it doesn't get any easier.

The conversion table is printed to scale. Make a photocopy, cut the scales out, and glue the two halves together, end-to-end, on the tongue of your framing square. I put clear packing tape over my conversion table to protect it. When it wears out, I just glue another one on. I learned this method in trade school from master carpenter Don Zepp.   — C.H.

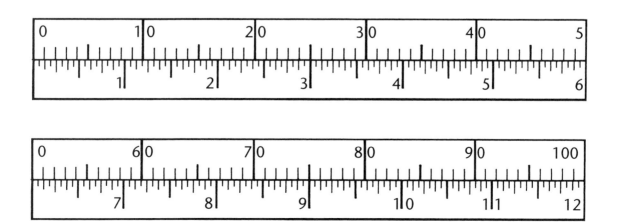

Case 1: Equal Wall Heights

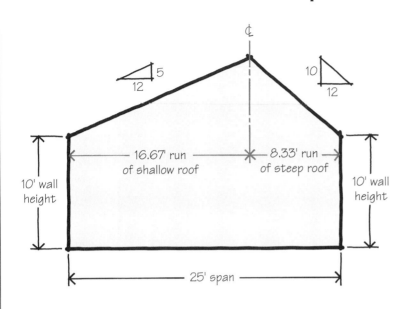

Step 1. Add the rise of the two different roof slopes: 5 + 10 = 15.

Step 2. Divide the rise of either roof slope by the sum of the two slopes from Step 1. For example, using the shallower slope, 5 ÷ 15 = .3333. (Use four decimal places for accuracy.)

Step 3. Multiply the span by the answer in Step 2: 25 ft. x .3333 = 8.33 ft. This represents the total run of the steep roof.

Step 4. To find the run of the shallow roof, subtract this value from the total run: 25 - 8.33 = 16.67 ft. You can check the answer by repeating the process using the steeper slope:
10 ÷ 15 = .6666
25 ft. x .6666 = 16.67 ft.

Both rafters can now be laid out as common rafters.

Case 2: Different Wall Heights

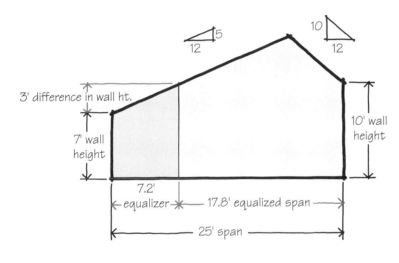

In this scenario, you first have to "equalize" the different plate heights:

Step 1. Divide the unit run of the shallow slope by the unit rise: 12 ÷ 5 = 2.4.

Step 2. Find the "equalizer" by multiplying the difference in plate heights by the value from Step 1. The difference in plate heights: 10 - 7 = 3 ft. Multiply by the value from Step 1:
3 x 2.4 = 7.2 (the equalizer).

Step 3. Subtract the equalizer from the total span of the roof: 25 - 7.2 = 17.8 ft. This smaller, "equalized" span is for a roof with equal plate heights.

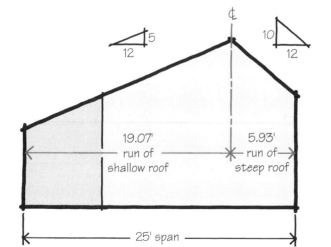

Using the "equalized" span, follow the four-step process from Case 1 to find the run of the steep roof:

Step 1. 5 + 10 = 15

Step 2. 5 ÷ 15 = .3333

Step 3. 17.8 ft. x .3333 = 5.93 ft.
This is the run of the steeper roof.

Step 4. To find the total run of the shallow roof, subtract the run of the steep roof from the span of the building: 25 - 5.93 = 19.07 ft.

Case 3: Steep Slope on Short Wall

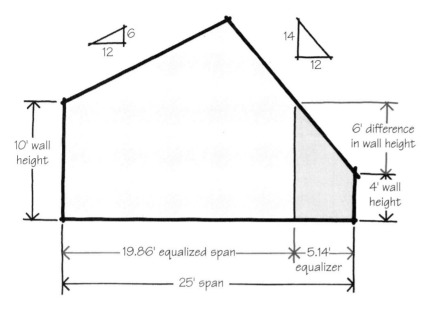

Find the equalizer by following the three steps from Case 2:

Step 1. Divide the unit run of the steeper roof by the unit rise: 12 ÷ 14 = .857.

Step 2. Multiply the difference in plate heights by the value from Step 1: 6 x .857 = 5.14 (the equalizer)

Step 3. Subtract the equalizer from the total span of the roof: 25 - 5.14 = 19.86 ft. This smaller, "equalized" span is for a roof with equal plate heights.

Find the run for common rafters:

Step 1. Add the rise of the two different roof slopes: 6 + 14 = 20.

Step 2. Divide the shallower roof slope by the sum of the rises from Step 1: 6 ÷ 20 - .3

Step 3. Multiply the equalized span by the decimal from Step 2: 19.86 x .3 = 5.96 feet. This is the run of the steep rafter on the shorter, "equalized" building span. To get the true run of the steep rafter, add the equalizer back in: 5.96 = 5.14 = 11.1 feet.

Step 4. To find the total run of the shallow rafter, subtract this number from the full span: 25 - 11.1 = 13.9 ft.

gable roofs. I'm essentially laying out common rafters for two different roofs. The trick is finding where the two different slopes will meet at the ridge. To understand the terms I'll be using, refer to the illustration at right.

Case 1: Equal Wall Heights

My four-step method results in the run dimension for each slope. In the first example (see Case 1), the bearing walls are both the same height, and I know the roof slopes and the overall span. After plugging the numbers into each of the four steps, I can use the results to lay out common rafters for both slopes.

Case 2: Different Wall Heights

Let's consider a more complicated scenario — an unequally-sloped roof with bearing walls at different heights. The slopes are the same as in Case 1, but the wall height on the shallow slope has been changed from 10 feet to 7 feet.

When the walls are different heights, I first create an "equalizer" using three easy steps (see Case 2). When the equalizer is subtracted from the span, I'm left with a roof that once again sits on walls of equal height. Then I can follow the four original steps to find the run for the common rafter of the steep slope. When laying out the common rafter for the shallow slope, remember to subtract from the original span (25 feet in the examples), not the "equalized" span.

Case 3: Short Wall, Steep Slope

Sometimes the steeper slope bears on the shorter wall (Case 3). Again, the trick is to find the equalizer — the point

Laying Out Unequally-Sloped Gables

Case 4: Saltbox

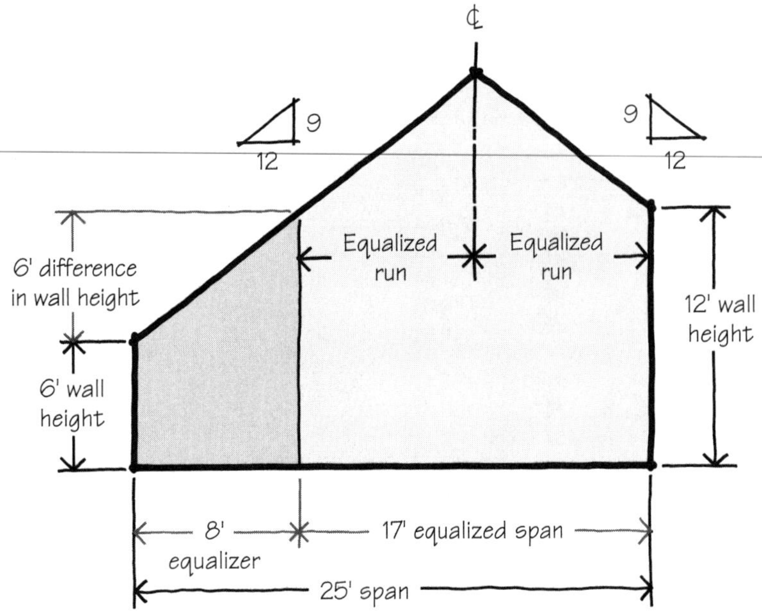

A Saltbox is a standard gable roof set on walls of unequal height. To lay out the rafters, first follow the three-step process in Case 2 to "equalize" the span.

Step 1. 12 ÷ 9 = 1.3333

Step 2. Difference in wall heights:
6 ft. x 1.3333 = 8 ft. (the equalizer)

Step 3. 25 ft. - 8 ft. = 17 ft. (equalized span)

Divide by 2 to get the run of the short rafter: 17 ÷ 2 = 8.5 ft. Add the equalizer back in to get the run of the long rafter: 8.5 + 8 = 16.5 ft.

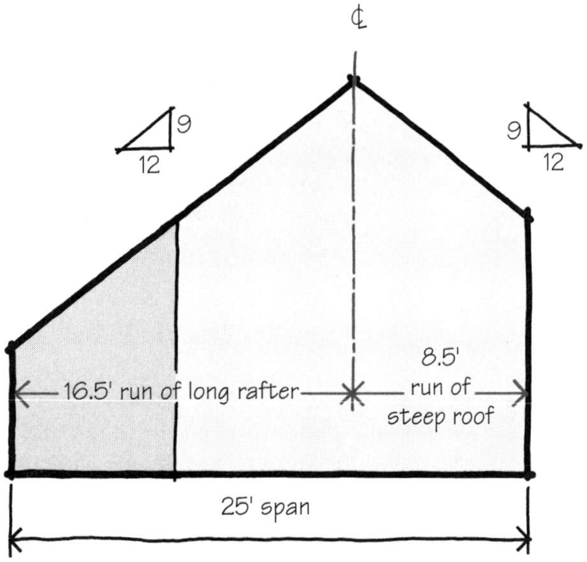

on the steep roof plane where a wall equal in height to the taller wall would meet the rafter. Follow the three-step process just described, but this time use the slope of the steep roof, because that's the side of the building with the short wall. The result is an equalized span for a building with walls of equal height.

Now use the four-step process described in Case 1. Remember, however, that in this case, you need to add the equalizer back in to get the run of the common rafter on the steep slope. Then subtract this number from the original span to find the run of the shallow common rafter.

Case 4. Saltbox Layout

A Saltbox is not a true unequally-sloped roof — it's really a standard gable roof set on a building with unequal wall heights. You don't need the four-step method described earlier, but you still have to "equalize" the span. To lay out the common rafters, find the run for the short rafter first by dividing the equalized span in half. To find the run for the long rafter, add the equalizer to the run of the short rafter.

—Carl Hagstrom

Joining Unequally Pitched Hips and Valleys

After struggling with hip rafters and talking to an old-timer or two, novice framers discover that joining two roofs isn't that tough — for hips and valleys you just cut all your bevels at 45 degrees, and use 17 instead of 12 for the run on the framing square.

Then it happens: That bid you submitted for the job with the adjoining, unequally pitched roofs gets accepted, and you've got to figure out a way to frame it.

One approach is to order twice the framing lumber required and keep cutting hips and valleys until you get it right. Another is to buy a fancy framing calculator and take a course in trigonometry. Or you can use the method I learned in trade school from master carpenter Don Zepp. It requires nothing more than a framing square and a pencil,

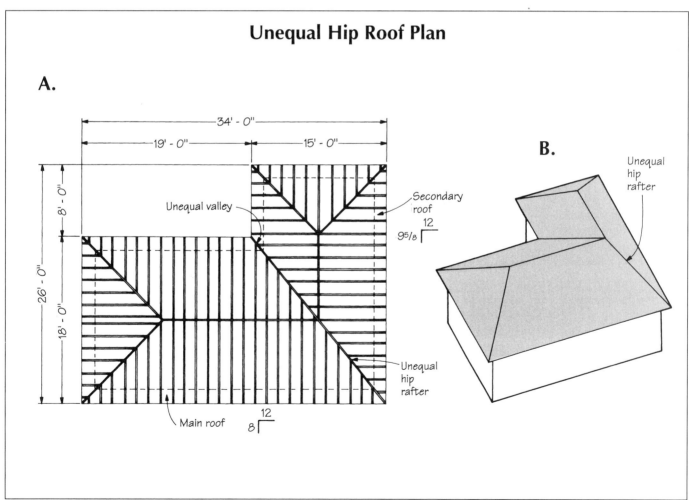

Figure 7. *The author begins any complex roof project by making a scale framing plan of the roof (A). This is useful for material takeoffs and as a job-site reference. The perspective sketch (B) shows the location of the unequal hip rafter.*

and I feel is the easiest method to understand and use in the field.

The key to the Zepp method lies not in fancy math, but in using the framing square to make scale drawings of the crucial angles and cuts, then transferring the measurements from sketch to lumber. The direct correlation between drawings and cuts makes this method accurate and reliable. Master it, and you can handle any of the numerous situations where two roofs of unequal pitches meet, whether at a hip or a valley. For the purposes of this section, I'll show how to frame the intersection of two hip roofs of unequal pitch.

A note before you start: You'll get the most out of this article if you read it with framing square and a scrap of plywood nearby, so you can perform some of the layout calculations.

Getting Oriented

The first step is to make a scale drawing of the entire roof, locating the exterior walls, ridges, and all rafters, as in Figure 7A. This drawing will serve as a map while you're putting the pieces together. It will also help you estimate the quantity and lengths of lumber you'll need. A simple perspective sketch (Figure 7B) helps give an idea of how the finished roof planes will look.

The L-shaped house in Figure 7 has both ridges at the same level. The overhang is 16 inches (measured horizontally), and the fascia runs at a continuous height around the entire house.

Joining these two roofs poses two main challenges: One is getting the two differently pitched planes to meet at the hip (or valley) rafter. The other is to get the fascias to line up.

Establish the pitches. First we should establish the pitch of both roof planes. We'll start with the knowledge that the main roof has an 8/12 pitch. How do we figure the pitch of the secondary roof?

When I use the word "pitch" I mean the ratio of rise in inches over 12 inches of run. From the plan, we see that the total rise of the secondary roof is the same as that of the main roof — 6 feet. The total run of the secondary roof is 7 feet 6 inches. To convert this to inches

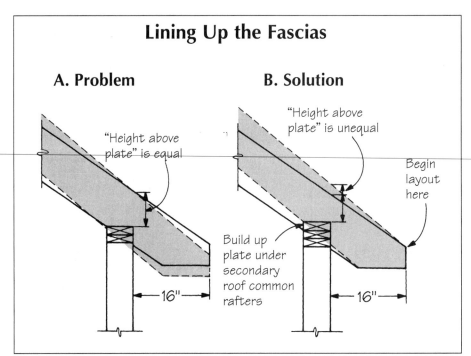

Figure 8. *Where roofs of unequal pitch join, it requires careful layout to create equal-width soffits with equal-height fascias. Drawing A superimposes the common rafters for the two roof pitches, and illustrates the problem that occurs when roof layout starts at the outside edge of the exterior walls. Drawing B shows how the author solves this problem by starting his roof layout at the ends of the rafters and raising the wall plates under the steeper-pitched roof.*

of rise per 12 inches of run, use this formula:

$$\frac{12 \text{ inches}}{\text{Total Run}} \times \text{Total Rise} = \text{Unit Rise in inches}$$

or

$$\frac{12 \text{ inches}}{7.5 \text{ feet}} \times 6 \text{ feet} = 9.6, \text{ or } 9^{5}/_{8} \text{ in. unit rise}$$

So $9^{5}/_{8}/12$ is the secondary roof pitch.

Lining Up the Fascia

Look at Figure 8A, which shows the two differently pitched common rafters. It would seem that lining up the fascias and cornices so you have equal overhangs on both pitches would be impossible, since the rafters of the two roofs extend beyond the plate at different angles. But it's really not that tough.

Making the fascia meet. To keep the overhang the same and line up the fascia, you must begin the roof layout from the inside top edge of the fascia, not from the outside of the exterior walls. Then you build up the wall plate under the steeper secondary roof, as in Figure 8B. The buildup equals the difference between the unit rise of both roof pitches. In this case, with unit rises of 8 inches and $9^{5}/_{8}$ inches, the difference is $1^{5}/_{8}$, or 1.6 inches. So the plate must be built up 1.6 inches per level foot of overhang.

Since the overhang is 16 inches, or 1.33 feet, the total buildup equals:

1.33 feet x 1.6 inches per foot = 2.13, or $2^{1}/_{8}$ inches

Rather than building the plate up exactly $2^{1}/_{8}$ inches, I use an extra 2x top plate ($1^{1}/_{2}$ inches thick) and increase the rafter height the remaining $^{5}/_{8}$ inch by adding it to the "height above plate" distance that I use to find the birdsmouth level cut.

Finding the Pitch of the Hip

Once you've temporarily supported the ridges with a few common rafters, you can lay out the hip rafter. I do the layout on a sheet of plywood. Using one corner, I start by drawing the angle of the hip and then develop this line into a full-sized plan view of the roof corner. From this drawing I can pull exact measurements for my bevel cuts.

Hip angle. Using the framing square's 12th scale, find the hip angle as shown in Figure 9A. Draw a diagonal line between 9 feet (the run of the main roof) on the tongue of the square and 7 feet 6 inches (the secondary roof run) on the blade. This gives the angle of the hip rafter.

Unit run of the hip. In a hip roof with equal pitches, the unit run of a hip or valley rafter is 17 inches for every 12 inches of run in the main roof. For example, in an equal hip roof with an 8/12 pitch, the pitch of the hip rafter would be 8/17. But here it will be different, since the hip rafter doesn't lie at 45 degrees to the commons.

The pitch of the unequal hip rafter can be expressed in relation to either the main roof or the secondary roof. Since you need only one set of numbers to lay out the cuts, let's work with the main roof, which has a unit rise of 8 inches.

Find the hips unit run as in Figure 9B. Lay your square in the corner of the plywood drawing, working from the side of the diagonal line that represents the main roof. Keeping the outside edge of the blade flush with the edge of the plywood, slide the square until the 12-inch mark on the tongue intersects with the hip. Make a mark there, then measure the distance from the mark to the drawing's corner — this is the hips unit run. It should measure $15^5/_8$ inches.

Unit run of hip = $15^5/_8$ in.

By the way, the unit run for the shallower pitch roof will always be less than 17 inches, while the unit run for the steeper pitch will always be greater than 17 inches.

So now we know the pitch of the hip rafter in relation to the main roof:

Hip pitch = $8/15^5/_8$

Laying Out the Hip

Working on the plywood drawing, next snap lines to define the soffit overhang, fascia, and wall framing at full scale, as shown in Figure 9C. Locate the centerline of the hip rafter not at the corner of the plates, but at the corner of the two intersecting fascia boards. Extend this centerline back to and beyond the wall plate. You'll notice the hip rafter doesn't sit on the corner of the two wall plates; it never will. In an unequally pitched hip roof, the hip will always fall on the wall plate supporting the steeper pitched roof.

With the full scale drawing completed, it's time to start putting some marks on a piece of lumber.

Hip tail. First lay out the tail end of the rafter, as in Figure 10. Using the unit run and rise of the hip, $8/15^5/_8$, draw a plumb line on both sides of the rafter end with the framing square. You can then take deductions for the sidecut angles directly from the full-scale draw-

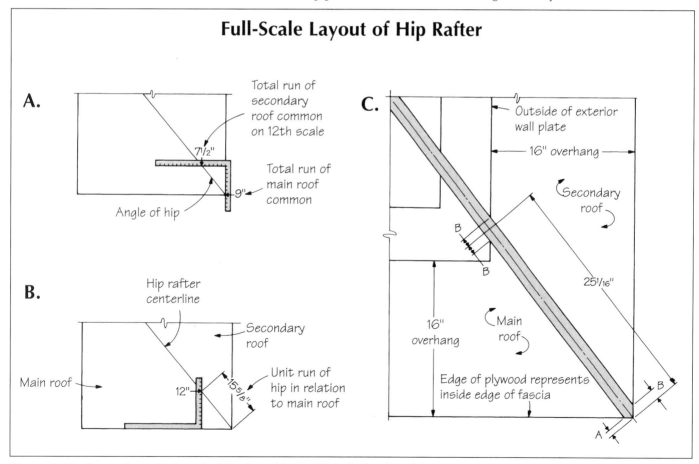

Figure 9. *The key to the author's method lies in making a full-scale drawing of the corner where the unequally pitched roofs meet, then transferring measurements from drawing to lumber. On a sheet of plywood, first establish the angle of the hip rafter, using the total runs of the two roofs (A). Next, derive the unit run of the hip rafter in relation to the main roof (B). Finally, complete the scale drawing with walls and soffits to establish critical dimensions for cutting the hip (C).*

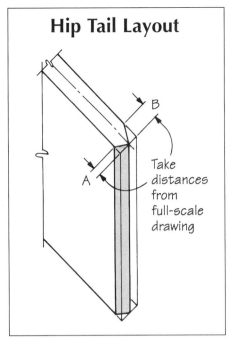

Figure 10. *To figure the bevel cuts at the hip rafter tail, take distances "A" and "B" from the full-scale layout drawing in Figure 9.*

ing — distances "A" and "B" in Figure 9C. Since you're taking these measurements from a plan drawing, *make sure you measure perpendicular to the plumb line* — not along the rafter edge. Hold off cutting the tail until you lay out the birdsmouth.

Birdsmouth. On the full-scale drawing, measure along the hip rafter centerline from the inside corner of the fascia to the point where the centerline meets the outside of the wall plate. This distance should measure about $25^{1}/_{16}$ inches.

Now use the framing square to step off this distance from the plumb line at the tail, just as you would with any rafter. The first step-off accounts for $15^{5}/_{8}$ inches. You'll have to add the remaining $9^{7}/_{16}$ inches by measuring perpendicular to plumb, as in Figure 11A. Mark a plumb line; this represents the centerline of the birdsmouth cut. Extend this line around the entire rafter.

Heel and seat cuts. Next, lay out the plumb and level cuts for the birdsmouth. Here again, we'll simply steal the measurements from our full-scale drawing.

To lay out the level cut, measure off the "height above plate" distance (see Figure 8B) along the birdsmouth centerline and draw a perpendicular line to mark the level cut. The "height above plate" for the hip rafter is the same as for the secondary roof common rafters, since the hip falls on the secondary roof plate.

To lay out the plumb cut, take measurement "B" from the full-scale drawing (Figure 9C) and transfer its dimensions to the rafter on both sides of the centerline of the birdsmouth, as shown in Figure 11B. Again, measure perpendicular to the centerline mark for the birdsmouth,

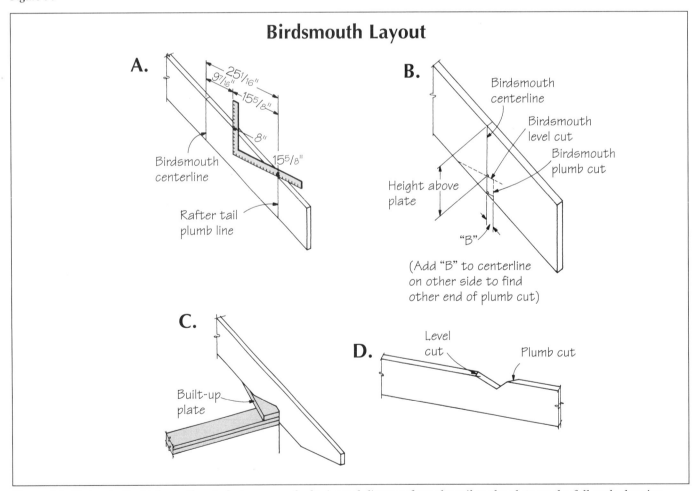

Figure 11. *To locate the birdsmouth cut, first measure the horizontal distance from the tail to the plate on the full-scale drawing (Figure 9C). Step off this distance, $25^{1}/_{16}$ inches, and draw a plumb line (A). The plumb line represents the centerline of the birdsmouth cut. Locate the level cut by measuring down the birdsmouth centerline the "height above plate" distance of the steep roof common rafters (B). To lay out the plumb cut, add and subtract distance "B" perpendicular to the birdsmouth centerline mark, as shown. Because the hip rafter doesn't cross at the corner of the walls (C), the birdsmouth plumb cut is cut at an angle (D).*

rather than along the length of the rafter.

Cutting to length. Once the birdsmouth is laid out, make a full-scale drawing of the hip-to-ridge connection, as in Figure 12. You'll use this to steal measurements for finding the length of the hip.

First, we need to step off the distance to the *theoretical* center point where the hip meets the ridge. Take the run of the main roof (9 feet), and using your square and the pitch of the hip in relationship to the main roof — 8 inches and 15⁵⁄₈ inches — step off this distance in the conventional manner. When you've stepped off 9 steps, draw a plumb line at that point and extend it completely around the rafter.

This is the *theoretical length* of the hip rafter. To obtain its *true* length, deduct distance "C" perpendicular to this plumb line. This gives the true length *at the centerline of the rafter*. To lay out the cheek cut, you must then add and subtract "A" perpendicular to the true length plumb line.

There you have it! Your unequally pitched hip is laid out and ready to cut. A few details remain.

Backing The Hip

As with any hip rafter, an unequal hip rafter must have its top edge "backed," or beveled, to match each of the converging roof planes. The question is how much to cut away.

Determine the backing for each roof plane as shown in Figure 13. Place your square on the side of the hip rafter that will face the main roof, using the hip pitch for that roof, 8/15⁵⁄₈. Measure distance "A" along the unit run and mark. This is the line of backing.

Repeat the same procedure on the side of the hip that faces the secondary roof, using the hip pitch for that roof and measuring distance "B". You'll have to find the secondary roof hip pitch the same way you found the main roof hip pitch, as in Figure 9B. The hip pitch in relation to the secondary roof measures out at 9⁵⁄₈/18¹³⁄₁₆. This is exactly the same ratio as the pitch for the other side, but I find it helps to work the two sides separately to avoid confusion, especially when it comes to laying out the jack rafters.

One important note: Wait until all your layout is complete and you've test-fitted the hip rafter before you actually bevel the top edge of the hip. Once the top is beveled, you will have lost an important edge of reference.

Placement of Jack Rafters

Before you nail up the hip rafter, be sure to lay out the position of the jack rafters. It's a lot easier on the ground than 20 feet in the air.

To determine the run of the jacks, draw in the *first two* jacks on the full-scale drawing of the hip rafter, as shown in Figure 14. We can then take all our measurements for the first jacks on each roof directly from the drawing, just as we did for the hip rafter. We can also derive the *common differences* so we can find the other jacks' runs.

With your square set on the main roof side of the hip rafter, step off the true run of the first hip jack using the pitch of the main roof hip, 8/15⁵⁄₈. Mark a plumb line; this is the centerline of the first jack. For each successive jack rafter, step off the common difference. Repeat the process on the other side of the hip, using the pitch of the hip for the secondary roof (9⁵⁄₈/18¹³⁄₁₆).

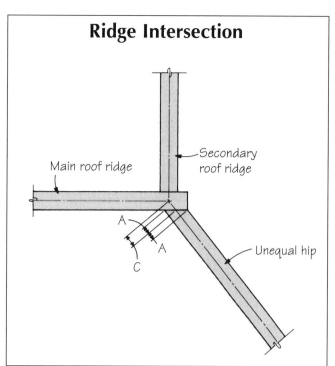

Figure 12. *To figure the cheek cut at the top end of the hip rafter, the author makes a full-scale drawing of its intersection with the main roof ridge, then transfers measurements.*

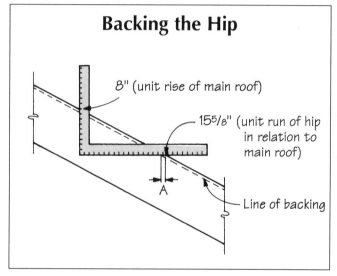

Figure 13. *"Backing," or beveling, the rafter ensures that the roof sheathing will rest flush on the hip. Use the framing square and the hip rafter pitch to determine the line of backing, as shown. Repeat the process on the other side of the hip, using distance "B."*

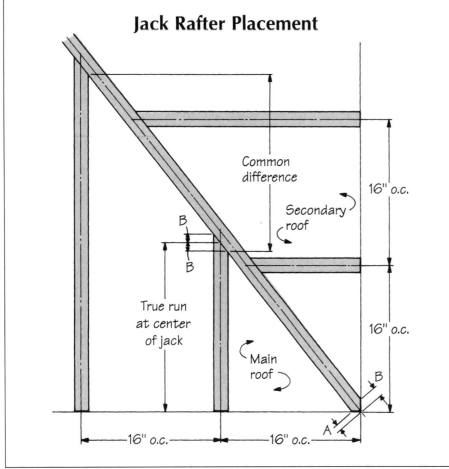

Figure 14. *To lay out the jack rafters, the author makes a full-scale drawing of the first two jacks on each side of the hip. From this drawing he finds the angle of the cheek cuts, the run of the first jack, and the common difference used to locate the rest of the jacks.*

Valleys

An unequally pitched valley rafter is laid out in the same way as the unequally pitched hip. The main difference between the two is that valley rafters sit on inside corners, while hip rafters sit on the outside corners.

Each roof will be slightly different. Sometimes, instead of backing the hip, I'll drop it in relation to the jacks. This still gives plenty of support to the sheathing. Also, I often double the hip rafter for greater strength and more nailing surface. In that case, the backing angle is the same but the line of backing will be lower. But however the details may differ, the Zepp method provides all the fundamentals I need to make sense out of a seemingly impossible situation.

—*Carl Hagstrom*
Thanks to Don Zepp for his assistance.

Stacking Supported Valleys

Recently while working on a beautiful Tudor home in the hills outside Medford, Oregon, I had the opportunity to cut the roof shown in the photo, and in plan on the facing page. From this roof plan, you'll notice the main roof is four large Versa-Lam hip beams converging to a near peak, with three smaller gable sections and supported valleys hanging from them.

This project contained every possible type of supported valley — a valley-to-hip (valley #1 on roof framing plan), a valley-to-valley (valley #2), and a "disappearing" valley (valley #3). My goal in this section is to explain how to calculate, lay out, and stack these three types of supported valleys. Since the birdsmouth cut for any supported valley is no different from that for a regular valley, I'll focus only on the connection at the top.

Valley to Hip

Whether you use a framing square, rafter tables, or my LL ratio method (see "Using Line Length Ratios," page 140), the first step is to calculate the correct length of the supported valley. Rafter lengths for supported valleys are figured for the full span and shortened by half the thickness of the supporting member. For example, the overall span for valley beam #1 is 286 5/8 inches and the roof pitch is 10/12, so its line length equals 235 1/4 inches. This length must be shortened by half of the 5 1/4-inch-thick hip, or 2 5/8 inches. Measure this shortening distance perpendicular to the plumb cut.

When stacking valley #1, locate its position by measuring up 235 1/4 inches along hip beam #1, starting from the point where the center of the hip crosses the outside wall plate line (Figure 16). Then, square across and plumb down to mark the center of the supported valley. The center of the valley is then aligned flush with the near edge of the hip it is hanging from.

Ridge cheek cuts. On this house, the hips are 5 1/4 inches thick and the valleys are 3 1/2 inches thick. To properly align the ridge that ties in here (ridge #1), with the centers of the hip and valley,

Roof Framing Plan

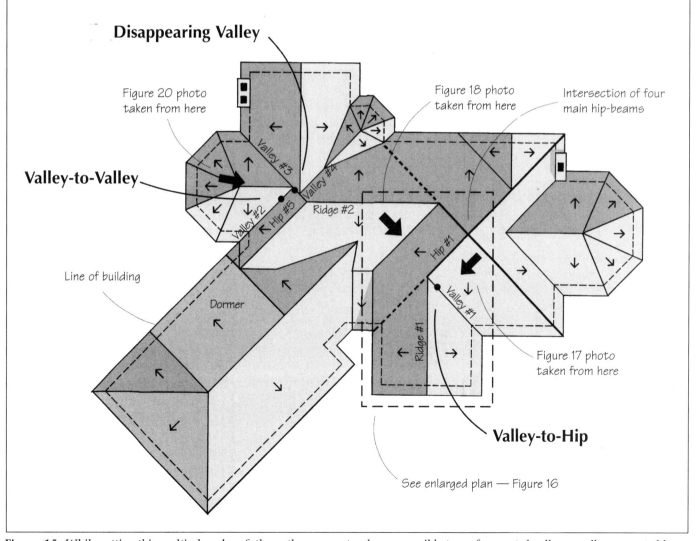

Figure 15. *While cutting this multi-planed roof, the author encountered every possible type of supported valley: a valley supported by a hip (valley #1); a valley supported by another valley (valley #2); and a "disappearing" valley that also serves as a hip (valley #3).*

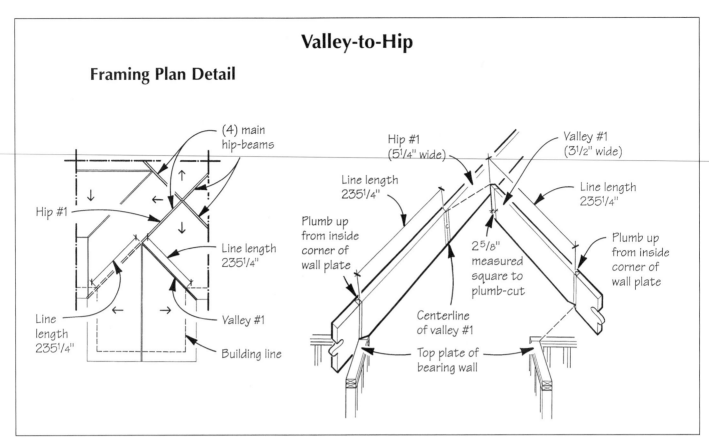

Figure 16. *To locate the intersection of the center of valley #1 with the supporting hip beam, mark the line length on the hip, measuring up the hip from the outside of the wall plate. Cut valley #1 to the line length, then shorten it by measuring perpendicular to the plumb cut half the thickness of the hip.*

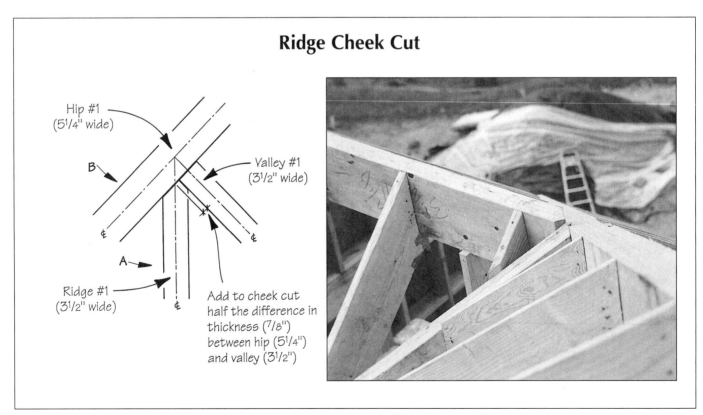

Figure 17. *A special cheek cut is required to align the ridge with hip and valley beams of different thicknesses (photo). First, lay out a standard double 45-degree cheek cut, then add 7/8 inch (half the difference between the two beam widths) to the side towards the thinner material (illustration).*

you must make a special cheek cut on the ridge (Figure 17).

To lay out this cheek cut, mark an equal double 45-degree cheek cut on the ridge, then add half the difference between the two beam widths ($5^1/_4 - 3^1/_2 = 1^3/_4 \div 2 = ^7/_8$) to the side towards the thinner material, as shown in Figure 17. This will center the ridge correctly.

The correct ridge height at this hip-valley connection is located where the side of the ridge nearest the hip (side A in Figure 17) planes in with the far side of the hip (side B).

The easiest way to stack this connection is to set a temporary ridge post at each end of the ridge. With this temporary ridge set to the correct height, everything will automatically line up if there aren't any other problems.

When filling in with jacks, the "continuation jacks" between the hip and the intersecting ridge must be held high to plane in with the far edge, as shown in Figure 18.

Valley-to-Valley

At the connection of valley #2 (supported) and valley #3 (supporting), the two inside edges of the valleys must align at the top (Figure 19), rather than as I've described for the valley-to-hip connection.

On a ridge that hangs between two valleys (as in the case of ridge #2 at valleys #2 and #3), the top edge must plane into the centers of the valleys on each side.

Disappearing Valley

Valley #3 is a special case. Notice in Figure 15 how this beam acts as a valley between the plate line and the intersection with valley #2. Then, for all practical purposes, the valley disappears, and reappears as a hip above the intersection with valley #4.

To make the upper section of valley #3 a hip, it must be backed (Figure 20). Back it on the right side (looking up the hip from the outside) from valley #2 up to ridge #2, and on the left side from valley #4 to ridge #2. A skilled craftsman would rip these angles on the ground prior to stacking. But as a production framer, I simply eyeballed the backed section and

Figure 18. *Because the main hip interrupts the gable rafters (above), the upper jacks will meet high on the hip (left) to stay in plane with the lower jacks.*

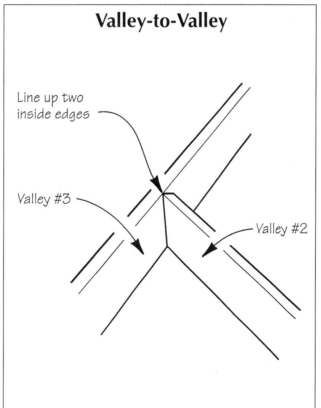

Figure 19. *At a valley-to-valley connection, the two inside edges of the valleys meet at the top.*

Stacking Supported Valleys

Using Line Length Ratios

Most roof framing books are far too complicated for me. I'm not a math wizard, so I've relied on a simple system of using ratios to calculate rafter lengths, which I call LL ratios. This method does not require a fancy calculator with trig functions and fractions, pages of rafter tables, or a computer program. A low-budget calculator with standard functions (+, −, x, ÷) will do.

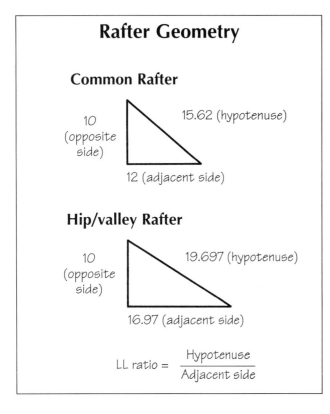

Figuring LL Ratios

LL ratios can be figured for any roof slope (see chart at right). For example, from the Pythagorean theorem we know that a triangle with a rise of 10 and a run of 12 has a slope, or hypotenuse, of 15.62 (see illustration). To find the LL ratio, divide the slope by the run: 15.62÷12 = 1.3017.

For hips and valleys, the LL ratio is figured using the hip/valley run: a diagonal distance of 16.97 for every 12 inches of run. For a 10/12 roof, divide 19.697 by 12 = 1.6414.

When I walk onto a job, I write all the LL ratios I'll need on the floor deck and on the prints where I can always see them. That way I don't have to memorize all the different ratios I might be using on one roof, or stop to figure them out.

Figuring Rafter Lengths

Use the following formula to figure any line length:

$$\frac{\text{Total span} - \text{Ridge thickness}}{2} \times \text{LL ratio} = \text{Line length}$$

When figuring this dimension, I prefer to subtract the ridge thickness out at the beginning, so I don't have to remember to shorten the rafter later while laying it out. (Note: This works for standard rafters, but for a supported valley, follow the shortening method described in the text.)

—W.H.

Rafter LL Ratios

Pitch	Common	Hip/Valley
1	1.0034	1.4166
1½	1.0077	1.4196
2	1.0137	1.4239
2½	1.0214	1.4294
3	1.0307	1.4360
3½	1.0416	1.4439
4	1.0540	1.4529
4½	1.0680	1.4630
5	1.0833	1.4742
5½	1.1000	1.4865
6	1.1180	1.4999
6½	1.1372	1.5143
7	1.1577	1.5297
7½	1.1792	1.5461
8	1.2018	1.5634
8½	1.2254	1.5816
9	1.2500	1.6007
9½	1.2750	1.6206
10	1.3017	1.6414
10½	1.3287	1.6629
11	1.3567	1.6852
11½	1.3850	1.7083
12	1.4142	1.7320
14	1.5366	1.8333
16	1.6666	1.9436
18	1.8028	2.0615
24	2.2361	2.4495

This chart lists Line Length ratios for roof pitches up to 24/12. Find the pitch of the roof in the first column, then select the ratio for either common or hip/valley rafters.

140   *Chapter Eight: Roof Framing*

Figure 20. *When a single beam serves as both a valley and a hip, it must be backed. The right side of this valley (valley #3 in Figure 1) is backed from the valley-to-valley connection up to the ridge, and on the left side from the ridge back down to the gable ridge (coming in from the left side in the photograph). These rips can be made on the ground or in the air after the roof has been stacked.*

cut it with a chain saw after it was stacked.

With valley #3 backed, the top inside edge will line up with the top edge of hip #5 and ridge #2. But the top center of the backed portion will be higher (Figure 21). Also notice that valley #3 and hip #5 must be cut full span to butt together at the end of ridge #2.

By Will Holladay, author of A Wood Cutter's Secrets *and freelance roof framing contractor.*

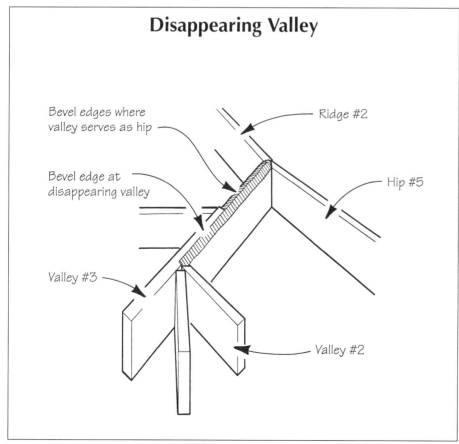

Figure 21. *The inside edge of the back cut on valley #3 lines up with the top edge of the hip and ridge; the peak of the backed portion, however, will be higher.*

Building Doghouse Dormers

One of the most economical ways to add living area to a house is to finish off the attic space, usually with a couple of dormers incorporated into the design. Dormers provide light and ventilation, and also add some visual interest, both inside and out.

But many first-time dormer builders make the mistake of underestimating the complexity of building even small dormers. On the exterior, you must contend with framing, roofing, fascia, windows, and siding. If you're also finishing the inside, add insulation, drywall, paint, and trim to the list. Along with these tasks, you're faced with protecting a gaping hole in your customer's roof from the weather.

Site Inspection

Before starting construction, examine the area where the dormer is going. Are there any wires or other mechanicals that need to be moved? Will kneewalls have to be repositioned? Are the existing rafters sized correctly? This inspection is easy if the attic is unfinished, but if the dormers are going in a finished space, it will take more time to locate potential problems.

Also, make sure you can get 4x8 sheetgoods to the interior, and look at the outside with an eye towards scaffolding — shrubs, porches, and chimneys can increase the difficulty of erecting staging.

Dormer Layout

The first step in laying out a dormer is to locate the centerline, and put a mark on the underside of the roof sheathing. (If the dormer is in a finished room, you'll have to remove enough drywall to provide access to the rafters.) On the rafter nearest the centerline, mark the height of the dormer ceiling framing. Since doghouse dormers are typically too small for anything but a flat ceiling, this mark establishes the top of the hole you're going to cut in the roof. To locate the bottom of the roof cutout, work backwards from the ceiling height mark: Subtract the height of the window header and the window, as well as the clearance between the window sill and the main roof.

Now work from the centerline to lay out the width of the dormer. Connect all these marks on the roof sheathing to create a rectangle: This represents the framed opening the dormer will require.

Strengthening the roof. At this point, before you get out your reciprocating saw, step back and take a good look at how your dormer cutout will affect the overall structure of the main roof. Doghouse dormers tend to be small, and typically require cutting no more than three rafters. How you support these interrupted rafters depends on the rafter size, and the roof pitch and span. I recommend checking with an engineer before cutting large holes in a roof structure. But in many cases, doubling or tripling the adjacent rafters will give sufficient support.

Adding full-length sister rafters is easy in an unfinished space. If the ceiling is finished, however, be sure to cut back enough drywall to add rafters. Slipping a rafter between the sheathing and the finish ceiling is not worth the trouble: It's almost impossible to do this without damaging the finish materials, and you can't get a solid connection at the inaccessible upper end.

Prefab the walls. Depending on the size of your crew and the complexity of the existing framing, you may want to frame the opening from the inside. You can also use your layout marks to build the triangular dormer sidewalls (Figure 22), and to precut your rafters. Performing all this work before breaking through the roof allows two carpenters to frame, sheathe, and shingle a dormer in less than eight hours.

Figure 22. *To reduce the amount of time the attic is exposed to the weather, frame the dormer opening from the inside and prebuild the walls. After cutting away the sheathing, pass the prebuilt walls through the opening (left), and install them in one piece (right).*

Opening the Roof

With the dormer hole framed, but the sheathing still intact, drive a 16d nail up through the roof at all four corners. From outside on the roof, enlarge the rectangle formed by the nails by adding the wall framing thickness (including the sheathing), plus one inch for flashing clearance. Remember, there is no wall at the top.

After marking out this rectangle with a chalk line, use a straightedge and utility knife to cut through the shingles (plan on using a half dozen blades). Remove all the shingles within the scored area. Then use the original nail holes to mark out the framed roof opening, and cut out the roof sheathing with a circular saw. In an unfinished attic, be careful not to allow the sheathing to drop through the first-floor ceiling.

Now remove any shingle nails that will get in the way of the step flashing for the sidewalls of the dormer. I use a slate ripper or flat bar to carefully lift the shingles, removing all nails within 6 inches of the opening. Be sure to get all of the nails at this point — trying to do it after installing the dormer walls is much more difficult.

Marking the valley. Instead of using a lot of math to figure out where the valley falls, I prefer to mark it by temporarily installing a few rafters (Figure 23). First, bring up your two prefabricated

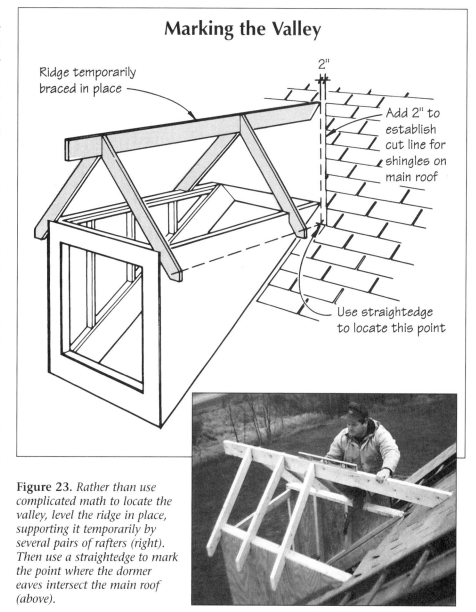

Figure 23. *Rather than use complicated math to locate the valley, level the ridge in place, supporting it temporarily by several pairs of rafters (right). Then use a straightedge to mark the point where the dormer eaves intersect the main roof (above).*

Building Doghouse Dormers

Figure 24. *Use a plate to catch the edge of the roof sheathing in the valley, and to provide solid nailing for jack rafters (left). Where the dormer eaves intersect the main roof, provide a rafter with a bevel cut that matches the main roof pitch (above).*

sidewalls and tack them in place, with temporary braces across the front to steady them. If you didn't precut the rafters earlier, lay out and cut four of them now. Tack one pair at the front of the dormer, and another pair back as far as the main roof will allow, letting the ridge cuts rest against each other. Then cut a ridge board about 6 inches longer than you will need, and cut the angle of the main roof on one end. Slip the ridge up between the rafters so it's flush with the tops and resting on the main roof. By holding a straightedge along these temporary rafters, you can mark the top and bottom of the valley formed by each roof plane of the dormer and the main house. This procedure will take some juggling if you're working alone, but two people can handle it easily.

Before you cut the shingles at the valley, disassemble your temporary rafters to get them out of the way. Also raise the valley layout line a couple of inches higher on the main roof plane, as shown in Figure 3. This allows for the thickness of the dormer roof sheathing, and gives you enough room to slip flashing up under the main roof shingles.

Cut through the shingles and remove them, using the same technique as before. Once again, it will be necessary to reach up and under the main roof shingles to remove nails that will get in the way of the valley flashing. This time, however, you'll need to remove all the nails 12 inches back.

Some builders avoid this "deep reach" by removing and tabbing back shingles around the valley area to provide access, and then reinstalling the shingles after the flashing is in. But if you can't salvage all of the shingles, you run the risk that the newer replacement shingles will vary slightly in color. The cut-back method confines any color variation to the dormer roof planes, where it tends to be less noticeable.

Framing, Flashing, and Roofing

After the shingles are cleared out, assemble the dormer rafters. Be sure to butt the last pair of common rafters snugly against the main roof to provide solid nailing for your subfascia (Figure 24). And always cut a rafter that will fall at the transition point between the fascia and the valley. Also install a plate at the valley to support the edge of the dormer roof sheathing. After the rafters are installed and the roof is sheathed, slip the valley flashing under the main

Figure 25. *When roofing the dormer, slide the new shingles up under the existing ones. This eliminates the need for a perfect cut on the dormer shingles and gives extra protection at the new valleys.*

roof valley shingles. I typically use 24-inch prepainted aluminum coil stock.

This job is best done by two people: one person to lift and hold up the shingles while the other feeds the flashing. If you missed any nails earlier when clearing the space for this valley flashing, now is when you'll find them. You can still remove a nail at this stage, but it's much more difficult with the dormer roof in the way.

With the valley flashing installed, start the dormer roof shingles at the gable end and work towards the valley. Slip the last dormer shingle in each course under the main roof shingles as far as you can (Figure 25). This eliminates the need for a precise cut at the valley and provides a measure of extra protection.

Start the flashing below the window with a piece of end-wall flashing, caulked at the corner (Figure 26). Flash the sidewalls with pieces of 7x7-inch aluminum step flashing, fastening each piece to the sidewall with one nail. Cut a tab into the first piece of step flashing and bend it around the corner of the end wall. Then weave step flashing into each course of shingles, running the step flashing up beyond the soffit line as far as possible.

Finally, staple tar paper or housewrap on the dormer walls, making sure it laps over the end-wall flashing and step flashing. You now have a weathertight dormer. If you have any time left in the day, you can install the window, and start the fascia and soffit work. Depending on the soffit material, securing the soffit where it meets the main roof can be a challenge. Vinyl soffit is no

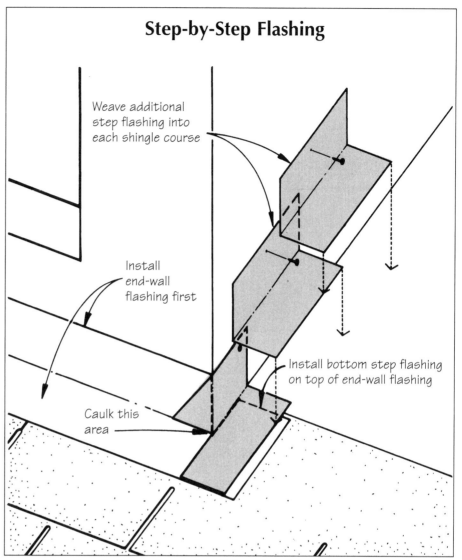

Figure 26. *When flashing a dormer, install the end-wall flashing first, then work up the roof with step flashing at each shingle course. Attach each step flashing with one nail in the dormer wall sheathing.*

problem — just slide the soffit pieces back through the F-channel and fascia. With a plywood soffit, however, you'll need to nail a block on the beveled dormer rafter at the valley intersection to keep the soffit from disappearing up into the dormer overhang. The pitch of the main roof will act as a wedge to press the soffit tight against this block. When it's wedged in tight, nail it off where you can reach it.

—Carl Hagstrom

Roof Framing with Wood I-Joists

While many builders have used I-joist floor systems, roof framing with wood I-joists remains somewhat of a mystery. But for certain kinds of roof configurations, wood I-joists may work better than dimension lumber — long-span cathedral ceilings are a good example. My purpose here is to clarify the differences between solid lumber and wood I-joists, and to provide some tips to make roof framing with wood I-joists easier.

I-Joist Roof vs. Conventional Framing

The first thing to understand is how a roof framed with wood I-joists differs structurally from a traditional stick-framed roof.

Figure 27. *I-joists make for a strong, flat roof, but you have to pay attention to the connections at the ridge and eaves.*

In a typical roof framed with dimension lumber, the rafters rest on the exterior wall top plate at the lower end and bear against a ridge board at the top. Continuous ceiling joists or collar ties span from rafter to rafter. There is no need for bearing posts under the ridge board, which is nonstructural. The roof loads are carried to the top plates of the bearing walls, where the floor joists, acting in tension, keep the rafter ends from spreading out. What you have here is essentially a truss, built on site. The strength of the roof system depends a lot on the connections between the joists and the rafter ends: As long as those nails are adequate for the loads and don't slip, the rafters are restrained from pushing out, the ridge board is compressed in place at the top, and the roof doesn't sag.

With wood I-joists, there is no practical way to make a strong shear connection between the floor joists and the rafter ends. Instead, a wood I-joist roof system is framed with either a central bearing wall or a structural ridge — a beam that carries the roof load to posts. The load from the top half of the roof is carried by the bearing wall or structural ridge; the bottom half is carried by the exterior bearing walls. The loads are primarily gravity loads, which push down, not out, on the bearing walls. So there is no need to engineer a connection between the floor or ceiling joists — if there are any — and the rafter ends.

In my work as a field rep for Trus Joist MacMillan, most of the wood I-joist roofs I see use a structural ridge beam rather than a center bearing wall (Figure 27). But whether you use a ridge beam or a bearing wall, there are two ways to support the joists at the upper end: with hangers or a beveled bearing plate (see "Wood I-Joist Details," page 148). The important point to remember is that no birdsmouth cuts are allowed at the high end of the I-joist. This would mean cutting through the bottom flange at the bearing point, which would damage the I-joist.

Using Hangers at the Ridge

The most common method is to use a face-mount hanger with a sloped seat (see Figure 28, and "Hanging an I-Joist From a Ridge Beam," page 150), such as the Simpson LSSU series or the USP (Kant-Sag) TMU. These hangers can be adjusted in the field to match the I-joist slope, and can be skewed side to side up to 45° for hip-and-valley jack rafters.

Web stiffeners. The sloped-seat hanger requires a beveled web stiffener on both sides of the I-joist to fill out the space between the hanger and the web. You can rip stiffener material out of plywood (the thickness depends on the I-joist size), then production-cut it on a chop saw to the right length. For larger I-joists, use 2x4s for the stiffeners. Make sure you check the manufacturer's literature for the proper stiffener size and thickness — it may vary from brand to brand.

Plywood stiffeners should be attached with three 8d nails with points clinched. For 2x4 stiffeners on the larger I-joists,

Figure 28. *Sloped-seat hangers can be adjusted in the field to match the slope of the roof. These hangers require a plywood or lumber web stiffener on both sides of the I-joist to fill out the space between the hanger and the web.*

Figure 29. *When supporting I-joists on a beveled plate, install metal straps across the tops of the butting joists for all roof slopes. When using hangers, straps are necessary for slopes above 7/12.*

Figure 30. *Sloped-seat connectors like the Simpson VPA or the USP TMP can provide bearing at the top plate if the roof loads are not too great.*

use three 16-penny nails. It's a good idea to drive two nails from one side.

It's important to install stiffeners with a gap at the top (we recommend 1/4 inch). This prevents the top flange from prying off the joist web under load.

Strapping for steep slopes. With a ridge beam and hangers, no additional lateral bracing is needed at the top end of the rafters. But for roof slopes greater than 7/12, you may need to install a metal strap tie, like the Simpson LSTA 15, across the top of each pair of opposing I-joists to resist the tendency for the joists to "slide" downhill.

Beveled Plate Details

The other method used at the ridge is to nail a double-beveled plate on top of the ridge beam (or bearing wall top plate) to provide the sloped bearing surface for the joists. (A single-beveled plate can be similarly used at the low end of the rafter.) Once the plumb cuts have been made, the I-joists are installed by butting them at the ridge and nailing them to the beveled top plate with at least two 10d nails.

Blocking. In order to provide lateral stability for the I-joists, you must also install blocking between them on each side of the ridge. This can be metal cross-bracing, dimensional lumber, or I-joist material (see illustration, next page). Probably the simplest and sturdiest is to use I-joist blocking, installed at the angle of the roof. This provides flange-to-flange support without the need for any filler pieces, and also makes a good shear block for transferring forces from the roof diaphragm to the ridge beam. In a cathedral ceiling, where continuous roof ventilation is needed, you can use narrower-width dimension lumber, metal cross-bracing, or engineered rim joist material notched to allow airflow.

Strapping required. Finally, metal straps should be nailed across the tops of the butting joists for all roof slopes (Figure 29, above). As an alternative, some manufacturers show a plywood gusset connecting the webs of the butting joists. Both methods will work, but metal strapping is faster.

Bottom Bearing Details

You can use a variety of details at the exterior wall plate, depending on the the roof profile you want.

Birdsmouth. The most common detail is to make a birdsmouth cut at the plate (see illustration, next page). But be careful: There's definitely a right and a wrong way to do this (see "I-Joist Mistakes," page 150). When laying out a birdsmouth, make sure the seat cut does not overhang the inside face of the bearing wall. The bottom flange must get full bearing on the plate. If this cut is not made properly, the joist's strength can be significantly reduced. With a birdsmouth cut, you'll have to use web stiffeners on both sides of each I-joist, as at the ridge.

Beveled plate. Another option at exterior walls is to use a beveled plate instead of a birdsmouth. This can save time, because there's less cutting to do and web stiffeners are usually not necessary (except in cases of very large roof loads). Using a beveled plate also provides more design flexibility, as the joists can cantilever up to one-third of the rafter span. The only possible drawback is the additional cost and availability of the beveled plates, although these can be ripped from dimensional stock on a band saw.

Sloped-seat connector. A third option at the low end is a sloped-seat connector attached at the bearing point (Figure 30). These metal connectors, such as

Roof Framing with Wood I-Joists

Wood I-Joist Details

Blocking

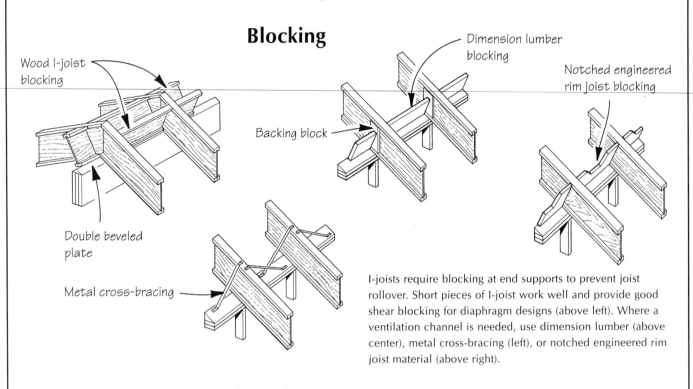

I-joists require blocking at end supports to prevent joist rollover. Short pieces of I-joist work well and provide good shear blocking for diaphragm designs (above left). Where a ventilation channel is needed, use dimension lumber (above center), metal cross-bracing (left), or notched engineered rim joist material (above right).

Birdsmouth vs. Beveled Plate

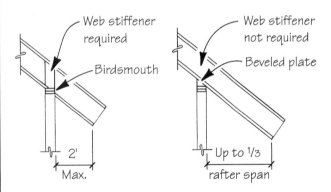

Though a beveled plate requires special fabrication, it has some advantages over cutting a birdsmouth.

Gable-End Overhangs

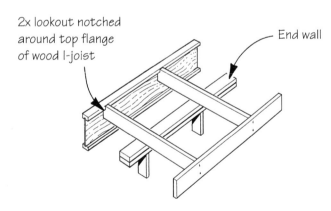

Frame gable-end overhangs with dimension lumber outriggers notched around the I-joist top flange. If the overhang exceeds the I-joist spacing, check with the manufacturer to see if a doubled I-joist is required.

Headers

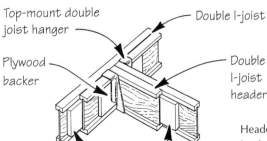

Headers at openings may require double joists, depending on the loads. Use filler blocks between the I-joists and a backer block to support the hanger.

Figure 31. *Many soffit profiles are possible with an I-joist roof system. Shown here are dimensional lumber rafter tails (top), a level soffit (middle), and an I-joist used as backing for fascia (bottom).*

Figure 32. *Sloped-seat hangers can be skewed up to 45° for hip and valley jack rafters. Compound cuts are not necessary; only the plumb cut needs to be made.*

USP's TMP or the Simpson VPA, provide a field-adjustable sloped bearing surface. Depending on the manufacturer and type of connector, the allowable slopes range from 1/12 to 12/12. Installation varies by manufacturer, so always check the instructions for the specific connector you are using. The advantages of these connectors are that no birdsmouth cut and no web stiffeners are needed. The disadvantages are that their loading capacity is somewhat limited (see manufacturer's catalog for specifics) and that installation can be time consuming.

Blocking. Regardless of the method you use at the low end, blocking or cross-bracing is required to prevent joist rollover.

Soffits and Overhangs

Soffit treatments seem to concern a lot of the contractors I meet in the field. "How can I nail my fascia to that skinny little piece of OSB?" is a question I hear a lot. The truth is, there are several details that work, depending on the roof profile you want (Figure 31). One typical detail is to sister on dimensional lumber rafter tails. You can also use plywood, an engineered rim board material, or even another I-joist for fascia backing. To frame a flat soffit, you can extend the birdsmouth cut to the end of the joist, then attach 2x4 blocking for a soffit nailer. Just about any traditional profile is possible with proper planning at the design stage.

The rules for overhangs are straightforward. All the manufacturers' design guides show many details. One point to remember is that if a birdsmouth cut has been made, the maximum allowable overhang for any of the details is 2 feet. If you want a longer overhang, use a beveled plate or sloped-seat connector.

Gable-end overhangs. Gable-end outriggers are framed by cantilevering dimension lumber across the gable-end top plate, similar to stick-framing techniques (see illustration, facing page).

Hips and Valleys

Hips and valleys are possible with wood I-joists, but the only practical

Hanging an I-Joist From a Ridge Beam

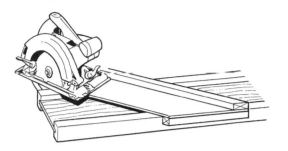

1. Cut the plumb cut at the top of the I-joist. Use a piece of plywood in the web or a cutting jig for the saw shoe to ride on.

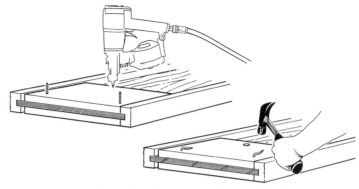

2. Attach the web stiffeners to each side with three 8d nails, leaving a 1/4-inch gap at the top. Clinch the nail points.

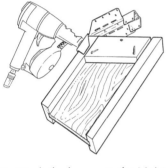 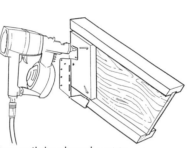

3. Attach the hanger to the I-joist. First, nail the sloped seat to the bottom of the rafter. Next, bend the hanger against the plumb cut and put the rest of the nails into the web stiffener and bottom flange.

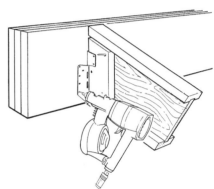

4. Lift the I-joist into place and nail the hanger to the ridge beam.

I-Joist Mistakes

The photos at right show examples of how *not* to frame a roof with wood I-joists. The bottom flange of the I-joist — the part that is most critical for carrying the bending forces — has no bearing, either at the top or the bottom.

It will not take a large load for this roof to fail. What will the failure look like? Most likely, at the upper end the I-joist webs will tear away from the top flanges. At the bottom end, the webs will split and the roof will come down.

COURTESY OF GARY LOZOWSKI

way to frame them is to use the field-adjustable hangers mentioned above. The techniques are identical to the methods mentioned earlier, except that when the hanger is installed to the beam, the hanger must be skewed (Figure 32). Although hangers can get expensive, one advantage of this technique is that compound cuts are not needed on the jack rafter ends. The only cut required is the plumb cut.

Header Details

Framing headered openings for skylights and dormers is also straightforward. As with dimension lumber, the size of the opening determines how many I-joists are needed to support the header. If the header hangs from a single I-joist, you nail a backer block (typically plywood) to the joist, then nail the hanger to the backer block (see illustration, page 148). Double I-joists require a filler between them — either plywood or dimension lumber — and a backer block for the hanger. For really large openings, it makes sense to use an LVL or Parallam beam to support the headers instead of multiple I-joists.

Keep in mind that all these connection details, whether for headers, hangers, or whatever, have been engineered by the I-joist manufacturer. It's critical that you read and understand — and follow — the application guide. If you have a question or a tricky installation problem, call the manufacturer for technical support.

By Curtis Eck, P.E., a technical representative for Trus Joist MacMillan in Seattle, Wash.

> **Sources of Supply**
>
> **Wood I-Joists**
> **Boise Cascade**
> Boise, ID
> 208/384-6161
> www.bc.com
>
> **Georgia Pacific Corp.**
> Atlanta, GA
> 404/652-4000
> www.gp.com
>
> **Louisiana-Pacific**
> Portland, OR
> 800/648-6893
> www.lpcorp.com
>
> **Trus Joist**
> Boise, ID
> 208/395-2400
> www.trusjoist.com
>
> **Weyerhaeuser Company**
> Federal Way, WA
> 253/924-2345
> www.weyerhaeuser.com
>
> **Hangers and Connectors**
> **Simpson Strong-Tie**
> Dublin, CA
> 925/560-9000
> www.strongtie.com
>
> **USP Structural Connectors**
> Minneapolis, MN
> 800/328-5934
> www.uspconnectors.com

Flat Roof Framing Options

Although there are several ways to frame a "flat" roof, none of them result in a truly flat roof. Even in mostly dry Southern California, where I have been production framing for over 15 years, you have to create a slight slope to get rid of rain water from the heavy, sudden storms we get. Otherwise, the water will pond and eventually cause leaks.

To give the illusion of a perfectly flat roof, many designers add a low parapet around the perimeter of the roof, which conceals the slightly pitched roof behind. The roof surface is designed to direct water to one or more drains behind the parapet wall.

Designers in my area commonly use two approaches to rooftop drainage (see Figure 33). One method is to use a two-part roof drain with a lower main drain and a higher backup drain, in case the low drain gets clogged. The other approach is to use a single drain with scuppers in the parapet as a backup.

Rip Strips for Tapered Rafters

On very simple, short-span roofs (over entryways, for example) where the rafter's strength isn't an issue, we rip a 1/8-inch-per-foot taper on the tops of the rafters to create the slope. As the span gets longer, we set the rafters flat, then attach long tapered "rip strips" to the top of the rafters to create the slope. A roof built with rip strips on top is quick and easy to frame because the ceiling underneath is flat, the wall plates are all one height, and the strips can be installed from above. We use a site-built jig to cut the rip strips with a circular saw (see "Quick Rip-Strip Jig," page 154).

Sometimes the plans call for two layers of plywood, one applied directly to the flat roof rafters, followed by rip strips and a second layer of plywood applied over them. With the approval of the engineer, however, we're sometimes able to apply the strips directly to the rafters and save one layer of plywood. Because we build in a seismically active area, the engineer has to be sure that there is adequate transfer of lateral forces from the shear walls to the roof diaphragm. Full-depth blocking over shear walls, anchored with Simpson A-35 clips, can often solve the problem (Figure 34).

Single-Slope Roofs

The simplest roofs have full-length rip strips nailed directly to the rafters to create a single slope (Figure 35). We try to limit the high end of the taper to 3 1/2 inches — the width of a 2x4. Tapering a 2x4 at 1/4-inch-per-foot slope allows a 14-foot run. We occasionally use 2x6s for slightly longer spans, but if the roof span gets much wider we'll create a low-pitched gable running to two sides of the roof, or a hip, which directs water to all four sides (more on hips below).

Other options for single slopes. Another way to frame a slope on a flat roof is to sister a sloping roof rafter onto each flat ceiling joist. This works fine for small roofs, but on large spaces, sistering wastes lumber and takes more time than ripping strips.

It's also possible to get custom tapered I-joists and trusses. We've used these on everything from small decks to large buildings. The only drawback with these is that they must be ordered far in advance, which makes careful job planning very important. If you are one joist short, it can cost you weeks of production.

The Rip-Strip Hip

Using rip strips, we can create a hip effect over a flat roof deck, directing water in four directions. With careful layout, the rip strips can even be installed before the roof joists are installed. Here's how we do it:

The outside rafter is installed flush with the outside face of the wall and has no rip strips attached (Figure 35, bottom). The second rafter gets a short

Figure 33. *A low parapet wall (left) conceals the slightly pitched roof behind, giving the impression that the roof is perfectly flat. A two-part drain provides for runoff. The main drain, protected by a leaf screen, is plumbed into the home's main waste line; the smaller drain will have an extension to provide backup drainage in case the main drain clogs. The author sometimes places a "cricket" in a corner, to direct ponding water back toward the drain (right).*

piece of rip strip — 15 1/4 inches for 16-inch on-center framing — tacked on each end, the second joist gets a 31 1/4-inch piece at each end, and so on. We then measure the height of the rip strips at their high ends and fill in with continuous flat strips ripped to that dimension.

We sheathe this low hip roof as if it were flat. Where the plywood crosses a hip, we snap a line and cut along the hip. The waste piece falls into place on the other side of the cut.

Double-Pitched Roof

We often build garage roofs that slope to a roof drain in one corner of the roof. To build the roof so it slopes to one corner, the trick is to first cut and set the four outside rim joists to form the slope, and then custom taper each rafter to fit inside the rim boards (Figure 36).

The two rim boards opposite the drain are set level and kept at full depth. The other two rim joists start out at full depth but taper toward the corner drain

Figure 34. *The plans often call for a fully sheathed flat roof to be covered with a second, sloping layer of sheathing (above). On smaller roofs, it is sometimes possible to save the second layer of plywood by nailing the rip strips directly to the rafters. In seismic regions, the engineer may specify full-depth blocking and metal clips over shear walls to ensure transfer of shear forces into the roof diaphragm (left).*

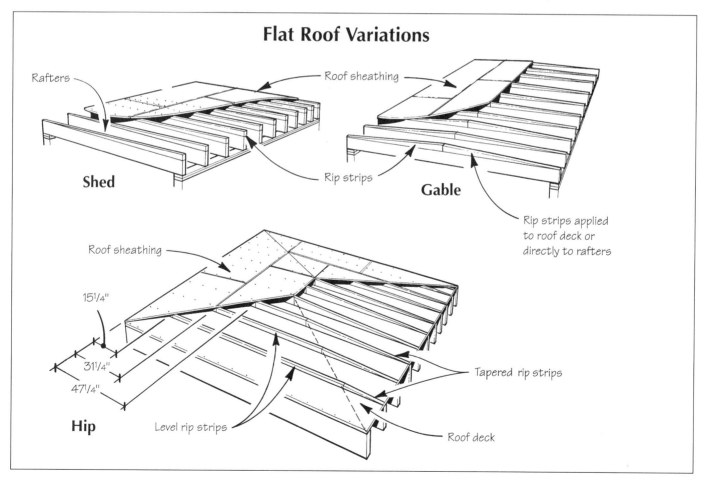

Figure 35. *On small and mid-sized flat roofs, the author applies rip strips to create a single-pitch shed roof, or a gable slope (top). On large flat roofs, a hip profile (above) often works better to direct water to multiple roof drains. Since there are no hip rafters in this approach, all the rafters run in the same direction, and the plywood can be laid as full sheets and cut along each hip.*

Flat Roof Framing Options

Quick Rip-Strip Jig

Measuring, snapping lines, and cutting rip strips one at a time is too time-consuming, so we devised a site-built jig that allows us to cut tapers quickly and accurately. Young guys can work right off the roof deck, but when we have a lot of ripping to do we usually work from a 2x12 set on sawhorses. Here's how to do it:

- Temporarily tack a straight 14-foot 2x6 to the work surface.
- Nail short scraps of 2x4 to the work surface at a 45-degree angle to the 2x6, with one corner touching the 2x6. This holds the material being ripped while preventing sawdust buildup.
- Nail a block at one end of the work surface to act as stop.
- Find a straight 2x6 and snap a line diagonally from corner to corner along its length.
- Nail another block to the work surface, this time spaced far enough away from the rip stock to allow for easy removal and replacement of rip stock. Cut a wedge that fits between this block and the rip stock. This secures the work and also removes any crown from the stock.
- With the chalked 2x6 wedged in place, nail a straight 2x8 on top of the 2x4 blocks to act as a saw guide. Set the 2x8 so the edge of the saw shoe will ride against it while the blade cuts the chalk line. (Put a couple layers of scrap building paper between the 2x4s and the 2x8; this acts as a spacer to allow the rip stock to slide easily under the 2x8.)
- To use the jig, rip the chalk line on the original 2x6. Remove the wedge, remove the two rips (which both taper from 0 to 5 1/2 inches in 14 feet), wedge a new 2x6 into place, and have at it.

— D.G.

A carpenter rips a long taper using the site-built jig.

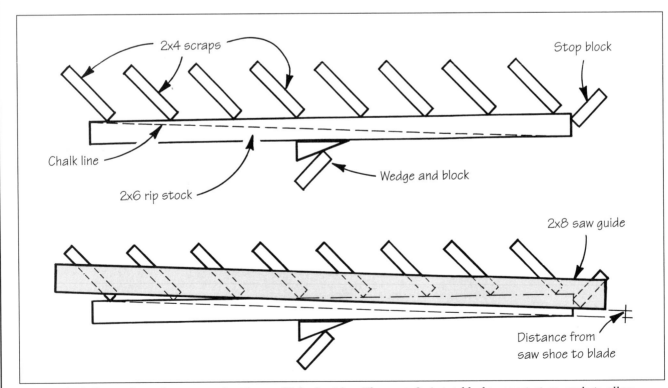

This site-built jig speeds up the process of cutting multiple rip strips. The scrap 2x4 stop blocks are set at an angle to allow sawdust to clear.

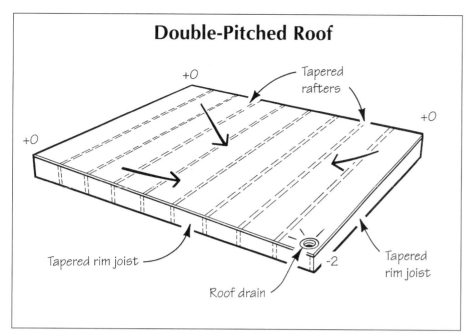

Figure 36. *For garages and other lightly loaded small flat-roof areas, the author sometimes frames a roof that pitches toward one corner. He begins by installing two full-size and two tapered rim boards, then scribes the rafters to one of the tapered rims. This results in a roof that pitches from all sides toward the corner drain.*

at the rate of 1/4 inch per foot. The tapered joists are laid out by simply measuring the height of the tapered band joist at the layout marks, snapping a sloping chalk line on each joist and cutting the taper.

When you lay the sheathing on a roof like this, don't be surprised if the plywood doesn't want to lay perfectly flat: The double slope creates a slight bow in the roof surface, but with a little weight, the plywood will conform.

By Don Gordon, of Gordon Fiano, builders of custom homes in Santa Barbara, Calif.

Chapter Nine
REMODELERS' SPECIALTIES

Roughing In for Kitchens and Baths 158

Tying into Existing Framing 162

A Second Story in Five Days 166

Case Study: A Retrofit Ridge Beam 169

Jacking Old Houses 172

Roughing In for Kitchens and Baths

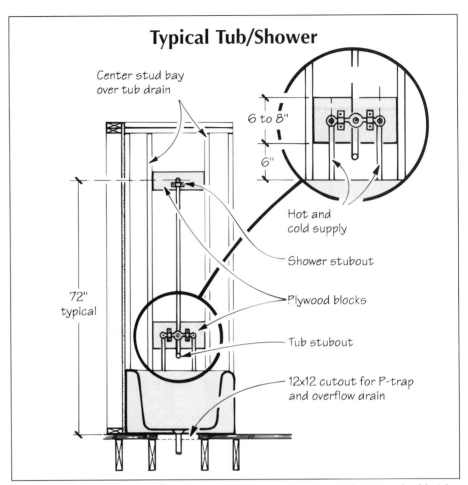

Figure 1. *Though most plumbers place the shower head and controls at standard heights, the author always consults with clients on this since many have individual height preferences.*

Having the right or wrong rough-in dimensions on a bathroom or kitchen remodel can make or break a job. Chipping out tile, hacking out drywall, butchering a cabinet, or yanking on Romex are sometimes the only ways to correct poorly located rough-ins.

Unfortunately, a bad rough-in usually remains undiscovered until near the end of the job, when the painter is finished and the tilesetter has collected his check. That's when the plumber finds out that the shower wall mortar was floated too thick and it's impossible to get the fixture knobs on. A tightly planned, two-week bathroom remodel can suddenly expand to a three- or four-week job — not good news to you, or to a client who has been living with one less bathroom.

Key Characters

There are several key elements that contribute to a successful kitchen or bath rough-in.

Lead carpenter. Invariably, a general contractor will take on full responsibility, but won't always have time to meet with the subs on the job. So information ends up getting passed through one or more workers. The result can be a disaster. To avoid problems, the contractor should assign one worker who has a well-rounded knowledge of all aspects of construction to the task of overseeing the layout and making sure subcontractors accurately place the rough-ins. As my company's lead carpenter, this responsibility falls to me. Spreading this responsibility over more than one person multiplies the potential for errors, and adds a lot of unnecessary communication.

Good plans. A detailed set of plans is essential to locating dimensions. If good plans don't exist, I draw them myself. If a drawing differs from the site dimensions by more than 1/2 inch, I get out a notebook and redraw the plan accurately, or make note of it on the plans. I've seen a fraction-of-an-inch error in a plan result in the need for a door casing to be scribed around a countertop or a switch plate to be cut down to fit next to a cabinet. This is not a pretty sight.

Spec sheets on site. Whenever possible, have the appliances and fixtures on site. At a minimum, have the manufacturers' specification sheets available for all appliances that must be built in. It's helpful with whirlpool baths, sinks, mixing valves, cooktops, and range hoods to have the fixture itself on site before rough-in begins.

Good client communication. Some rough-in locations are fixed according to the appliance or fixture requirements — those of a range, for example. No need for client input there. But the height of a shower head is usually of interest to the user. Asking your clients' opinions will make them happier with the job, even if you end up telling them how high you think the shower head should go, or how high everyone else plumbs their shower heads.

I've discovered that clients often have high expectations for under-sink storage. Unfortunately, under a three-basin sink with P-trap, garbage disposal, and instant water heater, the homeowners are lucky if they can get even a small wastebasket on the cabinet door. Make sure clients understand that the more appliances they have, the less storage space there will be.

Good subcontractor communication. I try to get all the subs together for one meeting before the job starts. It's helpful to have a contractor who does both heating and plumbing, since it isn't uncommon to have a DWV line conflict with a vent duct. Having the electrician there at the same time is helpful, just in case one of those ducts interferes with a recessed can light. Though hashing through these details with so many subcontractors at the same time is difficult, it's the best way to troubleshoot problems and come up with economical solutions.

Don't be shy about making things crystal clear to a sub. Post-it notes, ink markers, and spray paint are all helpful in directing subs.

Make a checklist. Once the demolition is done, the lead carpenter must visualize what the finished room is going to look like, even when the walls are nothing but open studs. To make this easier, I make a checklist of key rough-in dimensions. Making the checklist is a helpful way to consider each appliance location before it's too late. Referring to the list during the course of the job ensures that I won't have to call subs back to move things around.

Bathroom Rough-In

Here is a list of typical bath rough-ins, and some important considerations for each one.

Bathtub with shower. I don't attempt to precisely locate a 2-inch drain until after the bathtub is installed. But I do need a rough idea of the drain location when framing the floor. I usually cut a 12x12-inch cutout in the subfloor to give the plumber room to install the P-trap and the drain overflow, and this cutout must fall between joists. Because tubs are usually 32 inches wide, I often have to deviate from a standard joist layout if the joists run parallel to the tub.

For shower controls, I frame a standard 16-inch stud bay, centered on the

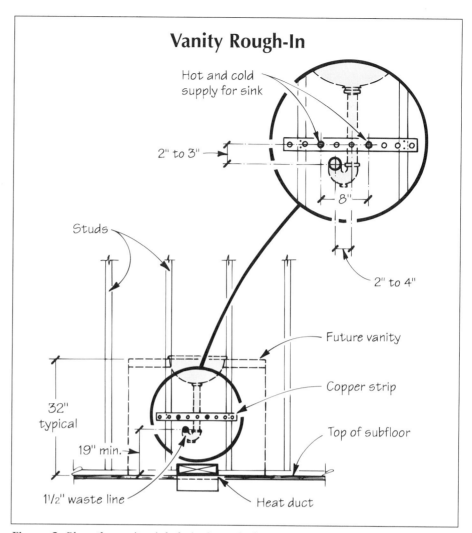

Figure 2. *Place the vanity sink drain 2 to 4 inches off center to allow room for the trap. Use copper strap to secure the hot and cold supply lines.*

width of the tub (Figure 1). I also install a piece of 3/4-inch plywood between the studs in this bay as a base for the faucet body. Plywood won't crack as easily as a 2x block with the many screws used to fasten the faucet body.

As I mentioned earlier, I leave the decision of where to locate the faucet body and shower head to the client. Before soldering in the faucet body, put all the escutcheons and handles on, and make sure the handles don't hit each other. I've discovered that the more expensive shower fixtures tend to be larger and require more space than economy models.

Deciding the depth at which to place the faucet body (and the plywood block) is largely determined by the thickness of the faucet body itself and the thickness of the wall surface. Most drain bodies give you at least 3/4 inch to play with in relation to the faucet. Many manufacturers have faucet extensions available, too, but don't depend on it. Grohe's (Bloomingdale, IL; 630/582-7711; www.groheamerica.com) newer faucets, which I install frequently, have an incredibly large range — close to 3 inches — that allows me to place the faucet just about anywhere within a wall, and not have problems getting the finishes on.

Sink. For a pedestal sink, the 1 1/2-inch drain must be centered exactly. For a vanity cabinet rough-in, jogging the P-trap over 2 to 4 inches to one side of center is desirable (Figure 2).

To quickly and accurately space the hot and cold supply lines the required 8 inches apart, I use copper straps with 1/2- and 3/4-inch holes at regular intervals. The copper pipes are then soldered to the strap for support.

Roughing In for Kitchens and Baths

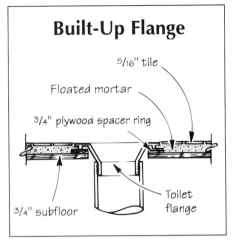

Figure 3. *Remember to allow room for the escutcheon plate when bringing the toilet supply line through the wall (right). For a tile floor, the author installs a 1/2- to 3/4-inch-thick plywood spacer ring to ensure that the flange isn't recessed too far below the finished floor (above).*

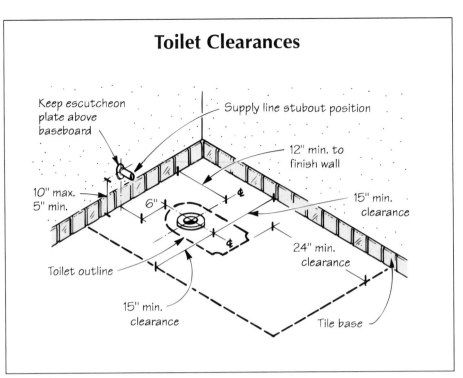

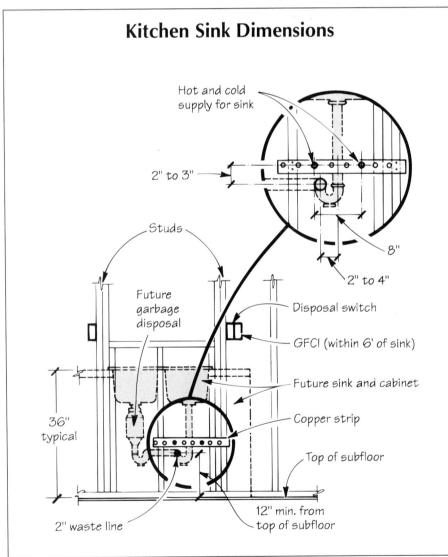

Figure 4. *For a double-basin sink with disposal unit, place the waste pipe 2 to 4 inches off center from the sink drain that doesn't feed into the disposal.*

Toilet. Most toilets have standard rough-in dimensions. Nearly all require the flange to be centered a minimum of 12 inches from the finished wall surface. With a typical drywall interior, I center the flange 13 inches from the framing, giving me 1/2 inch for drywall and 3/4 inch additional clearance. One plumber told me this additional clearance is essential for accommodating the inconsistencies in porcelain castings. If there is to be tile with a mortar bed on the wall around the toilet, another 1/2 inch is needed (Figure 3).

Code here requires a minimum of 15 inches of space on either side of the toilet, and 24 inches in front. I've found that additional space is helpful on at least one side for a wastebasket and toilet paper holder. I used to put the toilet paper holder on the back wall if the toilet rough-in had minimum side clearance, until an occupational therapist alerted me to the risk of back injury that occurs when twisting 180 degrees to reach for the toilet paper.

With a linoleum floor, I usually rest the toilet flange directly on the subfloor. If the floor has tile on a mortar bed, I build up the flange 1/2 to 3/4 inch with a plywood ring so that the flange isn't recessed too far below the finished floor. Otherwise you need two wax rings, which creates a risky seal. However, installing the flange too low is better than placing it too high.

If the cold water supply comes through the wall, it must be below the tank and above the baseboard, but not so close that the escutcheon is hanging half on the baseboard and half on the wall. I aim for the pipe to be 6 inches to the left of the toilet's center, and at least 5 inches above the subfloor. When it comes through the floor, place it 6 inches from the toilet's center and 2 inches from the wall.

Fans. I've rarely moved a ceiling joist to make room for a bath fan, unless it's a light/fan combination and the homeowner wants to center it in the room. I usually rough-in the bath fan near the bathtub/shower, but not inside. Make sure you position the fan with the louvers facing away from the bathroom entry, so that people can't see into the mechanical part of the fan when they come in the door.

Vanity lights. Symmetry is critical with vanity lights. In order to rough-in the electrical box, you often need to know the mirror or medicine chest height. I've often reframed a wall and even moved plumbing to make room for a recessed medicine chest. Strip vanity lights are nice because they usually allow you to rough-in the electrical box anywhere along the length of the fixture, as long as the height is accurate.

Outlets. Homeowners usually want outlets along the longest stretch of cabinet. Through-the-backsplash outlets require careful consideration of cabinet, substrate, tile, and backsplash dimensions. Set the outlet boxes far enough out to accommodate the thickness of the backsplash.

Outlet boxes need to be at least 3/4 inch above the backsplash so the plate doesn't hit the backsplash. If the wall above the vanity is tiled, I try to vertically center the outlet in a tile course. I usually don't worry about the horizontal placement of the outlet in relation to the tiles unless it's a high-end job. Then I'll either center outlets and switches in both dimensions of the tile or center them at the corner of four tiles. But this means you have to know the exact countertop placement, the tile dimensions, and the grout line width when the boxes are placed. If these aren't available (and they rarely are at the framing stage), I may let a wire run wild and cut in my boxes just before setting the tile.

Heating ducts. Whenever possible, I try to rough-in the heat through the kick of the vanity cabinet. Most homeowners don't like the look of a register on the wall or floor of a kitchen or bath. Standard 3x10 duct works best in a kick space, but getting a custom sheet-metal boot to make the right angle is often difficult. In my area, it is permissible to seal the kick cavity with caulk to make a wood plenum. Then I cut a simple 2 1/2 x14-inch rectangle out of the cabinet base for the register.

Kitchen Rough-In

Kitchens require fewer plumbing rough-ins than bathrooms, but have more appliance rough-ins. When laying out the rough-ins, the cabinets are the main consideration. With the subfloor swept clean, I mark out the location of the upper and lower cabinets. It's a good idea to use an optical level to locate the high and low points on the floor, to see how that will affect the height of the countertop. In general, the countertop will be the cabinet height plus counter thickness. Snapping a line or nailing a 2x4 at the top of the backsplash and the bottom of the lower cabinets is helpful for locating the rough-ins, especially the electrical outlets, on a framed wall.

Kitchen sink and dishwasher. A kitchen sink drain comes through the wall lower than a vanity drain to make room for the garbage disposal — even though the kitchen sink is nearly always higher (Figure 4). In a standard 36-inch-high base cabinet, roughing in the 2-inch drainpipe 12 to 15 inches above the subfloor works well.

In most cases the dishwasher will drain into the garbage disposal if there is one. The hot water supply to the dishwasher can be tapped off the sink hot water supply, but good plumbers will

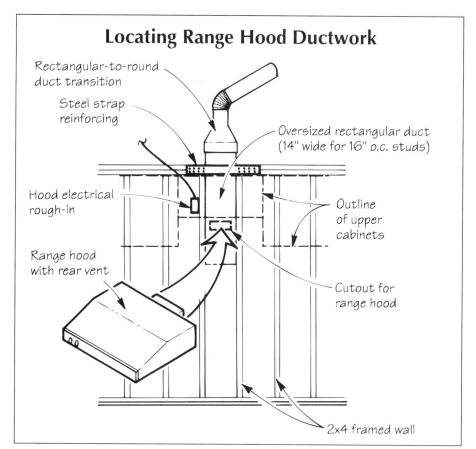

Figure 5. *One way to locate the duct for a range hood is to frame a full stud bay in the center of the hood location, then install an oversized duct before drywall goes up. After the cabinets are installed, cut through both drywall and sheet metal to connect the range hood duct.*

provide another supply near the dishwasher. This will keep the sink's hot water pressure from falling when the dishwasher is running.

Ranges and cooktops. Freestanding or slide-in ranges require accurately placed gas and electrical rough-ins. Most ranges have a back panel that allows space for the gas valve and electrical outlet. If not placed accurately, the electrical cord or gas line can keep you from pushing the range all the way back to the wall.

With a drop-in range, keep in mind that the cabinet space below is usable unless the range has a downdraft vent. Keep the outlet or gas rough-in just a few inches below the bottom of the countertop.

Hoods. In my opinion, ducted hoods are the most difficult appliances to rough in. There are two basic ways to duct a range hood: through the cabinet or into the wall.

A through-the-cabinet vent is the easiest to install. At worst, a ceiling joist will have to be moved to make way for the rectangular duct. Most clients still want to use what little space is left in this cabinet. It's worth finding out if you can get a piece of 1/4-inch plywood and the correct stain from the cabinetmaker so the duct can be hidden. Then there's usually room for a spice rack.

For an into-the-wall duct, which is preferable to many homeowners because of the storage that's saved above the hood, the rectangular sheet-metal boot needs to be precisely roughed in before drywall and cabinets are installed. If you're 1/4 inch off, it might be impossible to get an adequate seal between the hood and the duct.

I have found two solutions: One, I find the hood location to the best of my ability, using cabinet drawings. Then I make sure the duct rough-in falls in a stud bay with enough latitude to move either way a couple of inches. When the duct is installed, I ask the heating contractor not to fasten it in this bay, and to extend it a couple of inches above the plate line in the attic before turning a corner. This allows me to shift the boot a few inches vertically and horizontally after the drywall and cabinets are in place.

The second solution, suggested to me recently by a plumbing/heating sub, is to have the duct built oversized, then drywall over the entire duct. When it's time to position the hood, cut through the drywall and sheet metal at the correct location (Figure 5).

Either solution is better than removing a cabinet and tearing into a wall only to move a duct an inch or two. If you are wrestling with a 50-pound microwave/hood combination over an expensive countertop, you want the connection to be as painless as possible.

Downdraft cooktops are the toughest. Most subcontractors won't attempt to rough in the ductwork for these monsters without having them on site. Often, the duct must run through the back of the cabinet, underneath the subfloor, and through the exterior wall. If the downdraft cooktop isn't on site, a lead carpenter can spend hours trying to communicate to the cabinetmaker and subcontractors the best way to install it, and still end up modifying a cabinet to make it fit.

Lights. Architects often draw recessed lights centered between upper cabinet doors and above the front lip of the counter. This often means that ceiling joists have to be moved or headed off. Sometimes the top of the can light comes too close to a rafter, which is a more serious problem. In this case I've usually been able to convince the client to deviate slightly from the plans.

Outlets. Above-counter outlets are handled in a manner similar to that for bath vanity outlets. The refrigerator, microwave oven, trash compactor, and garbage disposal all need outlets. The garbage disposal also needs a switch. Most clients prefer it to be placed to the right of the sink.

Sometimes I photograph the open walls before drywalling, just in case an outlet gets accidentally covered. This saves me from trying to find the "lump" in the wall.

—*Jim Hart*

Tying Into Existing Framing

Building an addition is a lot like new construction except when it comes time to tie the new into the old. Although it may be difficult or impossible to uncover every important construction detail in the existing building beforehand, the more you have learned before you start the project, the better off you'll be.

Look Before You Leap

The best place to solve problems is in the design, estimating, and contract-writing phases of a job, where it can be done without the use of a nail puller. For instance, in a 20 x 28 two-story addition I recently bid on, the architectural plans specified 2x12 joists for both floors. While doing my cost estimate, I noticed that the existing house had 2x10 joists, making it impossible to match both the ceiling line of the first story and the floor of the second. One solution was to divide the span with a flush-framed steel beam and use 2x10 joists. We took this approach and produced the seamless look that is the mark of a well done addition. It added hundreds of dollars to the cost, but by catching the problem early, I kept the cost from being even higher.

Oversights of this nature are common in designing and building additions, and unless they are caught before construction begins, they will leave the designer distressed, the builder with an idle carpentry crew, and the owner unhappy with both.

In some cases, it's easy to point a finger at the architect; but similar mistakes turn up on the builder's end, and for the same reason — we make incorrect assumptions about the existing building. To reduce these costly errors, I have adopted the following as my number one rule regarding additions: "Understand the structure you are adding on to." This may not be good grammar, but it has proven to be good practice.

The best place to start is with a set of plans for the existing building, which the owners often have tucked away in the back of a drawer. While you can't assume everything was built as drawn, there may be valuable information that would be hard to come by otherwise. I recall an elaborate house where I could not fathom what route a soil stack took from a second floor bathroom to the basement. But the original plans showed the location of the stack at each floor, and a telltale bulge around the bell end of the cast iron pipe in the plaster wall nearby solved what was left of the mystery.

I also photograph the existing building for later reference. Two or three dozen 35mm photos take the place of all the sketches and half the notes I used to make when I went to look at an addition. I snap the exterior elevations, closeups of the trim inside and out, the interior wall and floor finish, and even the service panel and plumbing. Later, I can sit in my office and search the photos for the things I missed when looking at the building itself — for example, the fact that all the double-hungs on the first floor are one-over-one, but three-over-one on the second floor. The $20 I spend on photos for each project more than pays for itself, even if I only get one job in four.

What to Look For

While it's obvious that you should check the depth of the existing floor and ceiling joists, the size of the rafters, and the capacity of the existing electrical service panel, some things are more subtle. The exterior corner boards, for example, may at first glance appear to be 1x6 when they are $5/4$ stock instead.

Build Level or Not?

Some years ago, while working as a foreman for a design/build firm in New England, I supervised the construction of a one-story, shed-roofed addition to a two-story, wood-framed building more than a century old (parts of it were, anyway). I elected to go the plumb-and-level route, without considering any alternatives. When it was done, I was taken to task by the owner because the new shed roof accentuated the settling of the existing building. It cut completely across one row of clapboards. At one end the shed hit the sidewall just below the sill of the second-story window; at the other, the sill stood clear by 4 inches. Some artful "fudging" would have produced a roof that shed water just as well, and perhaps a happier client.

I say "perhaps" because, less than a year later, I was faced with an almost identical situation, except the addition was longer and the old building was even farther out of level. But I had learned my lesson. I took the extra time and effort to set the rafters so that the roof hit the second story wall an equal distance below the sill of each of six windows. I knew this created a twisted roof plane, but I assumed it was the preferable alternative. I was wrong. The new standing seam roof drew attention to the problem, and the owners were distinctly displeased with the result.

The moral of this story is communicate with your client (and the architect, if one is involved). Joining an addition to an older building sometimes leaves you facing a problem to which there is no "right" answer, only a choice of compromises. I don't suggest compromising structural integrity or failing to meet code. But if it's a matter of appearance only, inform your clients as fully as possible, and let them make the decision.

— D. S.

Unconventional practice. Unfamiliar construction techniques can easily be overlooked or mistaken for their modern counterparts. For example, in Kansas City, Kan., where I do most of my work, many homes have floor joists on 19.2-inch centers. Lining up the new joists in the addition with these old ones might simplify things for a mechanical sub who plans to run a trunk line from an existing joist bay into the crawlspace under the addition.

Level and plumb. Buildings more than a century old present a special challenge. Often, because of inadequate foundations or substandard framing or masonry practices, tying in to these old buildings means choosing between building plumb and level and making some accommodations for the older structure.

To the uninitiated, plumb and level is the quality way to go, but to get an acceptable result, you sometimes can't avoid matching an existing structure that's out of kilter (see "Build Level or Not?"). Allow for this in your budget and schedule because, while it's easy to build out of plumb and off level by mistake, to do so deliberately and precisely to achieve a particular result is much more difficult.

Opening Up

My first day on site rarely involves breaking ground. Instead, the lead carpenter and I pick apart the existing exterior siding and sheathing to reveal the framing or other load-bearing structure. If the layout of the wood framing is out of the ordinary, or if there are wires or pipes in the way, I make a story pole so that later we'll be able to locate the hidden features by transferring marks around the structure. We also remove enough siding and sheathing so the foundation work can be done.

Adding on usually means opening up a portion of load-bearing exterior wall, providing temporary support, and inserting new headers or beams. If the addition doesn't add to the load on the wall, I prefer to do this after the new work is closed to the weather. This lets

Tying Into Existing Framing

Figure 6. *Floor joists in a new addition may not match those of the existing building. To set the elevation for the new foundation, work backward from the height at which new and old floors must meet. In many cases, the top of the new foundation wall will be lower than the top of the old.*

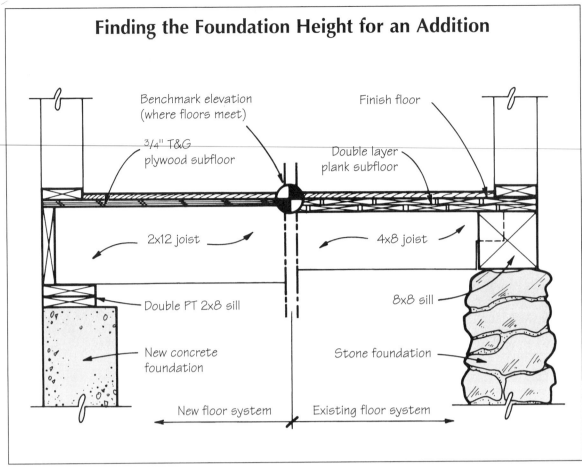

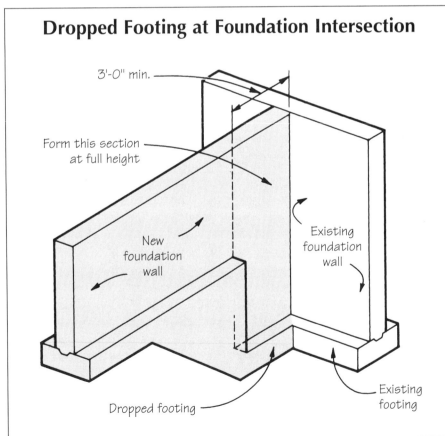

Figure 7. *To avoid increasing lateral pressure from a crawlspace wall against an existing full foundation wall, drop the new footing to the elevation of the existing footing for at least the first 3 feet of the new wall.*

the occupants go about their business undisturbed for several more weeks, and I don't have to make a stretch of wall good against the weather. But if the new structure bears on the old over the openings, I build the headers before we break ground. There is less shoring to do and less risk that I'll have to "remodel" my new work later.

Opening up long sections of load-bearing wall and installing a clear-span beam is not to be taken lightly, even in the case of a "simple" wood-framed house. The total load on, say, 12 feet of first-floor wall in a two-story house may require LVL or steel for the header. I don't hesitate to consult an engineer when I'm unsure of a structural scheme, even if it has been specified on an architect's plans.

Foundations

Complete the foundation phase of an addition as quickly as possible to avoid the mess that rain may make of foundation work. In the case of an addition, the work will take place in an established lawn or landscape, and water may find its way into the adjoining basement. It's a good idea to check the floor drains in the old basement: It's easier to clear them if you don't have to do it through 3 inches of standing water.

I try to expedite things for the foundation contractor by having all the head scratching done before he shows up. On anything but the simplest jobs, I make a large-scale drawing showing the old and new foundation walls and all related wood framing up to and including the finish floor. This means establishing a benchmark elevation for the new foundation. This is usually determined by where the new and old finish floors meet and may mean that the new and old foundation walls will not be the same height where they abut (see Figure 6).

If the addition is built over a crawlspace against a house with a full basement, the footings of the addition will be 3 or 4 feet up on the basement walls. According to a local structural engineer, unless the new wall meets the old at an outside corner, the new downward load of the addition can create lateral thrust in the bearing soil and cause the old wall to buckle inward. Our solution is to form the first 3 feet of the new wall at full 8-foot height, with the footing stepped down to the elevation of the existing footing (see Figure 7).

Framing

Few framing subcontractors have the skill and patience to deal with the difficulties of tying in to an old building, so my crew and I do our own framing on remodels and additions.

Floors. Ideally, an addition — particularly the floor joists — would be framed with reasonably dry dimension lumber, but in practice, this is very difficult to do. In my area, KD15 lumber is available only occasionally and sometimes not at all. The best I can hope for is a 19% moisture content, so I'm left with 2x10s and 2x12s that may ultimately shrink nearly 1/4 inch. This won't matter if the old and new finish floors don't need to meet flush anywhere (where there's a step up or down, for instance), and even if they do, carpet on underlayment can tolerate the movement that occurs after its installation.

But some finish flooring (tile, for instance) won't tolerate movement at all, and may be very costly to repair. In these cases, you should take some extra precautions. Bring the new joists in a bit higher than the old — say, half the anticipated shrinkage. Temporarily fasten the new subflooring with screws in the area where the finish floor will run continuously from the existing building into the addition (see Figure 8). Later, when the addition is closed in and the existing building opened up, remove the temporary subfloor and cut and remove enough of the old subfloor to allow the new subfloor to span the existing and new floor framing about half and half. This isn't a foolproof cure for finish flooring woes, but it tilts the odds in your favor.

Windows. If the new windows are the same type as the old ones, frame the heads of the new rough openings at the same exact height as the old. But if, for example, you're changing window types, lining up the rough heads may mean the trim won't line up, either inside or outside. If neither is aligned, it's particularly hard to explain to the client. To lay it out right, you need to have one of the new windows on hand or a head section detail from the catalog.

Roof framing. Allow extra planning

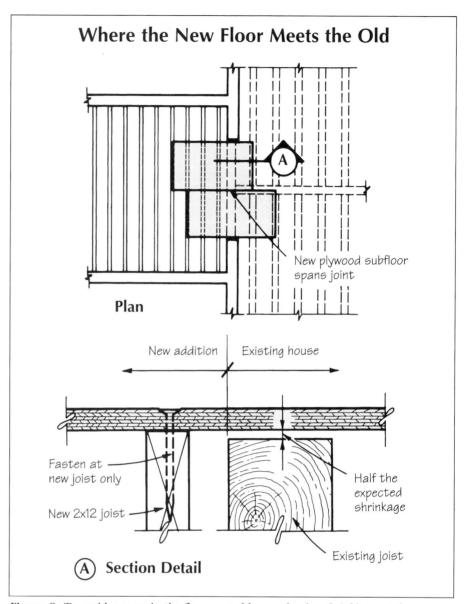

Figure 8. *To avoid a seam in the floor created by new lumber shrinking, set the new joists high by half the expected shrinkage and be sure to span the joint with the new subfloor.*

time before starting to cut rafters, even for a simple shed roof. Even when the new roof is independent of the existing one and the two don't actually tie together at all, the fascia and soffit will still need to match or closely resemble the existing.

For a complex roof — a hip and valley, say — I make a full-scale drawing of the eaves details on the deck or a sheet of plywood. Sometimes I go so far as to nail in a mockup of the fascia and soffit after the first two rafters are set. If the new roof has a different pitch from the existing roof, it is difficult to bring the eaves trim together.

Trusses. The very first time I ordered gable roof trusses to extend an existing roof, I carefully measured the rise of the existing trusses from top plate to peak and asked my truss builder if he would match this, even if it varied from his standard pattern for that pitch and span. He readily agreed and then sent me thirteen trusses, all 1 1/2 inches too low at the ridge. I had no shop drawing or paper trail to fall back on, and faced with a probable argument and a certain week's delay even if I won, I accepted the trusses. We made them work by shimming out the difference at the ridge over the first three trusses.

Now when I use trusses for roof framing, I won't authorize any to be made until I have a shop drawing for approval.

By David Schwartz, a builder and remodeler in Overland Park, Kan.

A Second Story in Five Days

Recently, our company added a second story to a three-bedroom ranch-style house. The owner was about to be married in June. Both he and his wife-to-be had children from previous marriages and they needed the extra space for their expanding family. This gave us about six weeks to convert the house to four bedrooms, which involved adding a new master bedroom and bath, plus an office, on the second floor.

What made this project unusual was that we used a crane to remove two-thirds of the existing roof, then replaced it atop the new second level a few hours later. The house was opened to the weather for less than a day (Figure 9). In remodeling, you rarely see so much progress in one day.

Adding on by going up with the existing roof meant we would not need to cover an opened house each night. On one job I'm aware of, a remodeling company had to pay thousands of dollars for damage caused when tarps failed to keep a rainstorm out of an opened building. Plus, demolishing a good roof, paying to have the debris removed, then rebuilding another roof 8 feet higher than the original did not seem to make sense. Even with the expense of the crane, reusing the old roof would save at least $2,500, and it would certainly shorten the length of the project.

The choice to add the second story on only two-thirds of the house was dictated by aesthetics and a zoning ruling that limited the space the owners could add. Also, for economy, we chose construction-grade 2x12s to make the beams used to support the roof during the lift; the longest ones available were 26 feet, which also limited the length of the roof section we could remove.

Day One: Preparation

Starting with a two-person crew, we began work on the house on May 1, a Friday. We met a lot of supply trucks that day, and did some framing work on the first floor to accommodate the new staircase.

Day Two: Demo

On the second day, we removed the lower siding and sheathing from the front of the house. We were planning to replace all the windows and siding on the front and gable end, so we didn't bother to be surgical with this part of the demolition.

Next, we made some structural modifications to the first floor. In order to carry the new loads from the second story, we had to double the first-floor joists in the front half of the house, where they cantilevered over the foun-

Figure 9. *Using a crane to remove and replace this 26-foot-long roof section saved about $2,500 and left the building open to the weather for less than a day.*

All in a Day's Work

❶ A few days before the remodel: The addition will go on the section of the house to the right of the front door.

❷ The crane lifted the roof with heavy nylon straps, swinging it over the house and setting it on the driveway. Four square holes cut in the roof provided access to two 26-foot-long lifting beams.

❸ After installation of the second-story floor joists, the crane lifted the prebuilt walls into place. Two-by-four bracing, run diagonally from the lifting points to the bottom plate, held the walls rigid.

❹ Carpenters nailed, straightened, and braced the walls, then the crane lifted the roof again. Workers on the ground steered it into place with guy lines.

❺ With the roof reattached, the author's crew relaxed for lunch. By the end of the day, they had finished the sheathing, tying the new structure to the old.

A Second Story in Five Days

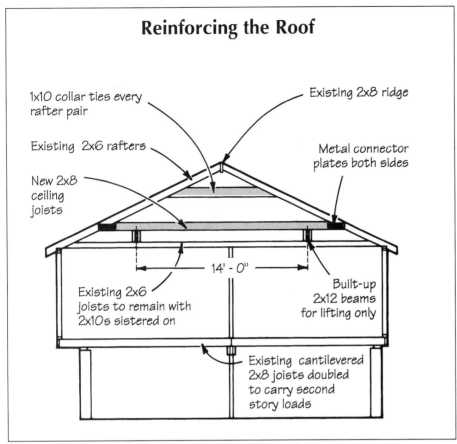

Figure 10. *To reinforce the roof for lifting, the crew added 1x10 collar ties and 2x8 second-story ceiling joists to every pair of rafters. Two 26-foot-long triple 2x12 beams supported the roof for the lift.*

dation. To do this, we removed the band joist and the basement ceiling, then slid in new 2x8s from outside and sistered them onto the original joists. In the rear, there was no cantilever, so no extra support was needed.

To convert the existing first-floor 2x6 ceiling joists into floor joists would require sistering on 2x10s. Considering how cramped the attic was, we decided to wait until the roof was off to add the 2x10s, when we'd have plenty of room to maneuver. In preparation, we cut a 14-inch square hole in the gable end and passed in the new floor joists. We also passed up the six 26-foot 2x12s that would become the beams for lifting the roof.

We also cut and removed the gutters and removed the top 10 feet of a small chimney that would be in the way of the crane.

Day Three: In the Attic

On the third day, the crew expanded to three. We worked in the very tight attic space, removing the cellulose insulation. We filled 68 30-gallon garbage bags with the old insulation — easily the most grueling part of the job.

After the insulation was out of the way, we reinforced the 26x26-foot section of roof that was to be lifted (Figure 10). We attached collar ties to every rafter pair and nailed together the two lifting beams using the 26-foot 2x12s. We used duplex head nails so we could dismantle the beams later. Working over the existing 2x6 ceiling, we set the laminated 2x12 beams in place, then laid what would become the new 2x8 second-story ceiling joists right on top of them. We scribed these new ceiling joists to the rafters at each end, cut them, and butted them up tight to the rafters, making the connections with large Simpson metal plates.

We also built two T-shaped strongbacks of 2x8 material and installed them between the beams opposite the lifting points, to prevent the beams from sliding towards each other.

Day Four: Prebuilding Walls

The fourth day I added four more carpenters to the job. We prebuilt the front and back walls of the second story in the front yard. These would give us all the support we would need to set the roof back on. We made sure we stacked the walls properly for the crane to lift them directly into place.

We left off the lower 3 1/2 feet of 1/2-inch sheathing from each wall. This we would install after the walls were in place to overlap the lower floor and tie the structure together. We braced the 2x6 walls for the crane lift with temporary 2x4 braces set diagonally down from the lifting points. We also prebuilt the headers for the gable ends.

To finish out the day, we tore out the old staircase, completed the inside structural modifications, and stripped the gable end of siding.

Day Five: The Roof Comes Off

Starting early, with yet five more crew members, we released the roof by cutting the studs and sheathing in the gable end, pulling the toenails holding the rafters to the top plate, and cutting a line through the roof sheathing and shingles from front to back. We then cut four 10-inch-square holes in the roof above the lifting points on the 26-foot beams.

The crane arrived at 7:30 a.m. The operator hung two 14-foot-long heavy-duty steel spreader pipes from cables. From these we hung four strong nylon straps, which we passed through the holes and around the two beams. At 9:30 a.m. the crane lifted the roof free of the house and set it on the drive at the end of the house. Nobody was allowed to stand under the roof at any time while it was in the air.

The engineer had predicted the roof would sag 6 inches in the eaves. This seemed unacceptable to me, so I had installed solid blocking along the eaves. This kept the sag to less than an inch.

With the roof off, we immediately started installing the second-story 2x10 floor joists. I had one carpenter cut back the 2x6 floor joists 1 1/2 inches at each end, using a chain saw. We ran the new floor joists all the way to the edge of the existing 2x4 walls to give them a full 3 1/2 inches of bearing, then installed 2x10 blocking between the new joists at the outer edge of the plates. While we could have installed a band joist more quickly, it seemed prudent to give the floor joists as much bearing area as possible.

We then installed T&G plywood subflooring along the front and back of the second floor where we would be setting the exterior walls. We left the rest of the floor open for the plumber and electrician.

The crane switched rigs, reached over the house, and lifted the front and back walls into place. We nailed off, straight-lined, and braced the two walls in place. Meanwhile the crane switched rigs again to lift the roof back in place.

It was tricky to get the roof back in its exact location. The crane operator lifted it into approximate position. A carpenter stood at each corner with a guy line. I stood on the untouched part of the roof shouting directions to these carpenters; another man stood next to me to shout directions to the crane operator. We got the roof in place, then installed hurricane ties on each rafter to thoroughly secure it. We broke for lunch about 1:30 p.m. as the crane operator packed up.

That afternoon we built the new gable at one end and extended the existing gable with 2x6s to the new floor. We finished the exterior sheathing, tying everything together. Then, since the weather was fair, with no rain in the forecast, we relaxed for the rest of day.

The following morning we installed housewrap and windows, then dismantled the lifting beam and strongbacks. Rather than move them outside again, we recycled them right there, ripping them with a worm-drive saw for 2x ceiling strapping.

The rest of the job went smoothly, as did the planned wedding. We completed the house on schedule, in time for the new family to move in.

By Chuck Green, owner of Four Corners Construction in Ashland, Mass.

Case Study: A Retrofit Ridge Beam

Our local health center had outgrown the 50-year-old, 1,250-square-foot house they had moved into 15 years ago. Working with the health center staff as a team, we developed a plan that fit their needs and allowed them to remain open for business every day throughout the project.

The existing building was a one-story, 24x36-foot Cape, with the ridge running along the longer east-west axis. The rough 2x6 rafters rested on the top plates of the first floor walls and met without a ridge pole at the peak. With 91 inches beneath the rafters at the peak, there was limited attic space. The concrete foundation was 10 inches thick and 6 feet high, with a poured floor of unknown thickness and a cistern that took up a third of the basement floor area.

The plan called for a 14x36-foot addition across the front of the building. This would give the center the space it needed downstairs, as well as keep the construction more or less out of the daily sphere of operation. But for the addition to work upstairs as well, we would have to tie into the existing attic space. This meant removing most of the

roof. How to do that without compromising day-to-day operations of the center took some thought.

Providing Support

It seemed that if we could support all of the rafters on the south side of the ridge with a load-bearing wall, those rafters would not have to oppose the rafters on the north side of the ridge, and thus the north-side rafters could be removed. Then, we could put trusses on the load-bearing wall and span the 26 feet to the new front wall of the building.

The engineer we consulted recommended that we abandon our load-bearing wall concept and instead use a large built-up beam, point-loaded down to the foundation. The point loading meant building posts into the two end

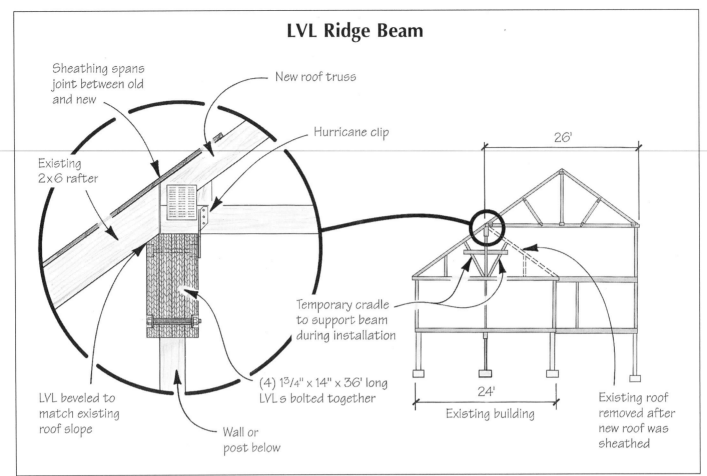

Figure 11. *The authors added a 900-pound, 36-foot-long structural ridge beam to support the new truss roof on this second-story addition. Instead of wrestling with the preassembled beam, the crew slid the four LVLs into the attic, then bolted them together.*

walls and at two points in between. This beam would carry the south face of the original roof and support the trusses that we needed for the new roof.

A Monster Beam

The engineer specified a built-up beam of four 1³⁄₄x14-inch x 36-foot-long LVLs bolted together. Available in up to 60-foot lengths, LVLs were perfect for this job because they can be put into final or near-final location one at a time, then bolted together (Figure 11).

To protect the health center below, we would have to leave the existing roof intact while getting the beam into place. Our plan was to feed the LVLs into the gable end wall from the lumberyard's scissor-lift truck.

We would need to notch out the existing rafters at the peak where the beam would eventually go. To preserve the integrity of the existing roof, we built two temporary kneewalls to carry each side of the roof (Figure 12). The kneewalls rest on a 2x4 plate we laid on the attic floor, centered directly under the peak and over the partition walls beneath. We tilted the two kneewalls outwards both to better support the old roof and to keep the center area free for the beam placement. Also, we laid out the kneewall studs so they wouldn't interfere with partitions and posts that would be built later on top of the same plate.

Assembling & Lifting

To support the LVLs, we installed some waist-high horizontal 2x4s across every fourth or fifth pair of kneewall studs. That way, we could slide the individual LVLs off the lift truck right onto this 2x4 cradle, where we could assemble the beam for final lifting.

Assembling the beam was a matter of clamping the LVLs together, drilling, and bolting. This took two men about three hours. The outermost LVL on the south side had to be beveled to meet the sloping underside of the existing rafters. We did this with a circular saw by crawling along the top of the beam.

We also installed vertical 2x4s between the horizontal ones and the rafters to prevent the beam from rolling as we lifted. This kept the raising of the now 900-pound beam safe, accurate, and calm. We used hydraulic jacks, one on each end, to slowly raise the beam. As the beam went up, we nailed additional temporary horizontal supports right below the beam as an extra precaution.

Once the beam was in its final position, we added posts down through the first-floor walls to the foundation. We had to pour a large pad in the basement for one of the posts, while the cistern wall and perimeter foundation walls supported the other posts.

Figure 12. *Before cutting away the peak of the existing roof, the authors built angled kneewalls on each side of the ridge. With horizontal 2x4s nailed across them, the kneewalls also served to support the built-up LVL beam while it was bolted together and lifted.*

Figure 13. *Plywood sheathing spanned the joint between the old roof and the new, tying the two sides together (above). Only a course or two of slates had to be removed to set the trusses (right), leaving the health center below protected.*

Bring On the Trusses

At this point, the beam was supporting the south half of the existing roof and was in position to accept the trusses for the roof of the addition. The original roof was still basically intact up to this point, as was always our priority, though the loading had been redirected. In the meantime, business had continued downstairs as usual, albeit with some noise overhead.

In preparation for the trusses, we had to remove only a course or two of slates and about a foot of board sheathing along the ridge, which was easily protected from the weather (Figure 13). We were able to set and brace all the trusses in one day, then apply enough plywood and felt paper to protect the junction of the new and old roofs. That day made all of the work in the attic worthwhile.

Only after the new roof was sheathed and shingled did we dismantle the old roof.

By Silas Towler and Rick Schneider, builders and remodelers in Addison County, Vt.

Jacking Old Houses

My company specializes in restoring turn-of-the-century and older homes in the Galveston Bay area of Texas.

Many of the structures we work on have been neglected for years in a very unforgiving climate: damp, cool winters and hot, dry summers punctuated by torrential rain showers and occasional hurricanes. Humidity hovers around 90% at night, and by midday it drops to around 50%. Intense sun and salty air also come into play.

Soil is also a big variable: It ranges from fairly stable sand at water's edge to a hard clay silt just a few feet inland. We have a heavy black humus known as gumbo, which is very unstable: It shrinks and heaves, turning from a soupy mess when saturated to a rock-hard mass when baked dry. Building movement on this type of ground is a given, so structural repairs must always be addressed before other work can begin.

Most of the houses I work on were built before ready-mix concrete was available. Many are supported on a grid of short brick piers spanned by 6x6 or 8x8 sills. In many cases, however, the sills were set directly on the ground. Even though they are made of bald cypress — a naturally decay-resistant wood — or creosoted yellow pine, after 50 or 60 years these timbers have rotted or been eaten by termites.

Using the Right Jack

The first step in a structural overhaul is to raise the house up above grade. This allows air circulation underneath, which in turn alleviates many moisture-related problems and also makes access to plumbing and other house systems easier.

I use three types of jacks: hydraulic bottle jacks, screw jacks, and hypoid gear jacks (see Figure 14). Most of my hydraulic jacks are rated at 10, 12, or 15 tons; the screw jacks I use have a 10-ton rating. I use the hydraulic jacks for lifting, backed up by the screw jacks for holding things up while I reblock the hydraulics. I have about 30 of each type. The hypoid gear jacks have either a 25- or 35-ton rating; they raise and lower with a 1-inch socket wrench. I have only a few of these, and use them at the heaviest points — usually at interior load-bearing partitions. The hypoid jacks are heavy and turn very slowly, so I use them only where I need their load capacity.

In very tight locations — behind a masonry fireplace, for example — where there's not enough room to operate a screw or hydraulic bottle jack, I'll use a Porta-Power Ram Pump (available from Northern Hydraulics, 800/823-4937; www.northernhydraulics.net). This is a small hydraulic ram used in auto shops that has a pump chamber like the one on a grease gun. You pump the hydraulic fluid through a hose to the ram piston, which may be set up a few feet away. I'll use the Porta-Power to lift until there is enough clearance to allow me to get cribbing and a standard jack underneath.

Setting Up

We start by distributing jacks around the perimeter of the house at intervals determined by the size of the structure. There are no hard and fast rules; you just have to properly estimate the weight of the building and place jacks accordingly. Obviously, a two-story structure will have jacks at tighter intervals than a single story. The corners usually hold themselves up pretty well, providing the framing is sound; most of the weight is in the middle of the building.

We usually place jacks alongside existing piers, if there are any, since these are already supporting the structure and provide a place for shims and blocking as the house goes up.

As we distribute jacks, we scope out the areas where we'll have to excavate. Since the jacks must bear directly underneath the sills, we usually have to do a fair amount of digging to accommodate both the jacks and the cribbing they rest on.

We take care not to position jacks or cribbing over any septic lines, gas pipes, water supplies, and so forth. When we're working on an unoccupied house, it's an easy matter to disconnect the plumbing and electrical before the job starts. If the house is occupied, we provide temporary wiring and plumbing to ensure uninterrupted service to the home while we work. We've learned by experience to watch out for live electrical wires lying in the dirt.

While setting out the jacks, we also look for rotted sills and joists — places where we'll have to scab on new framing

Figure 14. *The author uses three types of jacks (from left to right in left photo): hydraulic, screw, and hypoid gear. Steel angle slips, predrilled for attaching with duplex nails, spread the load over a larger bearing surface (right photo).*

or install "needle beams" (Figure 15). Needle beams are temporary wood beams set in from the edge of the house underneath and perpendicular to the joists. The needle beams stay in place for the duration of the job and are used for jacking until the house is high enough so that the sills can be replaced.

Once set up, the jacks stay in place throughout the job. Though it isn't a problem with hypoid and screw jacks, the hydraulic jacks can punch through a sill when pressure is applied because of their small bearing surface. To prevent this, I use 4-by, 6-by, and 8-by angle-iron "slips," predrilled for attaching to the bottom of sills with duplex nails. These also help prevent sills from twisting as uneven pressures build up when we initially lift the house, before we get a feel for how the structure is going to behave.

The slips, too, stay in place throughout the job. I've painted these pieces of steel bright green so they show up well under the house when we're cleaning up the job.

Lift Gradually

When all the jacks are in position with solid cribbing beneath, we first extend them until they are fully loaded — to the point where the cribbing stops pushing down into the soil. At this point, lifting can begin.

I prefer to work down the long side of the house — the eaves side of a typical gable house, for example. By sighting down the long sills and the center girder, I get a good idea of how the house is behaving as it is lifted.

It's ideal to have three people jacking and one going for material as they need it. The three men position themselves across the front of the house (if that's where we're starting from), one at each corner and one in the middle. They'll each raise their jack just enough to accommodate a $1/2$-inch plywood shim under the adjoining pier. Then they'll move down 6 or 8 feet to the next jacking point and raise that jack $1/2$ inch, and so on down the length of the house. A restrained approach like this eliminates a lot of consequential damage to interior plaster, doors, and windows.

We repeat the process until the fourth pass, when we pull out the plywood shims and insert a 2-inch concrete pad. Then the process begins again, adding plywood shims on top of the concrete pad until the seventh or eighth pass, when we insert a 4-inch concrete pad. We continue the process again, inserting

Figure 15. *After decades of settling into the ground, this house (left) is showing serious moisture damage. New floor joists with a needle beam underneath (middle) provide a sound structure for jacking. The framing nailed to the wall helps to pick up individual studs and the corner post. Note the new elevation of the porch rail relative to the concrete porch (right) as the house comes up.*

ture problems, as well as easier access. If the house gets set back down close to the ground, it's only a matter of time before the work has to be done again.

New Piers

Once we get the house jacked to where we want it, we pull string lines to lay out the new masonry piers. The piers are built plumb and square from either block or brick and allowed to cure. Then the house is lowered, in a reverse procedure.

Occasionally, if the client requests it, we'll excavate down a few feet and put in a half-cellar wall. We build them out of block; they're called chain walls in this area. I leave a ventilation hole by laying every fourth block in the top course on its side. In our damp climate, though, I prefer to leave the house up on pilasters and completely open underneath for good cross-ventilation.

While the house is raised up, I usually spread about a foot of bank sand underneath to avoid standing water and mosquito problems. I also slope the grade away around the outside.

Using Pipe Staging for Jacking

On a recent job we were able to use our pipe staging as a jacking platform. It was a carriage house with an apartment above, built around 1900 (Figure 16). It had been built on 6x6 sills directly resting on unstable soil. A previous owner had placed reinforced concrete, 24 inches thick in some places, on the floor and under the sills to try to stabilize the structure, but to no avail. Termites were eating their way up through the first-story framing to the second story. When we got there, the building was 19 inches out of level and 12 inches out of plumb.

We had to support the second-story apartment while we repaired and replaced the first-story walls. I used my pipe staging as a convenient jacking platform, setting it up two bays wide and three rows deep (Figure 17). Before using the staging for jacking, I called a local scaffolding distributor to find out how much weight it could carry. But rather than rely on the staging to carry the loads, I used 6x6 posts all the way to the existing slab under each jack. I put

Figure 16. *This turn-of-the-century carriage house was 19 inches out of level and 12 inches out of plumb before renovation. Termites had badly damaged much of its wall framing.*

an 8-inch pad, and so on, until the house is raised a few inches above the new specified grade, which might be anywhere from several inches to 4 feet, depending on the owner's wishes.

Occasionally an owner will ask me to lift the house just high enough to replace the rotted sills then set it back down only 3 or 4 inches higher than it was previously. That's okay if just one side or section of a house needs work. But if I'm going to overhaul the entire foundation, I prefer to lift the house at least 2 or 3 feet so I have room to work. Usually, once the house is raised up, I leave it up permanently. It saves me a lot of work, since I don't have to take the house back down, and the plumbers and electricians like it a lot better than working in the mud. It's also in the owner's best interest, since there will be fewer termite and mois-

Figure 17. Staging made a convenient platform for jacking the carriage house (top), but it had to be thoroughly braced all the way to the ground. In addition to continuous cross-bracing in both directions, the author set 6x6 posts under the critical jacking points (middle), and made stable bases for the scaffold legs (bottom) by drilling holes in 2x6s.

Figure 18. While the second story of the carriage house was suspended, the crew excavated, formed, and poured new footings (top), then repaired the wall framing and lowered the building onto new PT sills.

Jacking Old Houses

2x4 cross-bracing along the top of the staging, connecting all three rows together from the front to the back of the garage. I also stabilized the bottom by making bases out of 2x6s, with holes drilled for the feet. I then connected these bases and the bottoms of the 6x6 posts with double 2x4 braces running across the width of the garage.

In short, I braced the staging setup as thoroughly as possible. After all, the entire second story and roof would be essentially suspended for four to six weeks while people worked underneath.

Hanging the Needle Beams

So that we wouldn't have to wrestle with heavy beams high in the air, we hung the needles from the garage ceiling, perpendicular to the joists. We drilled through from the apartment floor above and used all-thread to snug them in place.

We used a transit, stringlines, and plumb bobs to keep track of the building's lateral movement as it was lifted. We also ran strings along the bottoms of the needle beams to ensure that we lifted the building uniformly.

While the building was raised, we excavated a trench for new footings, including jackhammering out the old concrete (Figure 18). We formed and poured a reinforced perimeter footing, with rebar pins sticking out on the inside to tie into the new slab. We poured the new slab on top of the old; its final elevation was 12 inches higher than the original slab.

The extra precautions we took setting up the job meant there were no problems later. Just to be on the safe side, though, I did increase my liability insurance. Because of the height and the extra measures involved, the carriage house job cost three to four times as much as it would have for a similar-sized building on the ground.

By Mike Shannahan, a carpenter in La Porte, Texas.

Chapter Ten
PICKUP WORK

Fast Fascia Techniques *178*

Framing Recessed Ceilings *181*

Framing a Simple Radius Stair *183*

Fast Fascia Techniques

On the West Coast, installing fascia is a specialized trade. "Board hangers," as they're called in California, methodically move through a tract development, hanging the fascia on house after house. The prices for this work are well established, and board hangers quickly learn that in order to make a profit, they must work efficiently.

Over the years, I've developed a method that allows me to work by myself and hang fascia faster than two carpenters working together. Using nothing more than a 6 1/2-inch wormdrive saw and the tools in my nail pouch, I'm able to hang 2x8 fir fascia boards up to 24 feet long.

Before climbing up on the wall plates, I lean the uncut fascia stock between the rafter tails. I position the boards so that a 2- to 3-foot-long section of each board projects above the roof plane. That way, I can easily cut the fascia to length right from the roof.

After marking the overhang at all four corners of the building, I transfer the overhang mark to each rafter tail with a chalk line, then mark a cut line on each rafter tail and cut the tails to length.

Where I start hanging depends on the type of roof. On a simple rectangular hip roof, I usually start in the middle of one side and work counterclockwise around the house. (I always work counterclockwise because of the way my wormdrive saw tilts.) On a more complex roof, I might start with a square cut at an inside corner or where a roof section meets a wall. For gable roofs, I usually start at a corner where the gable meets the eaves and work down in a counterclockwise direction.

Bent Nail Trick

I make the first cut while the fascia board is leaning against the wall (see Figure 1). Next, I pull the board up and lay it face up, across the rafter tails and about 4 inches from their ends. To keep the board from sliding off the roof, I tack a temporary 16-penny nail into the top edge of a rafter tail, about a third of the length from one end of the board. I then move to the other end of the board, coming in two or three rafter tails from the end. I position the board where I want it and make a V mark on the top edge where it crosses the left side of the rafter. Using this as a reference, I drive a nail down through the face of the fascia, 3/4 inch to the right of the V at an angle back towards me. I stop the nail when the point sticks out 1/4 inch at the top edge of the fascia.

Pivoting around the 16-penny nail I drove earlier at the other end, I slide the board out over the end of the rafter tail and line up the V with the left edge of the rafter. The protruding nail will touch at the middle of the rafter tail. I lift the board slightly and push the nail point into the rafter tail's top edge, then drive the nail home. At this point, I move over and remove the 16-penny pivot nail that was holding the other end. I slide the fascia out and over the rafter tails, turning the fascia board so it butts against the end of the rafter tails. The angled fascia nail that was driven earlier will bend as the board rolls over the ends of the rafter tails.

I like to leave the fascia partially nailed until the entire run is hung so I can sight and straighten the entire length before nailing it off for good.

Cutting to Length in Place

If I've started hanging where the fascia butts into a wall, as in the photos in Figure 1, I'll cut that end square first but leave the other end wild. Once the fascia's tacked up, it's easy to scribe the wild end to the last rafter tail, then cut it right in place. With my saw table set at 45 degrees, I run the saw down the outside face of the fascia, watching to make sure the blade follows the scribe line (Figure 2).

Another Bent Nail Trick

Successive boards are hung in a similar manner, using a different type of bent nail trick to support the end of the fascia as it's installed. I first cut the mating miter joint of the next piece while the board is still leaning against the wall. I then drive a 16-penny nail into the top edge at the miter cut before pulling the board up. As the fascia is lifted and set in place, this nail will support the end of the fascia, while the angle of the miter on the previously hung board holds the second board snug against the end of the rafter tail.

There are a couple of tricks I use to get a tight joint. Before lifting the second fascia board into position, I tap the back side of the first piece until there is a 1/8-inch gap between it and the rafter tail (the end should not be nailed yet). When I lift the second board into place, the gap allows me to easily position the mitered end far enough behind the first piece to get a snug joint. I drive a nail into a rafter tail approximately two-thirds down the length of the second fascia board, then go back to check that the miter is correctly positioned. The 45-degree cuts should be positioned slightly past each other; otherwise, the joint will spread open as you nail it and no angled or toe-nailed fastener will be able to pull it tight. Once I'm sure the second board is slid back far enough, I drive a 16-penny fascia nail straight through the lap into the tail.

When fitting the last board in a run on a hip roof, I lift the board into place, mark a line where it meets the outside corner of the hip rafter tail, and lay it back down on top of the rafters to make the compound angle cut. For gable roofs, the last board in a run is left running wild.

Hanging Barge Boards

On gable-style roofs, the gable fascia (called "barge boards" in my area) are fastened to lookouts that cantilever over a framed rake wall (Figure 3). I install these lookouts long, then cut them to length in place. I first snap a line between the overhanging ridge board and the wild end of the eaves fascia to mark the position. I scribe a line on the back side of the eaves fascia and, with the saw table held on the outside face of the fascia, watch the blade follow the scribe line. After cutting the eaves fascia, I start at the low end of the roof and cut

Figure 1. *To hang a long piece of fascia by himself, the author makes the first cut from the roof as the fascia leans against the building (top left). He then lays the board across the rafter tails, finish side down. Working near one end, he drives a nail at an angle until the point protrudes 1/4 inch from the back side (top right). He moves the fascia out to the edge and drives that nail into the top of the plumb cut of the rafter tail (above left). Moving to the other end, he swings the board down into place (above right) and nails it off.*

Figure 2. *To cut a mitered lap joint, the author scribes a line on the back side of the uncut fascia and, with the saw set at 45 degrees, makes the cut from the outside face of the board, making sure the blade follows the scribe line (left). An angled nail driven into the top edge of mating board provides a third hand for hanging the next piece (above).*

Fast Fascia Techniques

Hanging Barge Boards

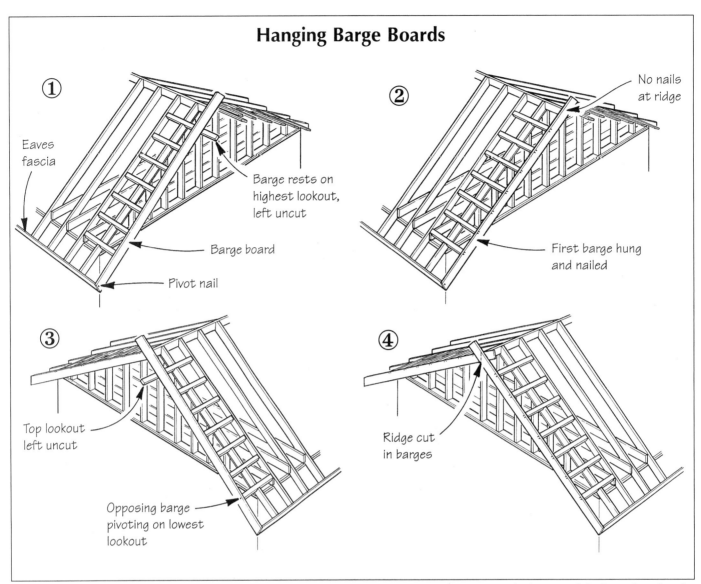

Figure 3. *The author first cuts the lookouts to length, except for the top one. He cuts the miter at the bottom of the barge board and supports it with one nail into the eaves fascia, resting the top end on the uncut lookout. He then moves to the ridge, cuts the lookout to length, and nails the barge to the lookout, but not to the ridge. He installs the opposing barge in the same way, letting it run long at the bottom.*

all the lookouts except the highest one next to the ridge. I cut the 45-degree miter on the eaves end of the barge board then pull it up, and start a nail at the top of the miter.

Next, with the top end of the barge board resting on the uncut lookout by the ridge, I drop the board over the edge and drive the started nail into the joint at the eaves fascia. Moving to the upper end of the barge, I hold the barge with my left hand and cut the lookout to length. I lower the barge into place and nail it off as I work my way to the lower end, making sure that I don't drive any nails into the ridge block. I then cut off the top of the barge about 1 inch past the ridge block, to provide a temporary resting point for the upper end of the opposing barge.

I repeat the process on the other side of the gable, again leaving the highest lookout uncut. With the upper end of the barge resting on the uncut lookout, I nail the barge with one 16-penny nail into the lowest lookout, letting the lower end extend beyond the eaves. Moving up to the uncut lookout, I lift the barge and cut the lookout to length, then carefully set the barge on the tip of the previously hung barge. I position myself above the ridge, lift the barge, and lower it until its top edge is flush with the top edge of the lookouts. Then, lifting the barge back up 1/8 inch, I make a vertical plumb cut. As I make this cut, the blade scribes a cut line on the barge board beneath (I hold the barge 1/8 inch high to compensate for the material removed by the saw kerf). Holding the cut barge up and away, I cut the other barge, following the scribed line.

By Mike Stary, of Stary Construction in Leucadia, Calif. Photos by Brent Emery.

Framing Recessed Ceilings

I build out on the West Coast, and most of the custom homes I frame contain one or more coffered ceilings. By "coffered" ceiling, I mean any angled ceiling recess — a design detail that's often called a "tray" ceiling in other parts of the country. By contrast, the traditional coffered ceiling — a gridwork of beams in the ceiling — is usually referred to out West as a "box-beamed" ceiling or a "library" ceiling.

With a soffit around the perimeter of the room and a crown molding around the top of the recess, this detail looks like a million bucks. But it's not expensive to build, as long as you frame in the proper sequence.

Design Details

The pitch and size of the angled recess depend on the designer's and owner's preference. Typically, the surrounding soffit starts at the top wall plate (at 8 feet) and the recess angles at 45 degrees or less to the finished ceiling at 9 feet or higher. The best design depends on the proportions of the individual room.

I always start by snapping out the perimeter of the soffit on the floor (anywhere from 12 inches to 48 inches from the wall) and the perimeter of the recessed ceiling, taking into account the area eaten up by the sloped sides. This helps to give the homeowners an idea of what the finished ceiling will look like, and it gives me a chance to "see" what they want. It's easier to change some chalk lines than rip out wood.

Soffit Box

It's usually best to frame the coffered-ceiling structure before the roof is built, so you won't have to dodge the rafters as you build. However, if the building has a low-pitched roof (6/12 or lower), the ceiling structure might run into the roof frame, so it has to be built afterwards. Sometimes, a coffered ceiling may be part of a floor system. In this case, I usually get to it during pickup. But more often I'm framing an independent ceiling structure, which is the procedure I'll describe here.

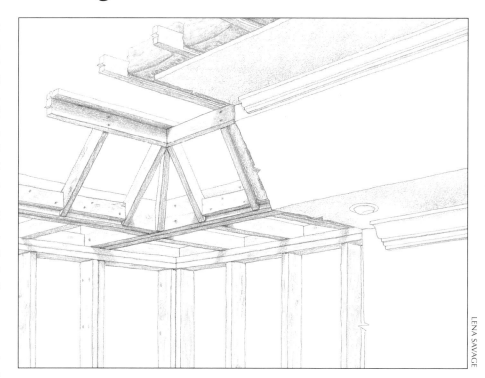

The first thing to build is the soffit box, which defines the soffit and carries the coffer rafters. The plans should call out the size of the soffit rim joist. If this is missing, use what would be required for a floor joist of the equivalent span, then double it up. If the ceiling is in a room that has exterior walls that will eventually support rafters, lay out the rafters along the wall plates first, so that the two systems don't conflict. Install the shorter soffit doublers first, then hang the longer ones from those (see Figure 4).

Before hanging the long doublers, make sure the short ones are straight, by eye or by stringing a dry line, then install 2x4 kickers to brace them in a straight line. These kickers run from the bottom of the nearby double top plate to the bottom of the doubler joist, so they'll be out of the way while the finger joists (the soffit ceiling joists) are

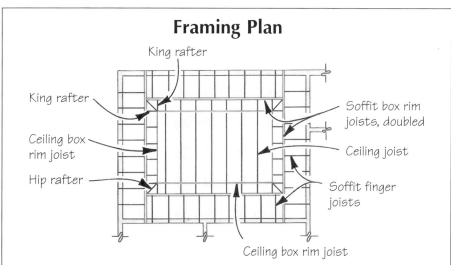

Figure 4. *The "soffit box" provides the main support for the coffered ceiling. Double 2-by carrying joists span from wall to wall across the short ends of the room; the long carrying joists hang from the shorter beams. Two-by-four "finger joists" provide a level soffit, while 2x4 coffer rafters slope up to a 2x6 "ceiling box."*

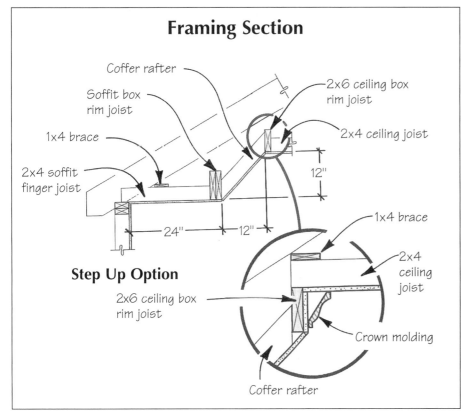

Figure 5. *The perimeter of the recessed ceiling can be either a simple angle or a vertical step-up trimmed with a crown molding.*

installed. Next, install the other sets of double joists, hanging them from double metal hangers, then line them straight and brace them.

To complete the soffit, cut and install finger joists (typically from 2x4 stock) all the way around the soffit box. The fingers not only create the soffit plane, but also prevent the coffer rafters from bowing out the doublers. Toe-nail the finger joists to the wall plates, then nail them to the rafters, installing blocking in every other bay to prevent twisting as the lumber dries. Set the finger joists 16 inches on-center for 1/2-inch drywall (you can use 24 inches for thicker drywall).

Coffer Rafters

With the soffit box and the finger joists installed, it's time to cut and install the coffer rafters. As with all my framing calculations, I use a Construction Master to calculate the diagonal measurement, or the length of the rafters. But you can use whatever technique you use to figure the length of common rafters. No shortening allowances are made — these are nailed directly to the ceiling box.

First, lay out the position of the king rafters on the inside of the soffit box. The run determines their placement. For example, if you have a 30-inch run, butt your tape to the inside corner of the box, measure out 30 inches, strike a line and mark an X on the side away from the corner. You should have 30 inches in the clear, from the inside corner to the start of the rafter. Lay out all four corners this way, giving you the positions for the eight king rafters. Then lay out the rafters in between, usually on 16-inch centers. Make sure the opposite sides match, just like rafters at a ridge.

On small areas like the coffer there isn't much forgiveness, so it's best to cut the rafters as precisely as possible. Once the rafters are all cut, nail up the eight kings, holding the bottoms of the rafters flush to the bottom of the soffit box. I nail these up with 8d nails (16ds just blow up the wood).

Ceiling Box

After the kings are nailed up, it's time to measure for the ceiling box, which is made up of four 2x6s that define the higher ceiling plane. Take dimensions from the layout on the soffit box, starting with the short side and measuring the distance from outside to outside of the two kings (check the opposite side to make sure it's the same.) Then measure the adjacent long side, from inside to inside of the two kings on that side, again checking the opposite side to make sure it's the same.

Install the short side first, fitting it in at the inside corner where the two kings converge. The rafters help put pressure on the board and hold it in place while you face-nail into one rafter then toe-nail to the other (this is easy with a helper). Flush the bottoms of each king coffer rafter to the bottom of the rim board.

After the two short rim boards are nailed up, install the remaining two. With the four ceiling rim joists fastened in place, I usually nail in a ceiling joist at midpoint to keep the long run of the box in line and use scrap 2x4s to temporarily kick the box straight.

Coffer Hips

Once you have straightened the box to satisfaction, nail up the remaining rafters and install the hips. The hips are just like the hip rafters for a roof and can be measured in place or calculated. Then put a cheek cut on both ends — two 45-degree bevels that allow it to fit into the corners. When installing the hips, align them so the drywall planes to the middle of the hip rather than to the edge, toenailing the bottoms to the soffit box and face-nailing to the ceiling box.

Ceiling Joists

After the rafters are nailed up, the work of filling in the ceiling box with the joists is straightforward. Crown the joists, and face-nail through the box into the ends of the joists using three 16d nails to keep them from twisting. On ceiling joists that are over 10 feet in length, I put a strongback down the middle to flatten out the joist plane. If you do this, pick a nice, straight piece of stock one size larger than the joist, roll it up on edge, center it, and toe-nail it to the tops of the joists.

Step-up detail. In many cases, the

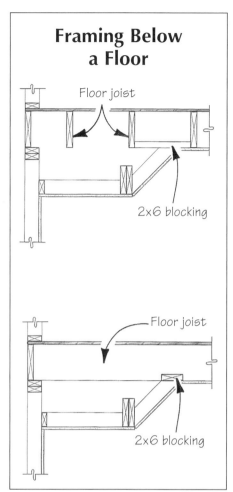

Figure 6. *When framing a coffer below floor joists, use a ledger at the wall to catch the finger joists, and 2x6 blocks between the floor joists to catch the coffer rafters.*

recessed ceiling joists sit on top of the ceiling box; this creates a step-up that can be trimmed with crown. In this case, the ceiling rafters have to plane to the inside face of the box, not the back. To accomplish this, you can use a speed square to guide the rafter into position, leaving the bottom of the plumb cut hanging just a bit below the ceiling rim board.

Then toe-nail the joists to the top of the ceiling box, and nail 1x4s flat across their top edge at each end to help prevent twisting (Figure 5). Also nail 1x4s across the tops of the finger joists. I usually catch this detail at the same time that I'm installing the recessed ceiling joists.

First-Floor Coffer

A slightly different procedure is used when building a coffer in a room where the ceiling is already built — the lower floor of a two-story house for instance. In this situation, the soffit rim joists and the finger joists are supported by a ledger board that is nailed around the perimeter of the room (Figure 6).

Since the ceiling is already in place, the rafters are attached to flat backing blocks nailed between the floor joists. This requires a level cut instead of a plumb cut on the end that attaches to the ceiling. The blocks also provide backing for drywall.

To install the blocking on the sides where the floor joists run parallel to the coffer rafters, first snap a line where the ceiling and coffer rafters intersect. To find this point, cut a sample coffer rafter, hold it in position, and make your marks. Then nail up the blocking, holding one edge of the block about an inch inside of the mark to provide perimeter nailing for the drywall. On the sides where the floor joists run perpendicular to the coffer rafters, place the blocks so they split the center of the rafter.

For a long run, it can be faster to install one long, flat backing board nailed to a supporting block at each end. To prevent this board from sagging in the center, I nail up a couple of flat 2x4 blocks near midspan. And sometimes I get lucky and the floor joist falls so that a couple of pieces of 2x4 will pad it out far enough to catch the ceiling rafter and provide the drywall backing.

When all the ceiling blocks are in place, snap a reference line for the coffer rafters, then cut and install them.

By Don Dunkley, a framer in Cool, Calif.

Framing a Simple Radius Stair

I'm a construction supervisor for a large framing contractor in New Mexico. Our ten crews frame about 15 custom homes each month, most of them high end, with 3,000 or more square feet of floor space. In this niche, radius stairs — curved stairs that arc as much as 180 degrees around a single point — are quite common. I've framed 40 or 50 of them in the last nine years and have developed a method that's both fast and accurate. In fact, the stair I'm about to describe took only 16 hours to frame.

Code Requirements

As with any stairway, the first order of business was to verify that the stair would meet code. A crucial dimension with curved stairs is tread width at the inside of the radius. The Uniform Building Code specifies a minimum inside tread width of 6 inches. Stairs that don't conform must be redesigned, which usually means laying them out with a longer radius. Fortunately this design met all codes, with an inside tread width of $8^{3/8}$ inches and an outside width of $16^{1/4}$ inches.

Planning

The structure consisted of a series of $3^{1/2}$-foot-long "pony walls," framed like little stud walls. Each pony wall was $7^{1/8}$ inches higher than the previous one, and each supported a tread. In general, the process involved lining up the pony walls to form steps, then fanning them out along two concentric arcs (Figure 7). The outside arc of this particular stair also defined a curved bearing wall for the second-story floor system; the inside arc followed the rise of the stairs and was left open for handrail installation by the finish carpenters.

This stair would also enclose a closet which meant that five of the pony walls, instead of being framed with studs, had headers to span the closet opening. The last three treads were framed in yet another way, which I'll describe later.

- Arc 1: 4 feet (the short radius)
- Arc 2: 4 feet 3 1/2 inches (the short radius plus 3 1/2 inches)
- Arc 3: 7 feet 6 inches (the long radius)
- Arc 4: 7 feet 9 1/2 inches (the long radius plus 3 1/2 inches)

I now had two 3 1/2-inch-wide arcs on which to align my curved floor plates. The total distance between the innermost and outermost edges was 3 feet 9 1/2 inches — 3 1/2 inches wider than the stair width — to allow for the curved bearing wall over the outside arc. The final stair width would be 3 feet 6 inches.

Rise and run. The next step was to calculate the unit rise and run. The stair had a total rise of 121 1/8 inches. That worked out to 17 rises at 7 1/8 inches each. The plans showed the stairs forming a 160-degree arc. I had 16 treads to place in this area, so I had to space them 10 degrees apart.

Next, I drew a tread layout on the slab. I placed a 12-inch speed square at the radius point, lined one leg up with my reference line, and marked off each 10-degree interval. I then snapped lines from the center of the arc through each 10-degree mark, extending the lines across the floor to arc number four. It took some work to get consistent treads. Having the 10-degree marks so close to the radius point meant that I had to

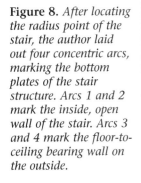

Figure 7. *This elegant curved stair conceals a frame structure (above) consisting of a series of stepped walls fanned out along an arc and tied together by plywood treads and risers.*

Layout

How smoothly a curved stair project goes depends on how carefully you do the layout. Working from the plans, I first snapped a line that I knew was square and parallel to the exterior walls of the house and that ran through the radius point for the stairway (see Figure 8). I sank a nail at my radius point, hooked the end of my tape measure over it, then drew four concentric arcs that began and ended at the reference line I had snapped on the floor. The arcs had the following radiuses:

Figure 8. *After locating the radius point of the stair, the author laid out four concentric arcs, marking the bottom plates of the stair structure. Arcs 1 and 2 mark the inside, open wall of the stair. Arcs 3 and 4 mark the floor-to-ceiling bearing wall on the outside.*

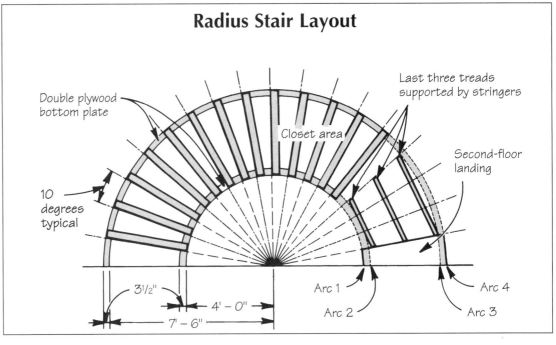

184   *Chapter Ten: Pickup Work*

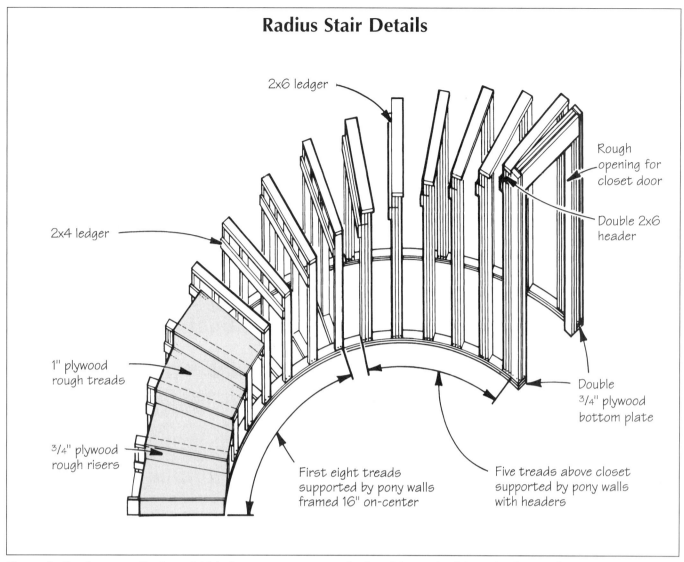

Figure 9. *Simple pony walls, framed 16 inches on-center, support the first eight treads of the stair. The next five pony walls include a header to span the closet opening. A 2x ledger fastened 6 1/8 inches below the top of each wall supports the back of each tread.*

constantly adjust my tread widths as I went along to make sure each matched the others.

Curved plates. For bottom plates, I used a double layer of 3/4-inch plywood. I cut the curve with a Skil wormdrive saw, setting the blade 1/2 inch deeper for every pass. This required three passes for each cut of the radius. After cutting all the plates, I placed them in position on the floor and transferred my tread lines across them. But I didn't fasten them until I had finished all the framing, to leave room for adjustment. I planned the curved wall so that a stud would fall just forward of each stair riser. This provided solid nailing to tie the stairs into later.

Building the Frame

We were now ready to build the stairs. Before framing a house, I typically spend several hours with the plans, and it pays off. When I hit the job I know exactly what needs to be done, with detailed cut lists for most of the house's components. On a curved stairway this includes dimensions for pony walls, headers, treads, and risers. With a good cut list, my crew can quickly build the pony walls and cut the rough treads and risers.

Pony walls. The first eight pony walls consisted of four studs, framed 16 inches on-center. As Figure 9 shows, the 2x4 bottom plate fit between the two curved plywood arcs; the top plate spanned the distance between the inside and outside of the same two arcs.

The 7 1/8-inch unit rise meant that the first pony wall would be 6 1/8 inches tall (adding a 1-inch-thick plywood tread would bring it up to 7 1/8 inches), the second 13 1/4 inches tall, and so forth. Each wall was fastened to the floor behind a 10-degree layout line. To support the back of the treads (the edge at the bottom of the next riser), we also nailed a ledger 7 1/8 inches below the top of each wall.

Closet headers. The ninth tread was the first clear span over the closet. Starting here, we cut four trimmers (two to a side) and a king stud for each tread, then bridged the distance between them

Framing a Simple Radius Stair

Finish Work

With the framing complete, the project was handed over to a local stair company. The stair crew spent close to 130 hours — a four-day job for four people — fabricating and installing the red oak stair parts. Even so, the job was relatively simple compared with other curved stairways. Instead of full-width treads and risers, this stair was mostly carpeted, with exposed oak treads and risers only along the inside edge.

Also, they chose not to use a finish skirt on the inside radius, instead dressing the exposed ends of the treads and risers with moldings. They simplified installation of the outer radius skirtboard by using 1/4-inch oak plywood backed up with 1/4-inch fir plywood instead of solid oak laminations.

The most challenging part of the job was the curved railing. The stair crew created the profile by sandwiching seven red oak strips between two premolded edges, then carefully laminating the assembly around a bending form. They made the rail in several sections, staggering the ends of the laminations to weave the sections together. But despite the care that went into lining everything up, the railing still required a good deal of hand-dressing after installation.

— JLC

Figure 10. *The author supported the last three treads with short 2x12 stringers to leave room for the closet door and a passageway beneath.*

with a double 2x6 header. The header was capped with a 2x6 plate which, in turn, was at the correct height to support the front of the tread.

Short stringers. The last three treads on this stair posed yet another problem. Because they extended above the closet door, they couldn't have a supporting wall beneath them. Our solution was to frame them like a conventional stair, using three short 2x12 stringers, as shown in Figure 10.

Using the closet door header for support at the bottom, we measured and cut one stringer so that it came tangent to arc number two at the middle tread. It had a unit run of 8 3/8 inches, which equaled the stair's inside tread width. We then cut a second stringer to form a chord of arc number three. This stringer had a unit run of 16 1/4 inches, which equaled the stair's outside tread width. With the inside and outside stringers in place, we used strings to lay out a third stringer for the midspan.

With the basic stair structure complete, we cut our rough treads and risers, using plywood templates to keep them consistent. We then glued and nailed them to our stair structure. To prevent squeaks, we glued and nailed every framing joint as we went along, including the one between the bottom plates and the floor. As extra insurance, I also ran an additional bead of adhesive around all completed joints.

By Robert Thompson, a construction supervisor for Framing Square Construction Company Inc., in Albuquerque, N.M.

Section Three
ENGINEERED MATERIALS

Chapter Eleven
TRUSSES

Installing Gable Roof Trusses 188

Installing Hip Roof Trusses 193

Truss Bracing Tips 197

Critical Bracing for Piggyback Trusses ...199

Making Room at the Top 202

Site-Built King Truss for Ridge
Beam Support 205

Installing Gable Roof Trusses

Occasionally we run across a builder who refuses to use roof trusses. His reasons usually include "They sent the wrong size trusses," "We nearly got killed putting them up," or "The roof was crooked and flimsy." These common complaints usually point toward lack of experience rather than any type of truss system failure. As with any new building product, there is a learning curve. Fortunately, costly errors can be avoided if the builder takes the time to learn how to order roof trusses properly, develops good site handling and erection practices, and follows simple installation, straightening, and bracing methods.

Usually we can set the trusses and sheathe a 2,500- to 3,500-square-foot colonial with a two-car garage in a day with five guys. In our experience, roof trusses are always more economical than stick-framing when you factor in the labor savings.

Begin With a Thorough Plan Review

We've had our share of incorrect truss orders. Although the mistakes weren't ours, we still suffered downtime waiting for replacements. The problem was always poor communication — usually the intervention of a salesperson or a misunderstanding on an "over the phone" order. To eliminate these pitfalls, we now create a "truss package" from our takeoffs.

Beginning with a clean set of prints, we become familiar with the roof design and ceiling details. Roof design may be obvious from the exterior elevations, but determining ceiling details requires close examination of the building section drawings. We highlight rooms with any details that need to be built into the trusses, like vaults, coffers, or trays. Some complicated details like barrel ceilings can't be efficiently fabricated into the trusses, but can be stick-framed after installation if accommodations are created in the trusses. We note critical items like wall measurements within rooms so it's clear where a vault begins and ends.

Color coding. Using colored pencils, we plan the preliminary truss layout by drawing lines across the floor plan on 2-foot centers (the typical truss spacing). Each different truss size and pattern gets its own color. Even on a basic colonial like the one shown in this section, we used several different types of trusses, including common flat bottom chord trusses, gable end trusses, vaulted ceiling trusses above the master bedroom, and attic trusses above the garage to create useable space in the future. There was also a girder truss, a small gable truss, and a "valley kit" for a gable bumpout on the front elevation.

Just to be sure there's no confusion about the length of the trusses, we make notes on the plan indicating the distance from outside face to outside face of the exterior walls the trusses will span. Often this information must be culled from several interior partition dimensions, and we double-check to be absolutely certain. The final design and layout will be done by the truss engineer, but our detailed notes create a basis for clear communication between all the parties involved.

Special notes. To complete our plan review, we make special notes on a separate page, including roof pitches, the widths of exterior bearing walls, the type of heel on the truss (top chord overhang, bottom chord overhang, or raised heel), and the live load requirements for the bottom chord. We also include the length of the overhang (projection of the truss beyond exterior wall). Twelve-inch overhangs are common in our area, but we spec $10^{1/2}$ inches. This leaves room for us to install a 2-by subfascia without trimming the truss tails (more on this later).

Roof-mounted equipment. Finally, we include information about the location and weight of any hvac equipment that will be mounted to the roof or hung within the attic space. The truss designer may need to beef up a couple of trusses to handle these loads.

These notes, together with the marked plans, become the package from which we make our truss order. On simple jobs, we just pass the package along to our building materials or truss dealer and await the engineer's design. With complicated truss roofs, we meet with the

Figure 1. *After the trusses are dropped on site, the crew first flushes up the stacks (top left), then marks important layout lines, including ceiling strapping (bottom left) and the exterior wall plates (near left).*

manufacturer's engineer to review our detailed plans and notes before he begins the design work. In either case, it's important to include a full set of plans because the truss engineer will trace any concentrated roof loads down through the house to the foundation. Sometimes we may need to beef up studs in a wall or install squash blocks to accommodate these roof loads coming down through the frame.

We always request a full set of truss design plans before actually placing the order. Careful review and a follow-up meeting with the designer give us a last chance to catch errors and omissions before fabrication and delivery.

Delivery Day

We plan the delivery date to mesh with our framing schedule. We want the trusses to arrive two to three days before we erect them. Receiving them any sooner causes problems. On small work sites, the bulky trusses get in the way. Also, the longer the trusses are stored on site, the greater the chance they may twist and warp from exposure to wetting and drying cycles and the heat of the sun.

On delivery day, we're always on site to ensure the trusses are dropped where we want them. The drop spot has to be flat and out of our way but within reach of a crane. Extra space will be necessary if we have a complicated truss package because we break up the truss bundles to organize the pieces. If the site is extremely constrained, we may have to split the delivery in order to have enough room to maneuver.

Layout on the Ground

We do some prep work to the trusses on the ground to speed erection and installation. First, we block up any low points under the truss bundles to prevent kinking and bowing of the trusses. Once the bundles are good and flat, we break the banding straps (see Figure 1).

Next, two crew members align the peaks and seats so we can mark them for precise installation. One guy works back and forth between the tails of the trusses tapping them in and out with a sledge hammer while the other eyes the peaks and directs the adjustments. We take care not to damage the trusses, using a block of wood to protect the trusses whenever we tap them.

Once the trusses are stacked perfectly flush with one another, we measure the length of the bottom chord and mark the center point on the top and bottom truss in the pile. Next we mark the inside edges of the exterior walls by measuring back toward each end half the truss span. We always measure the frame to get the actual field distance rather than going off the plan.

Using a straightedge, we then draw a line across the bottom chords of all the trusses between the marks on the top and bottom trusses (Figure 1). During erection, our tag line operator will extend the marks up the flat side of the bottom chord about 1/2 inch for easy visibility when the trusses are placed. We also use spray paint to mark one end of the trusses so they go up oriented the same way they were built.

Premarking the trusses this way ensures the peaks are perfectly aligned as they go up. This ensures a nice flat plane once the sheathing is applied.

Strapping layout. It's common practice in our region to strap the ceilings

Figure 2. *It takes a crew of five to set the trusses with a crane. A man on the ground attaches the crane hook and steers the truss with a tag line (top). A man at each plate positions the trusses to the layout lines, while two men above set the truss spacers (middle and bottom).*

with 1x3 furring. Laying out the strapping while the trusses are still on the ground is simple to do and saves a lot of time later. Starting from the mark we made indicating the inside face of the wall, we mark 16-inch on-center layouts and add an "X" on the furring side of each line.

Sheathing guides. We also make lines on the top chords for our first course of roof sheathing. We measure down from the peak to the last full 4-foot increment before the end of the truss tail. We add 1 inch for ridge vent spacing and $1/8$ inch for each of the gaps between sheathing panel courses. On a truss with a 19-foot-long top chord, for example, we would measure down 16 ft. $1^{3}/8$ in. (4 sheets = 16 ft. + 1 in. for ridge vent + $3/8$ in. for panel gaps). This operation saves us from having to measure and snap a line across the trusses once they're installed. When we sheathe the roof, we set the first course of sheathing above the line and continue to the ridge, then fill in the ripped sheet from staging at the bottom.

Prepping the gable trusses. The house featured in this article has standard gable ends, which we sheathe and trim out before lifting. But first we make sure that their bottom chords are straight. Occasionally they'll dip a little, which isn't a problem with regular trusses free-spanning between two walls. But because the gable trusses sit on top of the end walls, a dip in the bottom chord can raise the peak. So we snap a line along the bottom chord of the gable trusses and trim any excess off. This is permissible since the gable-end truss is supported its entire length by the gable-end wall. But no part of any other truss should ever be cut unless approved by the truss designer.

Presetting a girder truss. This house also had a small girder truss that supported one end of several of the main roof trusses at the gable bumpout. We chose to set this 12-foot-long truss by hand and install the hangers in advance to save time on crane day.

Organizing a Complex Roof

Complicated truss systems require a little extra attention. Difficult truss packages or hip roof configurations need to be restacked by hand in the order of erection so they get installed in proper sequence. The manufacturer marks each truss with an identification number; identical trusses are labeled with the same number and are interchangeable. The manufacturer's truss plan has each truss location clearly marked with those numbers. Following the plan keeps the truss reorganization in order.

With complicated roofs, we also use the truss plan to mark the top plates. The standard truss spacing is 24 inches but sometimes there are deviations, and they'll be noted on the plan. We write the truss number on the plate in place of the usual "X." This gives a fail-safe recheck during installation.

We prebuild the girder/jack portions of hip truss systems on the ground to save time and work safely.

Crane Day

Whenever trusses are relatively small (no longer than 24 feet with a 6/12 pitch or less), we lift them by hand, particularly on single-story homes. But manhandling larger trusses becomes a dangerous operation, especially on two-story houses. Besides personal safety, the biggest risk is damage to the trusses. Trusses are strong in their upright position, but are flimsy when handled horizontally. Sometimes truss plates can pop off or loosen enough to cause a future failure.

Most of the trusses we use are larger, so we hire a crane. It's safer for us and easier on the trusses since they're lifted evenly in an upright position. It also speeds the process so that it's cost-effective. While we have the crane on site, we also use it to lift second-floor interior studs, roof sheathing, and sometimes roof shingles up to the second floor deck.

We schedule truss-raising day well in advance. This gives us time to prepare for the arrival of the crane, which requires clear access to the site and a couple of spots to set up to reach the entire house. Occasionally on tight sites, we've had to move the trusses to put them in reach of the crane. Other building materials, trash containers, and debris piles also need to be out of the way. We mark out any underground installations: Cranes are heavy and can easily crush a shallow sewer pipe when the stabilizers are set. Tree and power line locations are also a consideration when planning the lift.

Staging. Good staging is essential for quick installation. We set up different types depending on the job's requirements — pump jacks if we'll be doing the siding, wall jacks if we're only doing the trim. We sometimes set staging up on the inside of the house. The bottom line is that we provide a working platform in every area involved rather than doing something unsafe like "walking the plates."

Bracing the walls. Just before the crane shows up, we line up and brace the walls. We also install the truss tie-down clips if there's time — they help hold the trusses in place until we nail them off. We then assemble the tools: long adjustable braces, levels, plumb bobs, tag line, strapping marked on 24-inch centers to be used as temporary braces, and our Truslock spacers

Figure 3. *Gable-end trusses, which are sheathed on the ground, are set first and temporarily braced to the deck below (left). As subsequent trusses are set, steeper, more stable braces are added (right).*

(Truslock, 2176 Old Calvert Rd., Calvert City, KY 42029; 800/334-9689).

Expanded crew. The crane operators we use are familiar with truss raising and have all the straps and rigging hardware we need. We usually run with an expanded crew of five on truss day (Figure 2). One crew member attaches the strap onto each truss and connects it to the crane hook; he also handles the tag line to keep the truss from spinning and in proper alignment with the house. Before each truss is lifted, he makes sure that the wall mark we made earlier is continued up the side of the truss at least 1/2 inch. Two members man the front and rear staging. They guide the trusses into their location on the top plate and set the position by the wall marks. The last two on the crew get the fun job of dancing the peaks of the trusses and bracing them as each one is set into place.

Start at the Gable

As the first gable-end truss is set in place to the locating marks we've made, the tag line person climbs a ladder and

Figure 4. *Temporary diagonal braces along inner webs, along with Truslock spacers on the top chords, keep the trusses from collapsing until the sheathing is installed.*

drives spikes through the bottom chord into cleats we previously fastened to the end wall top plate 1 1/2 inches in from the outside face. Before the truss is disconnected from the crane, we nail a long adjustable brace from the floor up to the peak of the truss (Figure 3). This brace is set at a steep enough angle to allow the bottom chord of the next truss to pass by. We roughly plumb the gable truss and send the crane for the next truss.

The first regular truss is positioned according to the layout lines and nailed on one wall only. The opposite end is left loose next to its tie-down until later. This allows us to straighten the walls later after the jostling they get while we're setting the trusses. The guys working the peaks space and secure the truss to the gable end, using temporary braces marked with 24-inch centers. Then we install a second adjustable floor-to-peak brace through the webs of the first regular truss to the gable truss. This brace isn't as steep as the first one and gives better support. Before lifting any more trusses, we accurately replumb the gable end using a long level and straightedge. Straightening the gable end at this point is easier than after all the trusses are in place.

The rest of the trusses are lifted in sequence, set, and nailed according to the marks at one end. Once we have a couple of regular trusses in place and have established our on-center spacing, we use the Truslock spacers, which automatically space the trusses at the top as they are set.

When we reach the opposite gable end, it is braced temporarily and the entire ridge is measured and adjusted to match the length of the building. We check the gables for plumb with a plumb bob all the way down to the floor deck just for accuracy's sake, and fine-tune using the adjustable braces. Next, we diagonally brace the top chords and vertical webs of the trusses to ensure stability until we can sheathe the roof (Figure 4). The process of lifting, setting, and bracing a typical house takes about three hours, while a complicated roof can take a full day.

Once all of the trusses are set, we check the strings and braces that are keeping the walls straight. We readjust if necessary before nailing down the truss ends we left loose during installation. Had we nailed off both ends during installation, we couldn't make any adjustments to the walls without using a cat's paw.

Sheathing, Fascia, and Bracing

While some of the crew goes to work sheathing (Figure 5), others work on the subfascia. We install a 2-by subfascia sized and sometimes ripped to match the end of the truss chords (Figure 6). The subfascia gives us a good base for attaching a 1-by finish fascia or for covering with aluminum coil stock. It also

Figure 5. *The sheathing is installed to premarked layout lines on the top chord. The layout is planned so that the installer begins with the first full sheet at the bottom and works up (left), leaving a ventilation gap at the top (right).*

Figure 6. *The authors prefer a rugged 2-by subfascia, which is installed at the truss tails before the last narrow piece of roof sheathing is filled in.*

Figure 7. *After the field of the roof is sheathed, the crew installs the valley kit for a small gable bumpout, leaving a gap in the sheathing below to allow for ventilation.*

eliminates flimsy installation of gutters by installers who don't bother to locate truss tails behind the trim. We shim between the trusses and the subfascia if necessary to adjust for deviations in the truss tails. Occasionally, we have to trim a couple of extra-long truss tails — made evident by a quick string line. We straighten the subfascia by eye in most instances; that gets us within 1/8 inch. After the subfascia is completed, we measure, rip, and install the bottom course of roof sheathing.

After the sheathing goes down, we install the valley kit for the small gable bumpout (Figure 7).

Permanent bracing. The process isn't complete until we install the permanent web bracing specified by our truss designer. We follow the provided details exactly, using the specified material. Omitting the permanent braces can, in extreme conditions, lead to truss failure.

By Rick Arnold and Mike Guertin, builders in North Kingstown and East Greenwich, R.I.

Installing Hip Roof Trusses

Many builders have an irrational fear of hip trusses. I've worked with hip roof trusses for 18 years and I currently run a truss installation company in Raleigh, N.C. In all that time, 90% of the problems I've seen were due to inaccurate placement of the trusses. But with good drawings and an understanding of how the pieces go together, a hip trussed roof isn't difficult and offers the same cost advantages as a standard gable trussed roof.

Good Communication

Ordering the hip roof trusses is the most critical link in the entire process. Any mistakes made at this stage will most likely go unnoticed until the trusses arrive, so I'm careful that the information I give to the truss manufac-

turer is accurate and a delivery date is agreed on. I typically provide the manufacturer with detailed drawings, and answer any questions that come up.

If the truss manufacturer is working from an architect's or designer's set of drawings, it's important that all three parties (the truss manufacturer, designer,

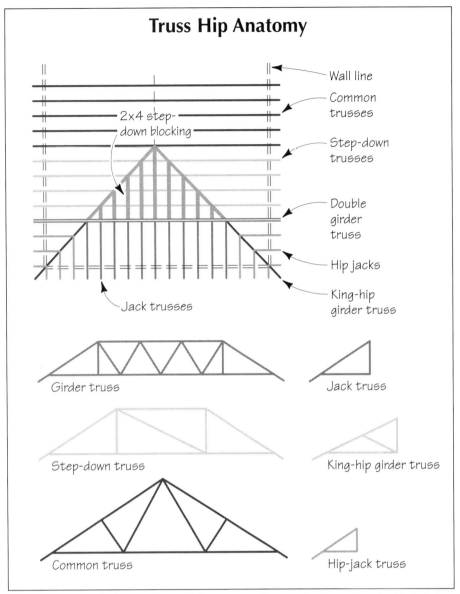

Figure 8. *In a typical truss hip roof, the common trusses step down to a doubled girder truss. The hip is then completed with a pair of king hip girders and jack trusses.*

Figure 9. *Framing truss roofs on the ground reduces dangerous work on the roof. The hip end shown here, part of a large commercial job, was completely framed on the ground (left), then craned into place (right). On residential work, the author usually installs the king hip girder and hip jacks on the roof.*

and contractor) are on the same page. The Wood Truss Council of America (WTCA) has established a set of guidelines that describes who is responsible for what. (Document WTCA 1-1995 is available through your local truss manufacturer, or can be ordered from the WTCA, Madison, WI; 608/274-4849; www.woodtruss.com.

I make a point to review and approve the truss design drawings and placement plan before the trusses are built. This reduces the chance of the trusses being incorrectly manufactured, and helps me become familiar with the truss plan.

It's Hip to Be Square

It's essential that the dimensions of the building's walls match those called out in the drawing, and that all walls are square, plumb, and level. If I arrive at the site and the wall plates are out of square or dimensioned incorrectly, I take the time to establish where an accurately framed roof system will rest on the plates.

Minor inaccuracies of an inch or less can generally be absorbed in the soffit, but major inaccuracies may require a meeting with the general contractor before work begins. In either case, the truss roof must be assembled accurately, or the roof sheathing and ceiling drywall will not break properly on the trusses.

Accurate Layout

Before the trusses arrive, I carefully lay out the truss locations on the wall plates, using the placement plan supplied by the truss manufacturer. These plans typically include a layout drawing that clearly indicates the location of each truss, and a drawing of each type of truss used in the roof assembly.

Ask your supplier to show you a sample layout drawing. Drawings that can't be deciphered won't do you much good the day the trusses arrive.

Starting on the Ground

In a hip roof truss system, the girder truss is set first, followed by the jack trusses (see Figure 8). To save time, I like to preassemble the girder and jack

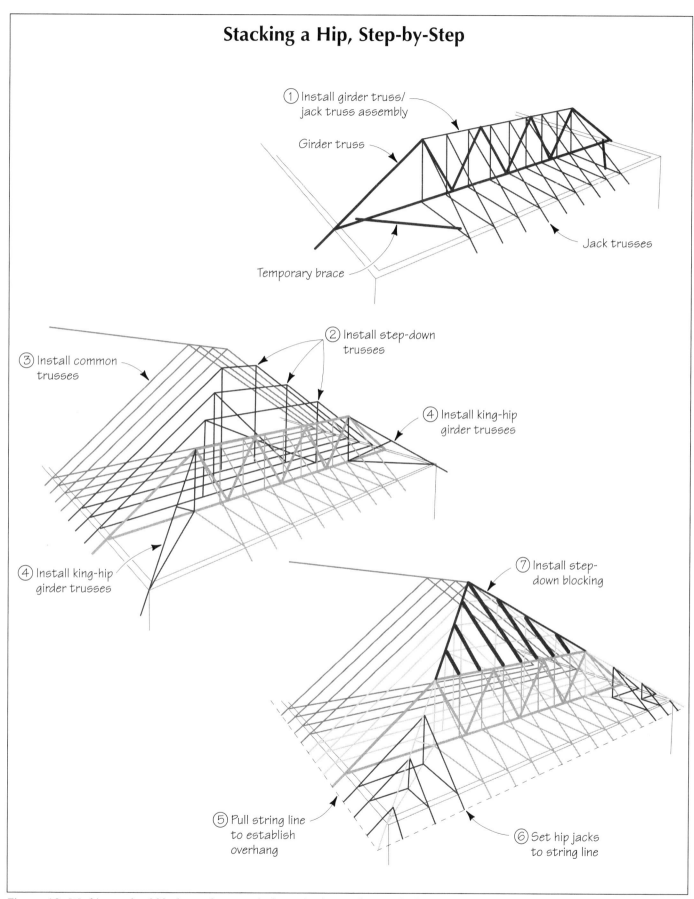

Figure 10. *Working on level blocks on the ground, the author's crew fastens the hip jacks to the girder truss, then stiffens the whole assembly with temporary braces. The assembly is then crane-lifted and carefully set on the top plates (top). Next, the step-down trusses are assembled on the ground and lifted into place, followed by the common trusses (middle). The king hip girders are then installed, but fastened only to the girder truss, not at the plate. Finally, a string is pulled around the ends of the jacks and step-down trusses to establish the proper overhang (above). The king hip is adjusted as needed and the hip jacks set to the string.*

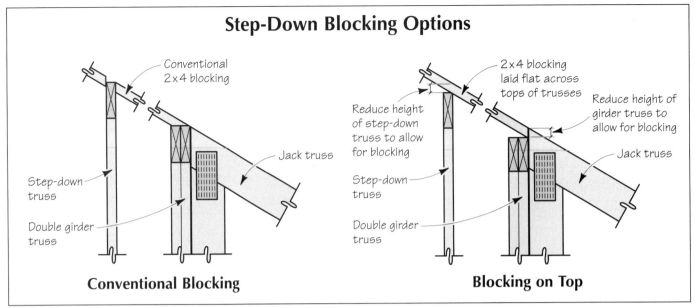

Figure 11. *Conventional step-down blocking requires two bevel cuts and a lot of installation time (left). Instead, at the design stage, the author reduces the height of the girder truss step-down trusses to allow 2x4 blocking to be laid flat across the tops (right). An alternative to individual 2x4s is to have the truss factory make a "blocking truss" (see photo, page 193), which can be dropped into place with a crane.*

trusses on the ground, and lift them into place as a unit (Figure 9). This eliminates the temporary bracing needed when assembling these trusses on the wall plates.

Using temporary sleepers to provide a flat assembly area on uneven ground, my crew of three men can assemble the trusses in less than 30 minutes. I don't worry too much about how the ends of the jack trusses line up while I'm assembling the trusses on the ground: I can adjust this after the assembly is set on the walls.

To crane the assembly over to the wall plates, I cinch two nylon straps around the top chord of the girder truss at quarter points. The entire assembly will be out of balance, with the ends of the jack rafters hanging much lower than the girder truss. To prevent the trusses from bending or distorting during the lift, I carefully brace the preassembled trusses while they're on the ground, and use the lift locations recommended by the manufacturer. I have the crane operator gently position the ends of the jack trusses at the layout marks on the plates, then slowly lower the girder truss into place.

After the preassembled unit is set in place on the walls, I check the positioning of the girder and jack trusses, nail them off, and install diagonal bracing (Figure 10). Bracing requirements for truss roofs will vary depending on the span, roof pitch, and other factors. I refer to the Truss Plate Institute's publication HIB-91 when deciding on bracing strategies ($7 from the Truss Plate Institute, 583 D'Onofrio Dr., Suite 200, Madison, WI 53719; 608/833-5900).

Step-Down and King Hip Trusses

After nailing and bracing the girder truss and jack truss assembly so it will stay rigid and straight, I install the step-down trusses, any common trusses, and the king hip girder. When I install the king hip girder, I fasten it securely to the girder truss (following the truss designer's specifications), but only tack it in place at the wall plate. This way, I can make adjustments later, when all the trusses are in place.

Before setting the hip jacks, I string a line from the ends of the common trusses to the end of the king hip girder, continuing around the corner to the ends of the jack trusses. Theoretically, this string line should run parallel to the walls, but in reality, I often have to adjust the end of the king hip girder to establish a parallel line. An improperly sized king hip girder can create a ripple effect that will cause problems when installing the roof sheathing, fascia, and soffit. If the king hip is long, I'll trim the end, and check to make sure that the top chord isn't above the roof plane. If the top chord crowds the roof plane, I use a skill saw to rip the top chord. If the king hip is short, I'll fasten a shim to the end.

It's important to remember that trusses are engineered products; unauthorized field modifications may reduce their strength or performance. If you're not certain what effect a field modification will have on a truss, check with the truss manufacturer before proceeding.

When I'm sure that my string line is parallel to the walls, I nail off the king hip girder at the wall plate.

Filling In the Corners

Using the string line as a guide, I set the remaining hip jacks. I position each hip jack over its layout mark (making sure that the truss is square to the wall plate) and check the end against the string line. If the hip jack is short, I slide it slightly off the layout mark toward the wall corner. If the hip jack is long, I slide it towards the main roof. By sliding the truss, I don't have to spend valuable time trimming it to fit. If a plywood edge will miss the relocated truss, I nail a scab to the side of the top chord.

When nailing the hip jacks to the king hip girder, I prefer to keep the bottom chords flush. If there is any difference in height where the two trusses join, I've found it easier to hide the discrepancy at the roof plane rather than at the ceiling plane.

I'm careful to keep an eye on the king hip girder as I nail off the hip jacks: It's easy to push the king hip girder out of line as you drive the nails home.

Step-Down Blocking

Typically, step-down trusses are designed so the roof sheathing is fastened directly to the top level chord of the truss. To provide a nail base along the hip, step-down blocking is installed from the end of the common roof's ridge down to the king hip girder. This type of step-down blocking requires a compound angle at both ends, and is very tedious to install. Intermediate step-down blocking is also installed between the step-down trusses to extend the nail base of the jack trusses all the way to the hip line.

To streamline the step-down blocking process, I have the truss manufacturer lower the top chord of the step-down trusses $1^1/_2$ inches perpendicular to the hip plane (Figure 11). This way, instead of fitting the blocking between the trusses, I can run continuous lengths of flat 2x4s on top. This technique cuts my blocking labor by 80%.

Another option, which we tried for the first time on a recent large job, is to have the truss manufacturer build a "blocking truss." All the step-down blocking is ganged together into one unit at the truss plant, and we set it in place with the crane. Since it's nonstructural, we can fine-tune the fit once the truss is set in place.

After all the trusses are set, I pull a string line from the end of the king hip girder to the ridge intersection of the hips and back down to the end of the other king hip girder. This string line defines the hip of the roof, and I use it as a guide as I install the 2-by on top of the step-down trusses.

By Paul Bartholomew, owner of PFB Company Inc., a truss installation company in Raleigh, N.C. Thanks to Jim Vogt, P.E., of the Wood Truss Council of America.

Truss Bracing Tips

Using roof trusses lets you close in a building quickly and gives you flexibility in placing interior partitions. But except for some data published by the Truss Plate Institute (Madison, Wis.; 608/833-5900), there is little written information on how to install and brace trusses. The few books that do discuss the issue typically show the first truss located directly over the end wall and braced diagonally to stakes driven into the ground.

However, other approaches often work better for at least three reasons. First, depending on the building's height and the slope of the site, the length of a diagonal brace from the peak of a truss to the ground can measure more than 40 feet — too long a span for strong, rigid bracing. Second, in the case of a hip roof, the girder truss usually sits at least 12 feet inside the end wall, where it's impossible to brace it diagonally to the ground. Finally, braces running from the roof to the ground clutter the site, slowing the work and creating safety hazards.

A Different Approach

To avoid these problems, I use a different technique for installing roof trusses.

First, I straighten and brace the exterior walls as usual. Then I beef up the bracing of one of the end walls — the end where I plan to start truss installation. But instead of starting directly above the end wall, I position the first truss 8 to 12 feet in, then brace it back to that wall. Here's how it works (see Figure 12).

Before plunging into the stack of trusses, I mark the truss spacings (usually 2 feet on-center) on the top plates of the bearing walls. While I mark the layout, I have another carpenter prepare plenty of two-by lateral braces by marking off the truss centers and starting a duplex nail at each mark.

Next, either with a crane or with lots of muscle, we raise the first truss and set

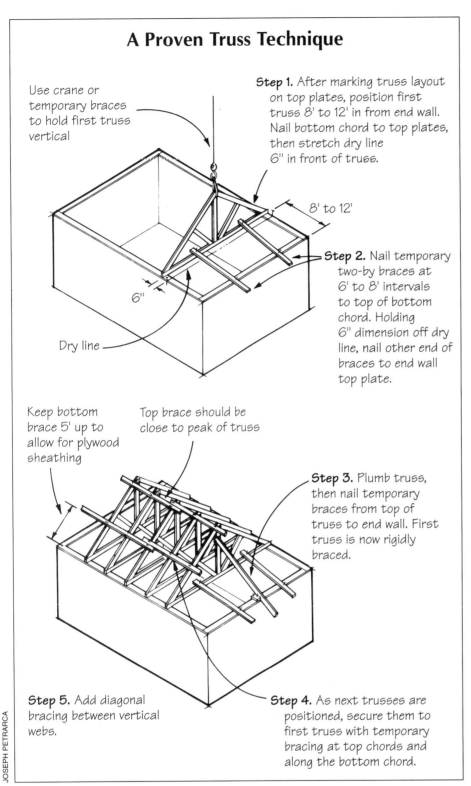

Figure 12. *Trusses are typically installed and braced with the first truss located directly over the end wall and braced diagonally to stakes in the ground, which can be problematic. The author's technique positions the first truss 8 to 12 feet in from the end wall, where it is then braced back to that wall.*

it in position on the first layout mark farther from the end wall than the height of the truss. For example, for 7-foot-high trusses, we would position the first truss at the fifth layout mark (8 feet in for trusses 2 feet on-center). This ensures that all the braces from the top chord of the truss will be at less than a 45-degree angle, making them much stronger than braces at steeper angles.

Then, with the help of the crane or with braces from the end wall or the deck below, we hold the truss plumb while we position it exactly on the layout marks and toenail it to the plates.

At this point we pull a dry line from the front to the back plate and 6 inches in front of the truss. We position braces at 6- to 8-foot intervals from the top of the bottom chord to the top of the end wall. Depending on the distance between the end wall and the first truss, I've used almost everything from 2x4s to staging planks. Never use 1-inch stock for bracing.

We first spike each brace into the top of the bottom chord. Then, as one worker measures, his or her partner adjusts each brace and spikes it to the end wall when the truss's bottom chord is exactly 6 inches from the line.

Next, with either the crane or workers on the deck or end wall still steadying the truss, I climb an extension ladder to the peak and plumb the truss, using a 6-foot level and straightedge or a plumb bob. (For lower-pitch trusses, staging or a tall step or trestle ladder also works for getting to the peak.) When the truss is plumb, I nail a brace from the peak to the top of the end wall. Then I work down the top chord at 6- to 8-foot intervals, straightening and plumbing it, then bracing it back to the end wall.

The Rest Is Easy

With the first truss in place, straight and plumb, the rest of the trusses go more quickly. We nail the second truss in place; then, using the premarked lateral bracing, we secure it to the first truss.

For a standard house truss, say 24 to 36 feet long, we install at least one row of lateral bracing along the peak, one along the attic walkway, and one along the midpoint of each top chord. Larger trusses require more.

To be sure the trusses will withstand a windstorm, we connect all the trusses to one another with diagonal bracing between the vertical webs. We begin installing this as soon as the fourth truss is up.

Once we've set a dozen or so trusses, another crew can start the roof sheathing to help stiffen them. We keep the lowest lateral brace at least 5 feet up from the bottom edge of the truss so we don't have to remove it until the first row of plywood is nailed off. Meanwhile, the first crew continues across the building, setting trusses and installing the diagonal and lateral braces. When the entire roof assembly is anchored together, we nail two continuous 2x6 braces from the ridge to the floor in the form of an "X" or an "A."

To finish the roof, we remove the end wall braces to the first truss and work back to the end wall. Finally, we install the permanent bracing and metal tie-downs per the engineered bracing plan from the truss manufacturer.

By Paul DeBaggis, an instructor at Minuteman Technical School in Lexington, Mass. A former builder, DeBaggis has installed thousands of roof trusses throughout eastern Massachusetts.

Critical Bracing for Piggyback Trusses

At a recent presentation I gave to 50 engineers on truss bracing, a structural engineer produced a set of photographs of a piggyback truss roof in a heap on the hardwood floor of a school gymnasium. The roof collapse occurred with only 5 inches of snow on the roof — there was no ice or any other load factors (fortunately, the gym was empty at time of the collapse).

In case you're not familiar with the term, piggyback trusses are used when the height of a required roof truss exceeds the width limit allowed by a state's transportation department for transport on the back of a truck. Depending on the state, the limit is usually around 12 to 14 feet. Say, for example, you need a 12/12 gable roof truss for a 40-foot span. The 20-foot ridge height will force the truss manufacturer to provide the truss in two parts — a bottom truss with a horizontal top chord and an upper truss that completes the gable tri-

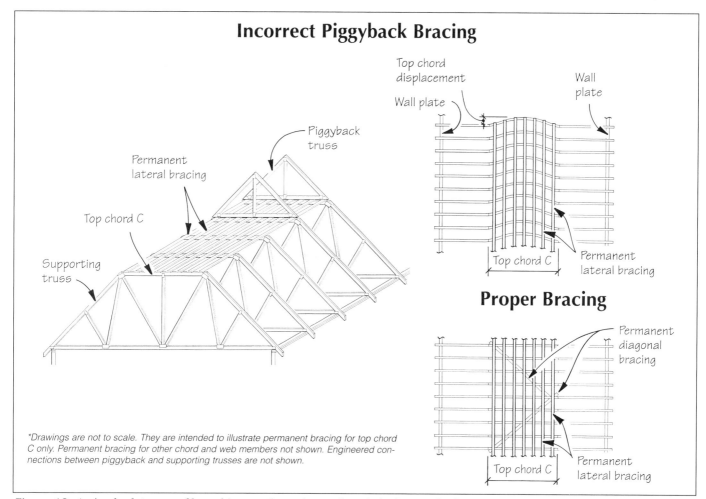

Figure 13. *A piggyback truss roof braced incorrectly as shown above is in danger of collapse. Because top chord C is a compression member, it may buckle under load if braced with lateral braces alone (upper right sketch). Though they are sometimes omitted, permanent diagonal braces (bottom right) are required to prevent this buckling action. Top chord connections between the piggyback truss and supporting truss are also required, but are not shown here.*

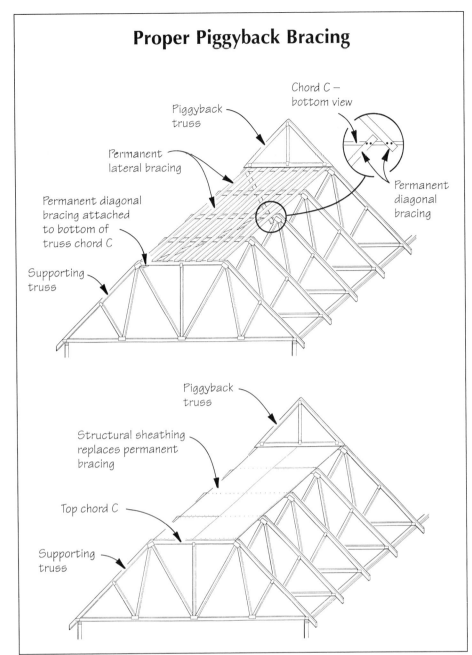

Figure 14. *Permanent cross-bracing installed on the bottom of chord C (top) will stabilize the lateral braces attached to the top of chord C. The diagonal braces should be installed at about 45 degrees to chord C. Structural sheathing (bottom) is also a simple and reliable method to provide permanent bracing for chord C. With the proper nailing, the sheathing provides diaphragm action, serving as both lateral and diagonal bracing.*

angle (Figure 13).

All the engineers who spoke up about the truss collapse at the presentation agreed that the sole cause of the failure was incomplete bracing of the piggyback system. Lateral bracing had been installed on the supporting truss as depicted in Figure 13. What was lacking was diagonal bracing.

Confusion About Truss Bracing

Temporary and permanent bracing of trusses are critical to the success of a wood truss system from the day of installation through the life of the roof. Yet confusion about bracing is common among builders and architects. The term "truss bracing" is ambiguous because it does not differentiate between temporary bracing and permanent bracing.

To help clear up the confusion, I suggest we avoid using the term "truss bracing" altogether, and refer instead to "temporary bracing" and "permanent bracing." Temporary bracing is the bracing used by the contractor to safely erect the trusses, and it may include elements of the permanent bracing system. Permanent bracing is the bracing required in the roof system to stabilize the trusses throughout the life of the structure.

For example, 2x4s used on the top chords of simple gable trusses to prevent the top chords from buckling during construction are temporary bracing. When the 2x4s come off and the structural sheathing goes down, the sheathing becomes the permanent bracing for the top chord. (Other permanent bracing may be required for other chord and web members.) When trusses are erected in sections on the ground with the plywood already attached, the sheathing serves both functions — temporary bracing and permanent bracing.

Contractors are responsible for determining the necessary temporary bracing, but not the permanent bracing. Permanent bracing should be specified by the building designer.

Back to the Gym

So what happened to the school gym trusses? Under the snow load (as small as it was), truss chord C in Figure 13 is in compression. Imagine what happens when you hold a yardstick on edge between your hands and push your hands together: The yardstick buckles. The same thing happened to chord C. Chances are the failure happened suddenly, without warning. When the chord buckled, the truss lost its integrity and fell in on itself. The piggyback section fell to the floor with the supporting truss.

Note the lateral braces in Figure 13. This is how some builders incorrectly brace chord C in a piggyback system, assuming perhaps that the roof sheathing on the rest of the roof will prevent lateral movement. This is not the case, however. As the top chord C displaces under load, the lateral braces simply move with the top chord. What is required are additional diagonal braces, nailed to the bottom of chord C, which act with the lateral braces to form a series of triangles stiffening the top chords. Without the diagonals, the lateral braces across the top chords are like a stud wall standing up without sheathing: Push on it and it racks easily; add a diagonal brace or sheathing and the stiffening effect is dramatic.

Proper Permanent Bracing

Figure 14 illustrates two ways to provide proper permanent bracing for chord C. (Note that this article is concerned only with bracing top chord C, because this is the most common problem area for piggyback systems. Other chord and web members also require permanent bracing.)

The top sketch shows how to use diagonal 2x4s to stabilize lateral braces like the ones used in the gym trusses. The cross-bracing is attached to the bottom of chord C, installed at about a 45-degree angle to chord C.

The bottom sketch in Figure 14 shows my preferred method of piggyback bracing — installing structural sheathing across the top chords of the lower truss (you may have to make some allowance for ventilation). After the sheathing is installed on chord C, the piggyback truss can be installed and attached to the supporting truss in accordance with the truss designer's specifications. It is the responsibility of the truss designer to specify the connection between the piggyback truss and the supporting truss. If this connection has not been spelled out on the truss design drawings, contact the truss manufacturer for the detail.

Retrofit Piggyback Bracing

Over the years, I have heard contractors comment that permanent bracing requirements appear excessive and that there seems to be as much lumber used for bracing as for the trusses. This is an exaggeration, of course, but it is an attitude that may lead builders to ignore the permanent bracing specifications.

So, what if you've built a piggyback roof that's braced like the roof in Figure 13? "If it hasn't collapsed yet, it's probably okay" is flawed logic. Whether that

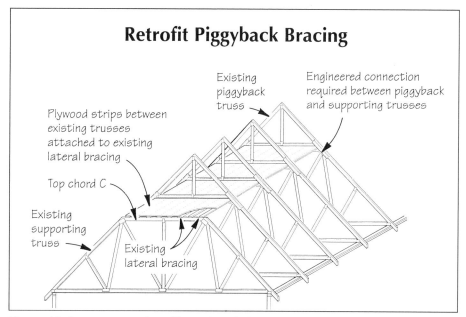

Figure 15. *In some cases, installing strips of plywood between existing piggyback trusses may serve to stabilize the permanent lateral braces. The building designer or engineer should specify the proper nailing.*

top chord C buckles or not is dependent on the compressive load level present in the chord under gravity load. It may take 50 years to get the load level needed to buckle the chord, but it might also happen next week.

It is possible to install bracing in a piggyback roof that wasn't braced properly when it was built. You should follow specifications provided by the building designer. The retrofit may take the form of cross-bracing, as in Figure 14, or you might be able to install plywood strips between the trusses, nailed directly to the lateral braces, as in Figure 15. Again, you should follow the designer's nailing recommendations, and you may have to provide for roof ventilation.

Summary

Using a piggyback truss system versus a standard truss system is not an incidental framing decision. The contractor should determine the temporary bracing to safely erect the trusses and the building designer should specify the permanent bracing needed to support in-service loads.

A point to remember is that the building designer cannot complete the permanent bracing drawings until he approves the truss design shop drawings used to manufacture the trusses.

It is not uncommon for an architect to specify roof truss bracing "per HIB-91." This specification should alert you to a problem because HIB-91, "Handling, Installing, & Bracing Metal-Plate-Connected Wood Trusses" (published by the Truss Plate Institute; 608/833-5900), is about temporary bracing, not permanent bracing. Any note by the building designer that does not tell you specifically what permanent truss bracing to install should also be a warning. The truss manufacturer may not be able to respond to your questions since they are not responsible for design of permanent truss bracing.

Do not install piggyback trusses unless the building designer specifies the necessary permanent bracing for the roof system. If you are both the contractor and the building designer, but not an engineer with truss bracing experience, you should contact an engineer familiar with permanent bracing design. Your truss manufacturer should know a local or regional engineer.

—*Frank Woeste, P. E.*

Making Room at the Top

Roof trusses save contractors time and money, and they eliminate the need for interior columns or bearing walls. The downside is that the attic space becomes a web-filled jungle, unsuitable for storage. But it doesn't have to be that way. Storage trusses can provide usable space in the attic at a very reasonable cost.

Case in Point

During the initial design of a three-story addition we built earlier this year, the clients requested that several large rooms on the top floor be free of posts. Trusses were the natural choice to clear-span the 24-foot-wide addition, so I put together an order for standard "W" trusses and set up a delivery with my truss supplier. A few days later, however, the homeowner changed the scope of work to include finishing off part of the basement in the new addition, which meant losing a large storage area.

Since my order hadn't been fabricated yet, we were able to change to attic storage trusses to make up for the lost basement space. Storage trusses provide the same clear-spanning ability, but reconfigure the interior webbing to provide an open area in the middle of the truss (see "Storage Truss Design," next page).

To provide enough headroom within the storage truss, the original roof pitch needed to be changed from a 6/12 pitch to an 8/12 pitch. This provided a storage area 6 feet high by 10 feet wide (see Figure 16). A standard truss would have come with a 2x4 bottom chord, but because of the additional weight these storage trusses would be carrying, a 2x6 bottom chord was used.

Bring in the Sky Hook

Deciding to use a crane to set these trusses was easy: The top wall plate on the third floor of this addition was more than 30 feet off the ground. The cost of renting the crane was $80 per hour with a four-hour minimum billing. Billable time included round-trip travel time, setup time, and shutdown time.

To minimize crane time, we lifted the trusses in bundles of three and stacked all

Figure 16. *These storage trusses span 24 feet and provide a usable attic space 10 feet wide by 6 feet high.*

the trusses on top of the third-floor walls (Figure 17). While the crane was set up, we also lifted all the plywood sheathing onto the third-floor deck. The total crane time on site was two hours. After the crane had unloaded all of the trusses, we tilted the trusses up one at a time with a push stick (Figure 18, next page). We used Truslocks to maintain spacing and as temporary bracing (see "Spacing and Bracing," next page).

Had this been a one- or two-story addition, we would have unloaded the trusses by hand. Be sure to line up a crew of four sturdy people before you try this. The storage trusses we used were slightly heavier than conventional trusses, but what made them especially difficult to handle was the higher center of gravity created by the increased roof pitch.

Figure 17. *To minimize crane time, the trusses were lifted in bundles of three and placed on the third-floor wall plates.*

Weathering In

Once the trusses were set and temporarily braced, we applied the plywood sheathing just as we do with conventional trusses. Our local code allows us to use 1/2-inch CDX plywood or 7/16-inch OSB on trusses spaced 24 inches on-center. For a stronger roof and a more substantial nail base, however, we use 5/8-inch CDX for 24-inch-on-center

---

Storage Truss Design

Probably the most familiar truss design is the Fink truss, named after the engineer who designed it. This style of truss uses minimal material to perform its task, but the W-shaped web design discourages any use of what would normally be attic space. A storage truss reconfigures the web to leave the center open.

Extra Loads

Most attic space is quickly and completely filled by the homeowners, so storage trusses must be engineered for heavier loads. Our truss manufacturer allowed for an additional 20 pounds per square-foot storage when designing the trusses, which required a 2x6 bottom chord. When the open area is used as living space, storage trusses are designed to support 40-pound live loads, and often require a 2x8 bottom chord.

Bracing requirements for storage trusses depend on how the truss is configured, and what the end use of the functional area is. In our case, the building inspector required double catwalk bracing on the bottom chord, just outside the storage area. Be sure the truss manufacturer supplies you with bracing infor-

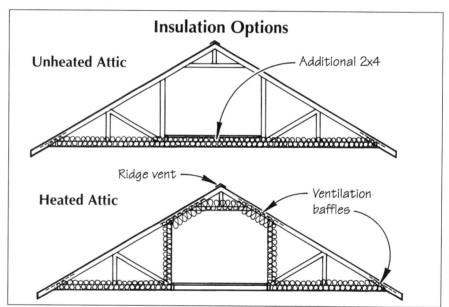

For an unheated storage area (top), the truss maker can add a 2x4 to the bottom chord to accommodate the required insulation. For heated storage (bottom), remember to use ventilation baffles to allow air circulation to the ridge vent.

mation for your particular truss design and loading situation.

Insulation

It's a little more difficult to insulate storage trusses than conventional trusses. If the storage area is unheated, you may need to raise the floor in the storage area to provide enough room for insulation (see illustration). If heated, then the walls and ceiling of the storage area will require insulation. With a vented ridge, be careful to maintain a ventilation space where the storage wall meets the top chord of the truss.

— G.R.

Spacing and Bracing With Truslock Tools

Truss fever: If the crew catches it, they think about one thing — staying ahead of the crane. Accuracy suffers, and mistakes are discovered when the plywood sheathing joints don't fall on the middle of the truss's top chord.

To help control the fever, I decided to purchase a set of Truslock spacing tools (Truslock, 2176 Old Calvert City Rd., Calvert City, KY 42029; 800/334-9689). The advertising claimed that these spacing tools automatically braced and spaced trusses during installation, saving time and improving safety. Knowing that my crew and I were going to be 30 feet above the ground, the idea of improved safety was very appealing.

Truslock tools work like a folding rule. After the first truss is installed and braced, the tool is fastened to the braced truss and "unfolded" as each subsequent truss is installed, rigidly clamping the trusses in place at their proper spacing. As trusses are set and the permanent bracing is nailed off, the tool can be folded up and moved ahead for the next series of trusses.

This tool eliminated the need for most of the temporary lateral bracing we used in the past to support and space the trusses until the sheathing was applied. The "guaranteed" spacing (within 1/8 inch per 100 feet, according to the manufacturer) allowed us to confidently nail off permanent bracing as the installation progressed.

Since the Truslock tools took care of the spacing and bracing, my crew was able to focus on the crane, avoiding any symptoms of truss fever.

— G.R.

The Truslock spacing tool unfolds like a folding rule to maintain proper truss spacing.

Figure 18. *The author's crew positions taller trusses with a push stick made from a 16-foot 2x4.*

roof framing. The thicker plywood costs more, but I've seen the edges of thinner sheathing telegraph through the shingles on a finished roof.

To quickly "weather in" the addition, we covered the roof with 6-mil polyethylene. Roofing felt installed at this stage will ripple and not lie flat for the installation of shingles, so the roofer ends up tearing it off and applying new felt anyway. Installing the plastic is quicker and, in this case, allowed us to frame interior partitions during a stretch of wet weather.

Extra Dollars Make Sense

The additional costs associated with the storage trusses were surprisingly low. The truss manufacturer charged an extra $8 per truss ($4 for the storage web configuration and $4 to increase the bottom chord to a 2x6). My roofing subcontractor added $2 per square (a dollar per square for each unit rise in roof pitch). So the total additional cost for upgrading to storage trusses was $160. That's a great buy when you consider that using the trusses added 300 square feet of storage space.

I was lucky on this job to be able to change the truss design late in the game. Next time, I'll decide on storage trusses early in the process.

By Gary Rowland, a general contractor in the Atlanta area.

Site-Built King Truss for Ridge Beam Support

The triangle formed by a pair of rafters and a joist is the most basic form of truss, and has been in use in one form or another for centuries. The two rafters can be envisioned as leaning against each other, just as you might lean a ladder against a wall. The vertical forces on the rafters are all transferred to the outer corners of the base of the triangle, which typically bears on the building walls below. The load transfer creates a horizontal force, which I refer to as rafter thrust. This thrust is restrained by the horizontal joist, which provides a restraining force to keep the base of the triangle (and the walls below) from spreading (see Figure 19).

The outward thrust of the rafters under load creates a tension force in the horizontal joist. As long as the connection between the rafter and the joist can carry this tension force (and assuming the rafters are correctly sized for bending), the roof will stand with no sagging at the peak. If the connections slip, as may happen under heavy snow loading, the roof may sag or even collapse.

It is important to realize that in this simple rafter-joist truss, the rafters get no vertical support at the peak. In many old buildings there may not even be any ridge member at all. All vertical loads are carried by the exterior walls.

Structural Ridge Option

Another way to carry vertical roof loads is to use a structural ridge: a beam that supports the rafters at their peaks and transfers loads to support points, which are typically posts at each end and along the span. Half of the uniform load on each rafter is carried by the structural ridge, reducing the load on the wall by one-half. Since the peak of each rafter is supported vertically by the ridge, there is no tendency to slide downward and no resulting outward thrust at the eaves. Structural ridges are thus useful where the designer wants to omit joists or collar ties — in cathedral ceilings or in a shed dormer addition where headroom is limited.

But sometimes due to floor plan constraints below, it's not possible to put a post under the end of the structural ridge. In this case, I sometimes use an enhanced version of the rafter-joist triangle described above instead of a post. By engineering the rafter-joist connections, it's possible to use a rafter-joist triangle to carry the point load from the end of the ridge beam out to the exterior bearing walls. I refer to this construction as a king truss.

Calculating the Loads

A structural ridge, like the one in Figure 20, must be sized to carry a total load equal to half the total load on all the rafters it supports. Of the total load on the structural ridge in Figure 20, half is carried by the king truss at one end, the other half by a post at the opposite end. We'll call this point load P; the calculations in Figure 20 show how to derive it. For the sake of this example, we'll assume a 30 psf live load and a 10 psf dead load; in many parts of the country the live load will be less. Remember, too, when calculating roof tributary areas to account for roof overhangs.

Once P is known, we can calculate the thrust, T, in the joist, then design the connection to resist it. I usually use bolts due to the heavy loads and the critical nature of the connection. Whenever possible, I use double-shear bolting, which means that one member is sandwiched between two others to which the force is being transmitted. Double-shear bolting can carry twice the force of single-shear bolting (where two members are simply bolted face-to-face).

In almost all cases, the limiting condition is the connection design. Rarely does the capacity of the rafter or joist for axial loading (tension along the length of the joist or rafter) prove to be a problem, although it should always be considered. The number of bolts is determined by dividing the thrust, T, by the capacity of the bolt to be used.

Table 1 (page 207) is intended as a guideline. It is based on use of S-P-F framing lumber, with the numbers of bolts based on single-shear applications. If double-shear bolting can be incorporated (that is, if you can double up either the joist or the rafter), the bolt count can be cut in half.

Note that as roof pitch decreases, the required bolt count increases dramatically. This is why we engineers like to see generous roof pitches, aside from water-shedding benefits. Note also that some of the bolt counts are marked "not recommended;" in these cases, the axial (tensile) strength of the joist or rafter might be at issue, as well as the practical difficulty of finding room for a large

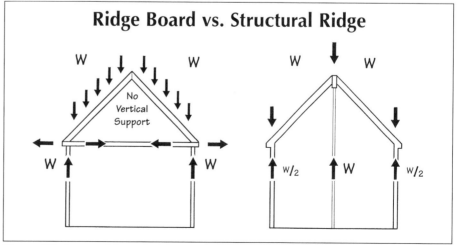

Figure 19. *In a roof framed with a nonstructural ridge board (left), the uniform load on the rafters creates an outward thrust at the top of the exterior walls. The ceiling joist acts in tension to resist this force. With a structural ridge (right) — typically a beam supported by columns — the outward thrust is removed.*

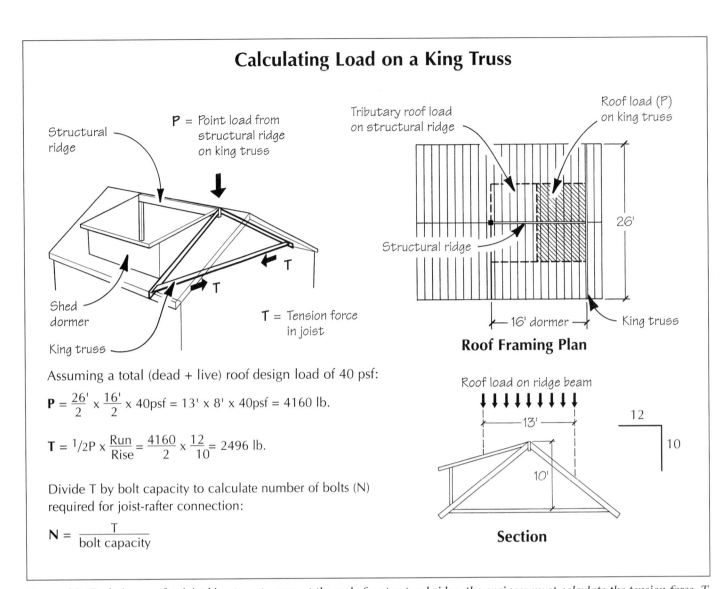

Figure 20. *To design a rafter-joist king truss to support the end of a structural ridge, the engineer must calculate the tension force, T, that will be induced in the joist. First, the tributary roof load flowing to the end of the ridge beam is calculated, then this point load, P, is used to calculate the tension force in the joist. The designer then specifies the bolts needed to carry this tension force at the rafter-joist connections.*

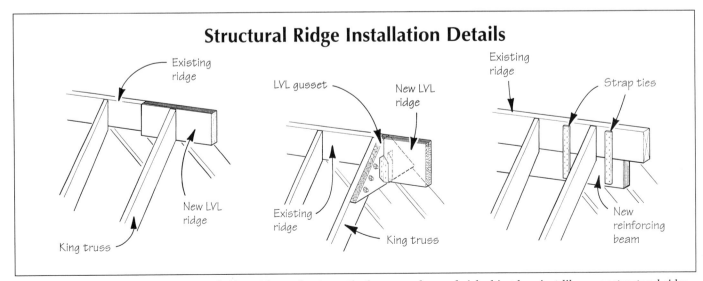

Figure 21. *When supporting a structural ridge with a rafter truss, the beam can be sandwiched in place just like a nonstructural ridge (left), supported by a hanger attached to a gusset plate (middle), or even hung by straps from an existing ridge board (right).*

Table 1. Number of Bolts Required for King Truss Joist-Rafter Connections

To use this chart, calculate the point load, P, at the top of the king truss (always round up), then find the number of bolts required for the roof pitch. Note that using seven or more 1/2-inch bolts is not recommended due to the weakening of the wood members by the number of holes. In some of these cases a smaller number of 1-inch holes will work. The number of required bolts assumes a single-shear connection. Reduce bolts by half for double-shear connections. The chart is based on use of S-P-F 2-by lumber (specific gravity = 0.42). In some, but not all cases, fewer bolts would be required if a stronger grade of lumber were used. Use of lower grades, such as redwood, eastern softwoods, S-P-F (south), western cedars, western woods, or northern species would in some cases require more bolts. When in doubt, the designer should refer to the NDS tabulated bolt design values, which were used as the basis for this chart.

Roof Pitch	Point Load (P) on King Truss				
	1000 lb.	2000 lb.	3000 lb.	4000 lb.	5000 lb.
12/12	2 / 1	4 / 2	5 / 3	7* / 4	9* / 5
10/12	2 / 1	4 / 2	6 / 3	8* / 4	10* / 5
8/12	3 / 2	5 / 3	7* / 4	9* / 5	12* / 6
6/12	3 / 2	6 / 3	9* / 5	12* / 6	14* / 7*
4/12	4 / 2	8* / 4	12* / 6	16* / 8*	20* / 10*
3/12	5 / 3	11* / 6	16* / 8*	21* / 11*	26* / 13*

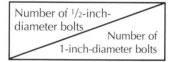

Number of 1/2-inch-diameter bolts / Number of 1-inch-diameter bolts

*Not recommended

Single-Shear Connection

Double-Shear Connections

Putting the King Truss to Work

This king truss concept becomes a part of many of my residential structural designs where either a cathedral ceiling or a large dormer prevents the normal use of regular rafter-joist triangles at common spacings. It also serves to carry the apex loads from hip or valley rafters in cases where an interior post is impractical.

For new construction, it's simplest to incorporate the structural ridge beam in the same way as you would a nonstructural ridge board, using a generous number of toe-nails through each king truss rafter (Figure 21). The downward force from the vertical ridge loads will cause a wedging action that will create a large friction force to lock the ridge in place.

For retrofit applications, the ridge can be hung below the existing ridge, or inserted between existing rafters, which might be sistered in some designs. The ridge beam might also be mounted with hangers to a suitably designed gusset fastened securely to a pair of rafters.

The connection of the regular rafters to the structural ridge also requires attention. These rafters have no thrust forcing them against the ridge, and therefore need a connection designed to carry half their tributary load. This may be accomplished by an appropriate number of toe-nails or, for enhanced confidence, a joist hanger, such as the field-sloped Simpson LSSU series or the USP (Kant-Sag) TMU series.

In a king truss design, it is very important that each connection be considered carefully and adequate fasteners provided for all calculated loads. This is an area that warrants consulting an engineer for a design review. In my experience, more field failures occur due to connection design deficiencies than from any other cause.

—*Robert Randall*

Chapter Twelve
Engineered Lumber

Wood I-Joist Fundamentals 210

Sizing Engineered Beams 211

Working With
Laminated Veneer Lumber 214

On Site With Parallam 217

Building With Glulams 220

Engineered Studs for Tall Walls 224

Wood I-Joist Fundamentals

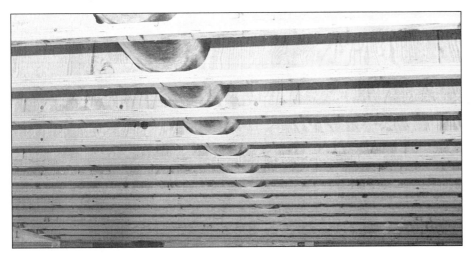

For builders accustomed to the shape and heft of conventional sawn lumber joists, the weight and appearance of a typical wooden I-joist often raises an immediate concern about its ability to handle the load. How could such a thin, lightweight material carry as much as a sawn 2x10 or 2x12?

Situations like the one in the photo above seem to defy common sense. How can you possibly remove so much of the I-joist web in the middle of a span? Yet the hole chart for that particular I-joist allows this. Can you imagine doing that with a sawn joist?

However, once you understand the kinds of loads present in a floor joist and how the shape of an I-joist accommodates these loads, you'll be able to use I-joists with confidence.

Stressed Out

There are two types of stresses that develop in a typical floor joist: shear stresses and bending stresses.

Shear. Horizontal shear is a tendency for the wood fibers in a loaded joist to slide past one another. Imagine a joist as a stack of individual wood strips, as shown in Figure 1. When a load is applied, the top of each strip in the imaginary beam shortens from compression while the bottom of each strip is lengthened by tension. Although the drawing greatly exaggerates the effect, the same thing happens to the wood fibers in a solid-sawn joist.

Shear stress develops as the beam resists the sliding action. The maximum shear stress is at the ends of the joist, and decreases toward the center of joist, becoming (theoretically) zero at the exact center of the span. Thus, there is practically no tendency for the wood fibers to shear apart near the midspan of the beam.

When selecting sawn floor joists, building designers need to check that the maximum shear stresses at the ends of the joist will not exceed the allowable shear stresses for a given species and grade of wood. It's also important for builders to reject planks with long checks at the ends.

I-joist web strong in shear. Solid sawn lumber is somewhat prone to horizontal shearing because of the orientation of the wood grain. By contrast, the plywood or OSB web of an I-joist can handle much greater shear stresses. This allows a relatively thin web to safely transfer the shear forces developed in typical floor framing applications. However, when selecting an I-joist, always check that the maximum shear forces that will develop do not exceed the maximum values specified for a particular I-joist. In some cases web stiffener blocks must be nailed on both sides of the web, over the support walls, to help transfer large shear loads down through the web. This prevents buckling of the web, as well as the "knifing action" of the edge of the web against the lower flange.

Bending Stress

In addition to shear stress, any simply supported beam under load (including the load from its own weight) also develops compressive stress in its upper fibers and tension stress in its lower fibers. Collectively, these tension and compression stresses are called bending stresses. They occur in any horizontal object that is supported only at its ends. Anyone who has ever tried to saw down through a log supported near its ends discovers this fact when the blade binds partway through the cut. Because the saw has removed material that was formerly transferring compressive stress, the saw kerf closes in on the top as the weakened log tries to reestablish this internal compression.

In a simply supported beam, bending stresses are greatest at the upper and lower edges halfway along its span (Figure 2). They decrease to zero at the center fiber of the beam.

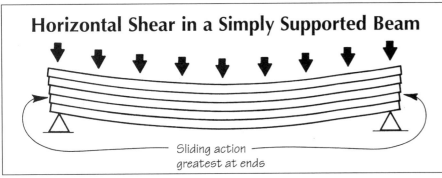

Figure 1. *To visualize horizontal shear, picture a beam as a stack of separate strips of wood. As the beam deflects under load, the strips slide in relation to one another as the top of each strip shortens in compression and the bottom lengthens from tension. The sliding action is greatest at the ends — the area of maximum shear stress.*

The spanning ability of a floor joist is, in part, limited by the bending stresses that the fibers at the edge of the beam can safely handle. These maximum safe stresses are called the Fb values. Fb values vary with different species and grade levels. A Select Structural Douglas fir 2x10 has an Fb of 1,667 pounds per square inch (psi) when used as part of a repetitive-member floor-framing system (value taken from 1991 NDS Supplement). A No. 2 spruce-pine-fir 2x10, in the same situation, has an Fb of 1,006 psi.

Optimization. By comparison, the laminated veneer lumber (LVL) flanges of a typical I-joist have Fb values in excess of 3,000 psi. This means that each square inch of LVL flange can handle 1.8 to 3 times as much stress as the same square inch of sawn lumber.

Simply put, I-joists are engineered to have the high-strength LVL material where the bending stresses are greatest and where it does the most good: in the flanges at the edges of the beam. If the same amount of LVL were placed near the center of the joist, where there is very low stress, it would add needless weight and cost to the joist without yielding any significant strength increase.

Likewise, the wood fiber near the centerline of a sawn floor joist also carries very little bending stress. It contributes to the weight and cost of the joist, but adds essentially nothing in strength. Its presence is simply a result of the rectangular sawn shape.

Cutting with Confidence

Once you understand how an I-joist works, it's easier to understand the manufacturers' rules for notching and hole-cutting. The high bending stresses in the I-joist flanges makes it imperative that no holes or saw cuts be made there. Even shallow kerfs in the lower (tension-carrying) flange can critically weaken the I-joist.

Likewise, the size and location of holes or slots in the I-joist web is tied directly to the magnitude of the shear stresses in the joist. That's why larger holes are generally allowed in the middle of a simple span, but not near the bearing points, where shear stresses are greatest. With that in mind, the photo of the duct chase that we started with should make a little more sense: The hole is at midspan, where shear force is minimal. Therefore, this particular manufacturer allows you to make a large opening in the web as long as you don't disturb the flanges, which are carrying the large bending stresses at that point.

As a final warning, don't make any holes until you refer to the manufacturer's hole charts. Also, no two manufacturers' charts are the same, so make sure you use the current literature that goes with the brand and series of I-joist you are using.

—*John Siegenthaler*

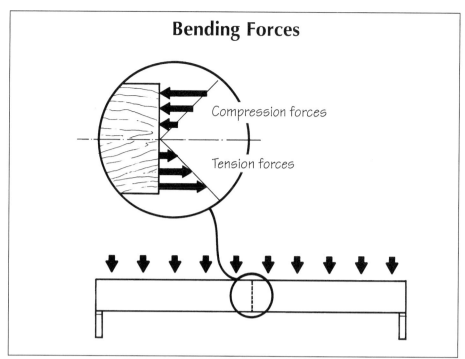

Figure 2. *Maximum bending stresses — tension and compression — occur at midspan in the top and bottom edges of a simply supported beam. They decrease to zero at the beam's centerline.*

Sizing Engineered Beams

In "Calculating Loads on Beams and Headers" (Chapter One), we discussed how to trace load paths and translate roof, wall, and floor loads into pounds per lineal foot of supporting beam. With this information, you can then choose the right beam size and material to carry the load. Here, we'll look at some uniform-load case studies, then compare the performance and cost of sawn lumber with several engineered-lumber products in meeting these different applications.

Sample House

The five applications and spans that I have selected as examples of beams and headers are arbitrary but common ones (see Figure 3).

The first step is to work out the uniform loads for each of the beams, following the procedure outlined on page 11 in Chapter One. Calculations are shown for two loading conditions. One design assumes a 50-pound snow load and the other is in a 20-pound non-snow climate (both loads are treated as live loads). The applications, as noted in Figure 3 and

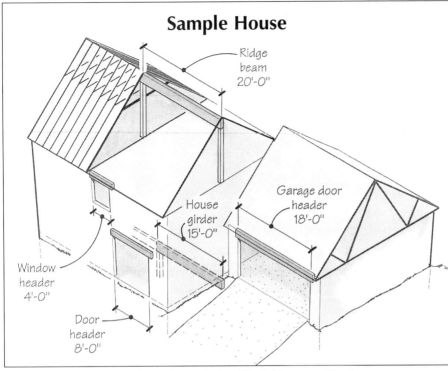

Figure 3. *This cut-away shows five typical residential beam applications.*

Table 1, are:

1) a structural ridge beam with a 20-foot span, as you might find in a cathedral-ceilinged master bedroom
2) a second-floor header with a 4-foot span, which picks up roof loads only, since there is no ceiling in that room
3) a first-floor header with an 8-foot span, which picks up roof, wall, and second-floor loads
4) a basement girder with a 15-foot span
5) a garage-door header with an 18-foot span, which picks up loads from half the truss roof

Based on the calculated loads, I sized and priced the required beams and headers five different ways, to compare the options (Table 2).

Beam Choices

There are many choices when it comes to specifying a beam material. To

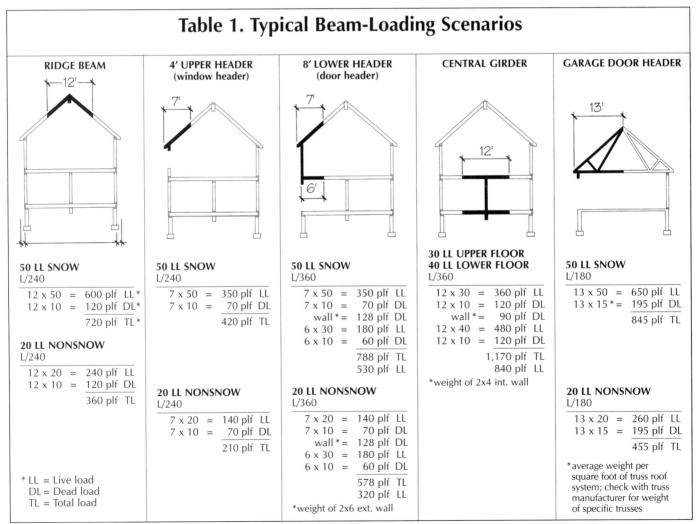

Table 2. Solid-Sawn and Engineered Lumber Options

		SAWN		LVL		POWER BEAM		PARALLAM		TIMBERSTRAND	
		Doug Fir-Larch Sel. str. or #2		F_b 3,100 E 2.0 other options avail.		F_b 3,000 E 2.1		F_b 2,900 E 2.0		F_b 1,700 E 1.3 F_b 2,250 E 1.5	
		Size	Cost	Size	Cost	Size	Cost	Size	Cost	Size	Cost
50 LL **RIDGE**	L/240	Sel. Str. DF-L Triple 2x12 max. span 9'3"	6.36 ($/ft beam)	Double 1³/₄x16 span 20'	11.22 ($/ft beam)	3¹/₂x16 span 20'	11.30 ($/ft beam)	3¹/₂x16 span 20'	11.76 ($/ft beam)	3¹/₂x18 span 20' 1.5 E	9.72 ($/ft beam)
20 LL **RIDGE**	L/240	Sel. Str. DF-L Triple 2x12 max. span 13'2"	6.36 ($/ft beam)	Double 1³/₄x14 span 20'	9.78 ($/ft beam)	3¹/₂x11⁷/₈ span 20'	8.25 ($/ft beam)	3¹/₂x14 span 20'	10.28 ($/ft beam)	3¹/₂x14 span 20' 1.5 E	7.55 ($/ft beam)
4'0" 50 LL **HEADER**	L/240	#2 DF-L Double 2x6	1.80 ($/ft beam)	Not practical		Not practical		Not practical		3¹/₂x4³/₈ 1.3 E	2.36 ($/ft beam)
4'0" 20 LL **HEADER**	L/240	#2 DF-L Double 2x6	1.80 ($/ft beam)	Not practical		Not practical		Not practical		3¹/₂x4³/₈ 1.3 E	2.36 ($/ft beam)
8'0" 50 LL **HEADER**	L/360	Sel. Str. DF-L Double 2x12	4.24 ($/ft beam)	Double 1³/₄x7¹/₄ or Single 1³/₄x11¹/₄	5.10 or 4.16 ($/ft beam)	3¹/₂x7¹/₄ (smallest available)	5.75 ($/ft beam)	1³/₄x11¹/₄	3.15 ($/ft beam)	3¹/₂x9¹/₂ 1.3 E	5.25 ($/ft beam)
8'0" 20 LL **HEADER**	L/360	Sel. Str. DF-L Double 2x10	3.54 ($/ft beam)	Double 1³/₄x7¹/₄ or Single 1³/₄x9¹/₄	5.10 or 3.35 ($/ft beam)	3¹/₂x7¹/₄ (smallest available)	5.75 ($/ft beam)	1³/₄x9¹/₄	2.52 ($/ft beam)	3¹/₂x8⁵/₈ 1.3 E	4.65 ($/ft beam)
GIRDER	L/360	Sel. Str. 2x12 max. span 7'4"	6.36 ($/ft beam)	Triple 1³/₄x14 span 15'	14.67 ($/ft beam)	3¹/₂x16 span 15'	11.30 ($/ft beam)	3¹/₂x16 span 15'	11.76 ($/ft beam)	3¹/₂x18 span 15' 1.5 E	9.72 ($/ft beam)
18'0" 50 LL **GARAGE HEADER**	L/180	Exceeds limit		Double 1³/₄x16	11.22 ($/ft beam)	3¹/₂x16	11.30 ($/ft beam)	3¹/₂x16	11.76 ($/ft beam)	Exceeds limit	
18'0" 20 LL **GARAGE HEADER**	L/180	Exceeds limit		Double 1³/₄x11⁷/₈	8.32 ($/ft beam)	3¹/₂x11⁷/₈	8.25 ($/ft beam)	3¹/₂x11⁷/₈	8.72 ($/ft beam)	3¹/₂x14 1.5 E	7.55 ($/ft beam)

Note: *This chart represents various beam choices and their respective costs.*

simplify this presentation, I chose one species of sawn lumber, Doug Fir-Larch, and four engineered beam options.

Sawn lumber has its limitations. It doesn't clear-span long distances (such as for the garage-door header), and select structural grades are not always available except by special order. Depending on the species and grade, its bending strength (F_b) values are often half to one third that of engineered products. But overall, for short spans, sawn lumber is the most economical choice. Note, however, that in the cases of the ridge beam and central girder, a triple 2x12 beam requires an intermediate post, which the engineered beams don't require.

Laminated veneer lumber (LVL) is strong, stiff, and versatile. It easily spans long distances, and is modular like sawn lumber: To increase load capacity, you can just bolt on a second member. Labor has to be factored in, but on the positive side, two or three workers can often assemble a beam in place that would otherwise require a crane to place. LVL typically comes 1³/₄ inches thick and ranges in depth from 7¹/₄ inches up to 18 inches.

Anthony Power Beam (APB) is a newcomer to the glulam market, and has positioned itself to compete with LVL and Parallam. It comes in 3¹/₂- and 5¹/₂-inch widths; depths range from 7¹/₄ to 18 inches. There is also a 7-inch-wide version available in depths up to 28⁷/₈ inches. APB comes as a full-size beam — that is, it requires no site lamination — so it is fairly heavy. For example, the 18-foot garage header for our sample house weighs in at 380 pounds.

Parallam, manufactured by Trus Joist MacMillan (TJM), is the only parallel-strand lumber (PSL) beam currently on the market. PSL is an assembly of long, thin strands of wood veneer glued together to form continuous lengths of beam. The wood fiber used is strong and stiff. Several widths from 1³/₄ inches to 7 inches are available in depths of 9¹/₄ inches to 18 inches. Like APB, Parallam comes fully assembled and is relatively heavy.

TimberStrand is a "laminated strand lumber" header material from Trus Joist MacMillan. It is made by upgrading low-value aspen and poplar fiber into high-grade structural material. The F_b and E values are certainly no match for LVL, Power Beam, and Parallam, but the performance of TimberStrand is still impressive. It worked for most of the applications in our case house, although the 18-foot garage-door header pushed TimberStrand beyond its structural limit. It comes 3¹/₂ inches wide in depths ranging from 4³/₈ to 18 inches.

Using Beam-Sizing Tables

Sizing information for the Doug Fir-Larch beam came from the *Wood Structural Design Data* (available from the

American Forest & Paper Association, 202/463-2700). Consisting primarily of tables that allow you to accurately size uniformly loaded beams and headers, this is a worthwhile book for any builder to have. Engineered-lumber manufacturers will gladly send you their sizing guides for free.

Engineered-wood span tables for uniform loads are used in much the same way as span tables for sawn lumber. The code allows reductions in live loads based on duration of load. Depending on which manufacturer's literature you refer to, sometimes these reductions are already applied and sometimes not. Typically, shear values are incorporated into the tables, and the required bearing length at the end of the beam is given. Tables for most products are limited to whole-foot spans.

Sizing begins with pounds per foot of beam. But with engineered wood, unlike sawn-lumber tables, you use both the live and dead load values: Live load determines stiffness; total load determines strength. To size engineered lumber:
- determine the total load and live load per foot of beam
- identify the type of load you are supporting (roof snow, roof nonsnow, or floor) and choose the appropriate table.
- pick the span you need
- match the total load and live load values to the values listed in the tables.

The thickness and depth of the member required to carry the load (or a choice of thicknesses and depths) will be listed.

No matter which product you specify, structural performance is controlled by bending strength (F_b) and stiffness (E). Just as sawn lumber of different species varies in strength and stiffness, the same can be true of engineered lumber. An LVL product that has an F_b of 3,100 can handle a greater load than an LVL product with an F_b of 2,400. So be careful to use the correct table when you compare products.

—*Paul Fisette*

Working With Laminated Veneer Lumber

In custom homes with large open floor plans, we've used steel I-beams, steel channels, flitch plates, Parallams, and laminated veneer lumber (LVL). The best choice varies according to the situation, but the product we've found to be most versatile and readily available is LVL.

LVL closely resembles plywood, both in appearance and in the way the two products are manufactured. Veneers are peeled from logs, dried, and then graded visually and ultrasonically. Adhesive is applied to the veneers as they are stacked, and the stacked bundle is fed through a roller press, then sized with a ripsaw.

New Material, New Rules

First-time users of LVL should carefully read the reference guides provided by the manufacturers. Because of LVL's ability

Table 3. Beam Sizing: Comparing the Options

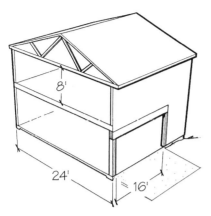

This chart compares various options for the 16-foot garage door header in the illustration at left. The header carries the weight of the walls above (10 psf), second-floor loads (40 psf), and roof loads. For each option, a beam size is given for both a 30-lb. and a 40-lb. snow load. The header is designed to meet an L/360 deflection criterion.

Material	E(psi)	F_b(psi)	Header/30-lb. roof live load	Wt. (lb/ft)	Header/40-lb. roof live load	Wt. (lb/ft)
Microllam LVL (2.0E ES)	2.0×10^6	2,925	2 LVL@1¾x18	18.2	3 LVL@1¾x18	27.3
Versa-Lam LVL (1¾")	2.0×10^6	2,400	3 LVL@1¾x18	24.3	3 LVL@1¾x18	24.3
Versa-Lam LVL (3½")	2.0×10^6	2,640	1 LVL@3½x18	16.2	1 LVL@5¼x18	24.3
Versa-Lam "Plus" LVL	2.0×10^6	2,800	1 LVL@3½x18	16.2	1 LVL@5½x18	25.4
Gang-Lam LVL (So. Pine)	2.0×10^6	3,100	2 LVL@1¾x18	18.02	3 LVL@1¾x18	27.04
Gang-Lam LVL (Western)	1.8×10^6	2,650	3 LVL@1¾x18	23.63	3 LVL@1¾x18	23.63
Gang-Lam LVL (Western)	2.0×10^6	2,950	2 LVL@1¾x18	15.75	3 LVL@1¾x18	23.63
G-P Lam LVL (1.8 E)	1.8×10^6	2,250	3 LVL@1¾x18	24.6	3 LVL@1¾x18	24.6
G-P Lam LVL (2.0 E)	2.0×10^6	2,850	2 LVL@1¾x18	16.4	3 LVL@1¾x18	24.6
Parallam	2.0×10^6	2,900	3½x18	19.7	5¼x18	29.5
Glulam	1.8×10^6	2,400	5x16½	22	5x17⅞	24
Glulam	2.0×10^6	3,000	5x16½	22	5x16½	22
A36 Steel	30.0×10^6	22,000	W8x31 W10x22 W12x19	31 22 19	W8x35 W10x26 W12x22	35 26 22

Note: *design values vary among brands and types of LVL, so it is not possible to substitute one brand of LVL for another without checking the specific manufacturer's design guide.*

CALCULATIONS COURTESY RANDALL ENGINEERING, MOHEGAN LAKE, NY

---

to support long spans, the supporting posts at the end of LVL beams often carry considerable loads. Double trimmer studs are almost always specified, and in some situations, three or more trimmers are required (see Figure 4).

It's essential that plumbing, electrical, and hvac systems are considered during the design stage. Holes, chases, or notches should never be cut in LVL beams unless previously approved by the structural designer, and all subtrades should be made aware of this fact. When LVL is used as a flush beam, these cutting restrictions can present challenges when routing piping and ductwork.

Before ordering LVL, it's important to note the grade, species, and manufacturer specified by the designer. Douglas fir LVL is marginally weaker than yellow pine LVL, but I have yet to encounter a situation in which the species affected the sizing of the LVL beam. Before changing species or manufacturer, however, get approval from the designer.

LVL is available in lengths up to 60 feet, but think twice when ordering long lengths of the material. I learned a painful lesson when a delivery truck carrying 60-foot lengths couldn't make the swing into the building lot from the street. Carrying the 400-pound LVL to the job site was a challenge.

Lead Time Is Important

When using LVL, it pays to plan ahead. While many lumberyards stock a limited supply of the product, only the largest typically stock all the lengths or

Working With Laminated Veneer Lumber

Figure 4. *LVL headers may require three or more trimmer studs to support the increased loads they're capable of carrying.*

Figure 5. *Clamps help with alignment when nailing or bolting together a built-up LVL beam.*

LVL Manufacturers

Boise Cascade
Boise, ID
208/384-6161
www.bc.com

Georgia Pacific Corp.
Atlanta, GA
404/652-4000
www.gp.com

Louisiana-Pacific
Portland, OR
800/648-6893
www.lpcorp.com

Tecton Laminates Corp.
Hines, OR
541/573-2312

Trus Joist
Boise, ID
208/395-2400
www.trusjoist.com

Weyerhaeuser Company
Federal Way, WA
253/924-2345
www.weyerhaeuser.com

widths required for a large project. I place my orders two weeks in advance so the material will be delivered when I need it, and I include any required special connectors in the order.

Connection Hardware

Connectors for just about any situation are available from Simpson Strong-Tie Co. (Dublin, CA; 925/560-9000; www.strongtie.com). However, when assembling LVL hips, valleys, or other angled connections, we've sometimes found it impossible to properly install premanufactured hangers. In these situations, we have custom saddles fabricated, and we through-bolt all connections. It's a good idea to line up a fabricator before you begin a complicated project, so any discrepancies between the drawings and job-site conditions can be resolved quickly.

Weather Protection

In the past, if LVL beams were exposed to the weather, the veneers tended to absorb substantial moisture and were prone to cupping. In one instance, a built-up LVL was exposed to wet autumn weather and cupped so badly we had to through-bolt the entire beam.

Manufacturers now coat their LVL with a waterproof treatment that protects the product from moisture. I prefer the black, waxlike material that Truss Joist MacMillan uses on its Microllam LVL, even though the coating rubs off on hands, clothing, and tools. Be careful: These waterproofing treatments make LVL slippery stuff to handle. The members will often slide in unpredictable ways as they're maneuvered into place.

Despite the waterproof coatings, LVL will still cup when exposed to moisture (although not as severely as in the past). We store LVL well off the ground and cover it with a tarp. It's important to inspect all LVL material when it's delivered but before it's unloaded. More than once, I've rejected cupped LVL that was stored improperly at the lumberyard.

Ganging LVL

With few exceptions, we fasten individual LVL members together to form built-up beams. In dry conditions, when cupping isn't an issue, I hand-nail these built-up beams using five rows of 16-penny hot-dipped galvanized nails, spaced 16 inches on-center. The crew would rather use a nail gun for this chore, but I've found that hand nailing does a better job of drawing the members together.

To keep the edges flush during assembly, we begin nailing at one end of the beam and use a clamp to align the edges as we nail off the beam (Figure 5). If you plan to use a nail gun, make sure you use a strong compressor and a gun with lots of power. LVL (especially the yellow pine variety) is hard material.

On complex projects where LVL will be exposed to extended wet weather before "dry in," I bolt the individual members together to ensure that cupping won't be a problem. Predrilling the bolt holes allows us to assemble heavier built-up beams in place — an important advantage when you consider that a 20-foot, three-member LVL beam can weigh in at 400 pounds. To ensure that the predrilled holes will line up during assembly, we clamp the members together on sawhorses, and drill all of the holes in one session. In the case of flush beams, be careful to lay out any joist framing before drilling, so the bolts won't interfere with joist placement.

Species can make a difference. I recently noticed one of my carpenters fastening two LVL members together using a cordless drill and 3-inch coarse-thread wood screws. My first reaction was to be amazed by cordless technology, but after closer inspection, I realized that the plies used in the LVL were Douglas fir, not the yellow pine I was accustomed to. The Doug fir LVL is easier to fasten, and because it's 10% lighter, easier to handle as well.

—*Ned Murphy*

On Site With Parallam

As builders of custom oceanfront homes, we use a lot of long structural beams. Whenever we build on pilings, as we often do, we use girders to transfer loads to the pilings. Also, increasingly, the architect-designed custom homes we build feature large open spaces and complex roofs that require heavy beams. Anytime the structure of a house requires something stronger, shallower, or longer than we can accomplish with three or fewer 2x12s, our beam of choice is Parallam.

I first used Parallam five years ago in a house that I designed and built for myself. The job went well and I was satisfied with how easily I could integrate Parallam into the frame (Figure 6). Not long after, I built a house where the architect specified that all the beams were to be multiple LVLs (Microllams) spiked together on site. After framing was finished, we noticed that the main girder in a flush-framed floor was sagging. The side-loading from the floor joists was causing the outside LVL in the laminated beam to separate from the other LVLs, putting too much load on the one member. So we had to go back in and install through-bolts to pull the LVLs together. This wasn't a big deal, but from that time on I've requested that architects specify Parallam instead of LVL (Figure 7).

Why Parallam

We like Parallams for many reasons. They are stronger, stiffer, straighter, and longer than nailed-up 2-by beams —

Figure 6. *Unlike steel, Parallam integrates easily into a wood-frame structure, requiring only standard carpentry tools.*

with the added advantage that they don't have to be nailed up on site. On many of the homes I build, I'm able to set three or four main beams, each up to 40 feet long (not clear span), in place of 30 to 50 separate pieces of 2x12. This is a tremendous timesaver, although a crane or lift is a must for beams longer than 16 feet (Figure 8).

Strengthwise, Parallam beams are equivalent to multiple LVL beams, with the added benefit that you don't have to worry about side-loading issues or the expense of bolting. Also, we've often had to return LVL that is cupped or warped, whereas Parallams don't have these problems.

Compared with glulams, which have comparable strength, Parallam is somewhat more available in our market. For

Figure 7. *For a ridge beam, a single piece of Parallam takes the place of a multiple-member LVL beam.*

Figure 8. *A Parallam beam is heavier than an equivalent-strength steel I-beam. The author's crew lifts beams up to 16 feet by hand; beyond that, they use a crane.*

Figure 9. *Parallam comes pressure-treated for exposed conditions. Recently treated beams may vary in size, but will eventually shrink back to the same dimension.*

Both beams and columns are manufactured in lengths up to 66 feet, although in our area, lumberyard trucks can handle only up to 32 feet and we've never been able to get anything longer than 48 feet via flatbed.

Availability is not usually a problem: We can get most sizes in a week or less. Also, Parallams are essentially modular: Two $3^{1}/_{2}$-inch-thick pieces can be nailed or bolted together to replace one 7-inch-thick piece. This can be a timesaver compared with waiting for a back-ordered size. TJM's specifier's guide gives schedules for nailing and bolting, depending on the load.

Although the glue used is waterproof, Parallams are intended for protected use. For outdoor use, CCA-pressure-treated Parallams are available (.4- and .6-lb. retention); we use these extensively in piling foundations (Figure 9).

Design and Planning

It's not uncommon for us to use 20 to 30 Parallam beams on a large (4,000 to 6,000 square feet) house. For these situations, we plan ahead and order early. Usually we give our list to the yard and the beams arrive precut. Because the lumberyard sells in 2-foot increments, we can often save money by having them cut two odd-length sizes from one longer piece.

Before ordering Parallams, we always check the engineer's drawings, looking for places where we can increase efficiency by ordering one longer beam instead of two shorter ones. For example, say the plans show a $3^{1}/_{4}$x$11^{7}/_{8}$ beam and a $5^{1}/_{2}$x$11^{7}/_{8}$ beam arranged in a straight line and ending on the same post. In this case, it's more cost-effective to increase the size of the smaller beam to $5^{1}/_{4}$ inches and set one long continuous beam instead of two shorter simple-span beams. If the beams are in a 2x6 wall, then there's also less furring out to do.

When ordering Parallam, don't forget to order the steel connectors you'll need (Figure 10). The specifier's guide references Simpson connectors, but other manufacturers also make connectors sized for engineered lumber. Make sure the capacity of the hanger is adequate —

an exposed beam, however, architectural-grade glulam is still the best option.

Parallam is more expensive than steel. Sometimes we use steel because of its greater strength, but Parallams are far simpler because they can be cut on site, drilled, and nailed with the tools found in every carpenter's toolbox. In my experience, steel almost always involves tedious extra work to integrate it into the wood frame. If a piece of steel is cut slightly too long, you have a problem; with Parallam, you get out a saw.

Manufacturing and Availability

Parallam currently is the only available parallel-strand lumber on the market. It is manufactured by Trus Joist MacMillan at two plants — one in British Columbia and one in Georgia. The manufacturing process starts with strands of veneer cut from small second-growth trees. These strands are aligned, coated with glue, then cured with a combination of pressure and microwave "cooking." The patented process helps the glue bond with the individual fibers by heating the mass more evenly than is possible with radiant heat methods. The result is a very condensed wood-fiber beam that is stronger and stiffer than solid-sawn lumber.

Parallam is made in widths of $2^{11}/_{16}$, $3^{1}/_{2}$, $5^{1}/_{4}$, and 7 inches and in depths to match the most common wood I-joists — $9^{1}/_{2}$, $11^{7}/_{8}$, 14, 16, and 18 inches. Parallam columns are also available, in a variety of sizes to accommodate standard wall framing — $3^{1}/_{2}$x$3^{1}/_{2}$, $3^{1}/_{2}$x$5^{1}/_{4}$, $3^{1}/_{2}$x7, $5^{1}/_{4}$x$5^{1}/_{4}$, $5^{1}/_{4}$x7, and 7x7 inches.

you'll typically need to carry heavy concentrated loads.

Parallams are perfect for flush-framed floor beams. Single-member beams can be side-loaded with no problem; with laminated beams, make sure the nailing or bolting pattern is specified. Joist hangers typically do not need to be top-flange types, but make sure you install all the nails or bolts specified.

Beware of too many flush-framed floor beams, however: They can be a nightmare for your electrician, plumber, and hvac sub. Our rule on site, and the recommendation in the manufacturer's guide, is "no holes." In truth, within very strict criteria, a very few small holes are allowed, but too many flush-framed floors can greatly complicate electrical and mechanical installations.

If you're using the specifier's guide to make your own Parallam selection, be sure to allow enough bearing length for each end of the beam. The large loads carried by the beams will of course concentrate at the bearing points. Too little bearing surface area will result in crushed grain and the beam will settle.

Handling and Installation

As with any lumber, it's possible to mistreat Parallam. If you leave an untreated Parallam in a mud puddle, it will turn gray, absorb water, and eventually show some signs of delamination. After the beams arrive on the job site, they should be protected from the weather and stored flat. After installation, Parallam will show normal signs of weathering. This is nothing to worry about, but if you notice any delamination of fibers, ask to have the manufacturer's rep inspect the beam.

We have noticed some size variation with CCA-treated Parallams, depending on how recently the beams came from the treatment tank. On a recent job, some "14-inch" beams ranged in size from 14 to 14 1/2 inches. TJM engineers assured us that in a few months the oversized beams would all shrink back to 14 inches. Within a week or so, the discrepancy had narrowed to about 1/8 inch, so we set all the beams even at the bottom, assuming continued shrinkage would fix the problem.

For very long lengths, you will need a crane or lift to unload and set the beams, although we have successfully handled lengths up to 16 feet with four men.

Cutting Parallam is about the same as with sawn lumber. Circular saws work fine, although you may need to cut from both sides and finish the cut with a handsaw. On jobs where we have to do a lot of cutting, we usually rent a 10-inch circular saw so we can cut most sizes in one pass. Drilling and nailing are no different than for solid lumber.

Manufacturer Support

Technical assistance with the sizing of Parallams is readily available from the manufacturer. The specifier's guide is easy to understand and is adequate for most standard applications. If your plans have not been stamped by an engineer and you think there's a problem, you can get help directly from TJM's staff engineers. Call and explain your application or fax a plan, and they will advise on sizing, spans, connection hardware, drilling, and other details. Design software is also available, but generally is only needed by engineers, architects, and high-volume users.

By Patricia Hamilton, owner of Boardwalk Builders in Rehoboth Beach, Del.

Parallam vs. Glulam & Steel

On a recent custom home, the author used Parallam for a grid of 18-foot clear-span beams. The chart shows how Parallam stacks up against equivalent-strength steel and glulam.

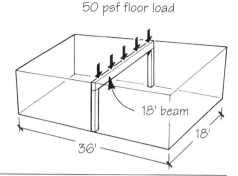

	Size	Weight (plf)	Cost
Parallam	7x14	30.6	$389
Steel	W12x19	19	$303
Glulam*	7 1/4 x 14	26.8	$347

*Southern pine 26F-V4, Industrial-grade

Figure 10. *Steel connector manufacturers carry special hardware sizes to match the dimensions of engineered wood products like Parallam and I-joists. To avoid delays, be sure to order the connectors you need when you order the beams.*

Building With Glulams

Though wood I-joists and laminated veneer lumber (LVL) will do for many long-span applications, glued laminated timbers, or glulams, are the only substitute for large, exposed solid timbers. In this section, I'll share some of what I've learned from ten years of using glulams for headers, joists, cathedral ceiling rafters, and exposed trusses.

Why Glulams?

Glued laminated timbers are made by face-laminating dry lumber of 1- or 2-inch nominal thickness under controlled temperatures and pressures. Glulams can be made to nearly any width, length, depth, or strength. And because their laminations are oriented horizontally, the fabricator can make them with nearly any curve or camber. As with any beam, most of the stresses on a glulam are concentrated along its top, or compression edge, and its bottom, or tension edge. (There's little stress along the beam's centerline.) But unlike I-beams, which derive their strength from their wide top and bottom flanges, glulams derive their strength from the higher-quality lumber used in the top and bottom laminations.

A glulam is a good choice when you can't get sawn lumber or an LVL in the size needed for a given span; when the structure doesn't give you enough depth for a solid sawn beam or girder; when you need a high-quality finished appearance; or when you have to support concentrated loads at midspan (a task that requires special engineering).

Ordering Glulams

I buy glulams only from manufacturers who belong to the American Institute of Timber Construction (AITC), since members must comply with stringent manufacturing standards. Beams from these companies are stamped with a label that outlines their structural characteristics. Glulams made by nonmember firms may or may not comply with these standards, so it may be impossible to determine their true strength.

Grades and materials. Several "appearance" grades are available. (Appearance is unrelated to strength.) From worst to best, these are Industrial, Architectural, and Premium Grade. To save money, use the most economical appearance grade acceptable in each location, and prefinish only those members that will be exposed in the completed building (see Figure 11).

Douglas fir is the standard glulam material in the western United States. Southern pine is the most popular species in the East, though other species, including Doug fir, may also be available. Some fabricators make pressure-treated members for exterior use.

Stock glulams. On most residential jobs, the builder estimates the headers and beams needed as if they were sawn beams, then orders stock glulams of equivalent strength from the

Figure 11. *Use Architectural or Premium Grade glulams for exposed beams (left), and less expensive Industrial Grade for concealed beams (right).*

lumberyard. You can't go wrong by using an equivalent size, but since glulams are stronger than solid wood, you can probably use a smaller member. Consult a glulam span table, available from AITC (see "For More Information," at end of section). Straight stock glulams come in widths of 2 1/2 inches, 3 1/8 inches, 5 1/8 inches, 6 3/4 inches, 8 3/4 inches, and 10 3/4 inches. Depths are multiples of the lamination thickness. Thus, a member with eight 1 1/2-inch laminations (typical of straight beams) will have a total depth of 12 inches.

On a big job that requires a lot of glulams, it's a good idea to use the same size throughout whenever practical. If the plans show several glulams with slightly different sizes, ask the architect or engineer if you can use the largest size everywhere. This will raise material costs somewhat, but I've found that it saves money and time because it makes things a lot easier for the workers on the site.

Custom glulams. Stock glulams are fine for most simple headers and joists. But if you need beams that will carry heavy loads over long spans (a heavily stressed girder, for example, or a set of flat roof joists that will carry a heavy snow load), it's better to send your plans to the fabricator, whose engineers can determine if you need a custom member. Unlike stock glulams, which are cut from perfectly straight, 60-foot lengths, custom members are made with a subtle camber. The camber approximates the anticipated long-term deflection, so that the beam will straighten, rather than sag, over time. Another advantage to getting a beam custom made is that the fabricator can make any custom connectors, and can predrill bolt holes and precut birdsmouths for rafters. Glulam hip rafters can even have their top edges beveled to receive roof decking — the shop will do a better job than you could on site. (Even if you would rather bevel the rafter yourself, it's a good idea to tell the fabricator what you're doing. They can place some additional high-strength laminations on the top of the beam so that you can bevel it without weakening it.)

Before making custom glulams, the fabricator will send you a set of shop drawings showing the overall size and shape of each member, as well as any cuts, bolt holes, or hardware. The drawings will also list the appearance grade and species, the glulam's design values, the adhesive used, and the finish or treatment.

The disadvantage of ordering custom is that you'll pay more and wait longer to get your beams. (I've seen lead times of up to three months.) It's best to get quotes from different suppliers, and to place your order before digging the foundation hole.

Handling and Assembly

Putting up one or two headers is no harder than with dimensional lumber. But if you're building a complex roof or a glulam addition (a solarium, for example), then good planning is crucial.

Make sure that the delivery truck can access the site, and find out whether it's a boom truck. If not, you'll need to schedule a crane to unload the beams. When lifting a glulam with a crane, always use slings to keep from scratching it (Figure 12).

Glulams normally come wrapped in a housewraplike material (Figure 13). Take

Figure 12. *Unless your delivery truck has a boom, you'll need a crane to unload and place large glulams. Use slings to protect the beams.*

Building With Glulams

Figure 13. *Glulams come to the job site wrapped in protective material. To prevent uneven weathering and stains, it's best to unwrap the whole beam at one time.*

the wrapping off the whole beam before assembly. Otherwise, sunshine and moisture may cause uneven weathering between covered and uncovered areas. A partially open package also invites moisture to get inside. At the job site, all glulams should be stored on blocks and separated from one another with wood stickers. Of course, it's best to schedule the job so as to keep the beams out of the weather completely, which means the best time for a set of glulam rafters to arrive is the day you're ready to install them. The roof should be sheathed and papered immediately after the glulams are placed.

On complex jobs, carefully check all anchor bolt, column, and base plate layouts, making sure that everything will fit together the way it's supposed to. And make sure you understand the proper assembly sequence. On some glulam trusses, for example, the hardware used makes it necessary to assemble the small web pieces before finishing the chords. Make sure that the designations in the drawings match the marks on the glulams themselves. (These marks should be located so that they'll be concealed in the finished building.) Each connector and glulam should also be given the same mark as in the shop drawings.

Glulam frames must be braced during assembly and until the structural system, including the decking, is complete. Arrange temporary braces so that their fasteners don't mar the glulams' exposed faces. If there will be a lot of traffic during construction, protect the corners of posts and headers by taping on temporary guards made from wood strips or heavy cardboard.

When sheathing glulam rafters or joists, snap the centerline of each glulam on the sheathing before installation and keep your fasteners on these lines. This will keep the fasteners from splitting the edges of the glulam. When the job is finished, remove dirt and dust from exposed surfaces with a detergent that's appropriate for the particular finish (check with the fabricator).

Making Connections

Part of a successful glulam job is knowing what connections to make where. Glulams can be fastened to each other and to other parts of the structure in a variety of ways (Figure 14). Connector manufacturers make joist and beam hangers, column caps and bases, and saddle hangers specifically for glulams. Most of these fall into one or more of the following categories:

- **Direct bearing connections**, the most common type, are those in which one structural member bears directly on another — a beam on a column, for example. Fasteners merely hold the members together until the walls, floors, and roofs are framed and sheathed. Direct bearing connections are inappropriate for joints that must resist lateral loads or uplift, such as the eaves of a roof (T-plates at beam-column intersections are often used for this purpose).

- **Indirect bearing connections** are those in which a steel connector supports one member and transmits the load to the other member. These include joist hangers, saddle hangers, and column caps where the plate supporting the beam is larger than the top of the column. An oversized column cap is useful where a heavily loaded floor beam

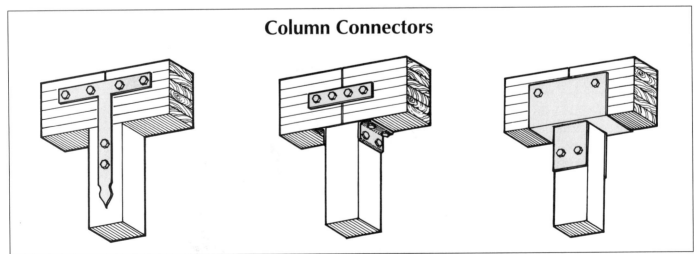

Figure 14. *T-plate connectors (left) provide alignment and resistance to uplift. Where two heavily loaded beams meet over a column, steel angles provide extra bearing while straps tie the ends together (center). Where a heavily loaded beam sits on a post, a welded steel U-plate spreads the load and prevents the beam's wood fibers from crushing (right).*

Figure 15. *In metal side plate connections, such as the ones at the top of this glulam post (left), the bolts must be adequately sized to transfer the loads to the metal plates. Cantilever connectors (right) are commonly used to join glulams end to end.*

sitting on a 6x6 column might get crushed at the point of bearing. The column cap's plate prevents this by spreading the load over a wider area.

- **Layered connections with wood side plates** are those in which overlapping wood members are fastened through the zone of overlap — for example, where a jack rafter meets a hip. With large members like glulams, these connections should be fastened with through-bolts or lag screws, not nails. When the loads get very high, consider using split rings or shear plates. These spread the loads over a larger area than bolts alone. These connections must be engineered; they're made by the glulam fabricator using precision cutting tools.

- **Layered connections with metal side plates** are very common (Figure 15). The side plates are bolted in place through the glulam, and transfer loads from one glulam to another. Metal side plates are generally $1/4$ inch or $1/2$ inch thick, which makes the overall assembly relatively slim. This type of connection is particularly useful when a number of glulams converge at a single joint.

- **Timber joinery** is seldom used with glulams, because of the additional labor and engineering required. Also, the loads that dictated that you use glulams in the first place are generally too high to be handled by traditional mortise-and-tenon construction.

Always use galvanized fasteners with glulams. Even an interior fastener could see excessive moisture during construction; if it rusts, it may stain the glulam.

Detailing for Moisture

Glulams are subject to many of the same moisture problems as sawn lumber. Here are some tips on how to compensate for them.

Moisture-related expansion. While glulams won't twist like solid timbers when exposed to moisture, they will expand, especially in width. Therefore, the pockets provided for them in steel connectors should be slightly oversized. An extra $1/8$ inch to $1/4$ inch is usually enough. This doesn't weaken the connection, since the bolt, not the plate, is what provides the connection's strength.

Never overtighten a bolt, or else it will crush the side grain of the member when the wood expands with moisture changes. Unless steel side plates are present, use washers on all bolts to prevent the bolt head and nut from crushing the wood fiber.

Drying-related checking. Most glulams pick up some moisture during construction. As they dry, they have a tendency to check; the bigger the member, the worse the checking. To minimize this, dry the building out as slowly as possible, especially if you have large exposed members. You might want to turn the heat up gradually by a few degrees each day, starting around 30°F.

Moisture absorption. Like sawn lumber, glulams will wick moisture from any damp surface they touch. Always leave at least $1/2$ inch between a glulam (particularly its cut end) and any masonry or concrete. Steel shoes, column bases, and other connectors that might collect water during or after construction should have weep holes drilled in them.

When you cut a glulam, you should immediately seal the end grain with a water repellent. To prevent mistakes here, it's best to have as much of the drilling, cutting, and sealing as possible done in the shop.

By Eliot W. Goldstein, an architect in West Orange, N.J. Mr. Goldstein serves on the American Forest and Paper Association committee that writes guidelines for engineered wood structures.

For More Information

The following organizations publish technical information on glued laminated timbers.

American Forest and Paper Assn.
Washington, DC
202/463-2700
www.afandpa.org

American Institute of Timber Construction
Englewood, CO
303/792-9559
www.aitc-glulam.org

APA — The Engineered Wood Assn.
Tacoma, WA
253/565-6600
www.apawood.org

Southern Pine Council
Kenner, LA
504/443-4464
www.southernpine.com

Western Wood Products Assn.
Portland, OR
503/224-3930
www.wwpa.org

Engineered Studs for Tall Walls

To strengthen stud walls for increased loading or to reduce deflection in tall walls, you can decrease stud spacing, increase stud thickness, or use doubled or tripled studs. With ordinary dimensional lumber, however, achieving a stiff stud wall that is also perfectly flat is easier said than done, particularly when longer studs are needed. Typical workarounds include nailing together two studs with crowns opposed, sistering steel channels to individual studs, or, in extreme cases, fabricating a steel moment frame.

New alternatives. A better answer may be found in the latest crop of engineered wood materials. If you're framing a tall wall, or a wall with a lot of large openings in it for doors and windows, or if you just need to make sure the wall is perfectly flat (to accept cabinets and countertops, for instance), consider using one of three types of engineered studs: finger-jointed, laminated strand lumber (LSL), or laminated veneer lumber (LVL).

Although all three types of engineered studs can be used throughout a building's frame, the higher cost of these materials has led some builders to confine their use to specific applications for which the superior strength, straightness, and stability, together with the availability in long lengths, make engineered studs a good choice (see "Design Values," opposite page). For example, window walls are often underdesigned to resist buckling and shear forces from winds. Since these walls must be balloon-framed, you need long, stiff material. Engineered studs may fit the bill, depending on the number of studs that can be fit into the wall and on whether or not the window layout allows you to run continuous structural studs from plate to plate. Long continuous studs won't fold at the "hinge" created by platform framing, and the added strength of the engineered stud material will counter the tendency of taller balloon-framed studs to flex.

Engineered studs, especially the LVL type, are considerably stiffer and can carry significantly more weight than dimensional lumber, making them a good choice for fitting between window openings. The higher design values give you the ability to make a stiff frame in a relatively small area.

Don't forget, however, that window walls must also resist shear, either with lateral and diagonal bracing or with structural sheathing. Windows reduce the area on a wall that can be used for shear bracing. If there's insufficient area for shear panel bracing, you may need to rethink the design or use a moment frame.

Finger-Jointed Studs

Currently manufactured by Louisiana-Pacific and Georgia-Pacific, finger-jointed studs are available in 2x4s and 2x6s. The studs are made by end-glueing comparatively short pieces of lumber using a Type I polyvinyl acetate

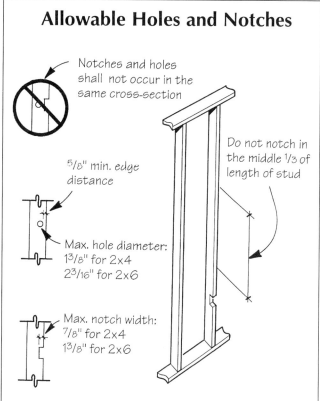

Allowable Holes and Notches

Notches and holes shall not occur in the same cross-section

5/8" min. edge distance

Max. hole diameter:
1 3/8" for 2x4
2 3/16" for 2x6

Max. notch width:
7/8" for 2x4
1 3/8" for 2x6

Do not notch in the middle 1/3 of length of stud

Certain restrictions may apply with regard to drilling holes and cutting notches, as shown in this illustration, which is similar to one supplied with Timberstrand studs. Whenever you use engineered lumber, consult the manufacturer's specs for design values, load tables, and nailing requirements.

glue. (L-P uses stud-grade lumber in three species mixes: S-P-F, Western Woods, or Douglas Fir/Larch; G-P uses three grades of SYP: stud, No. 1, and No. 2). Unlike the finger joints used for glulam headers, which are oriented parallel to the short side of a board's cross-section, structural studs for bearing wall applications have the joints oriented parallel to the long cross-sectional dimension. The quality of the glue joint is graded by the Western Wood Products Association (WWPA), and the material is stamped "Stud Use Only," which limits use to vertical end-loaded applications. The glue is water resistant, but these studs should be protected from prolonged exposure to the weather. The glued joints meet design requirements for short-term bending loads for lateral forces such as wind, seismic, and impact.

Finger-jointed studs aren't any stronger than sawn lumber of the same

Table 4. Design Values for Engineered Studs

	Finger-Jointed					LSL	LVL
Manuf.	Louisiana-Pacific		Georgia-Pacific			Trus Joist MacMillan	Louisiana-Pacific
Grade	Stud	Stud	Stud	No. 2	No. 1	Timberstrand	Gang-Lam
Species	S-P-F	Doug. Fir/Larch	Mixed So. Pine	Mixed So. Pine	Mixed So. Pine	Aspen/Poplar	Southern Pine
E	1,200,000	1,400,000	1,200,000	1,400,000	1,500,000	1,300,000	2,000,000
F_b	675	675	775	1,300	1,450	1,700	3,100
F_t	325	450	450	775	875	400	2,300
F_v	70	95	90	90	100	150	290
$F_{c\parallel}$	675	825	950	1,650	1,650	1,400	3,180
$F_{c\perp}$	425	625	565	565	565	300	1,020

Abbreviations
- E Modulus of elasticity, psi
- F_b Bending, psi. Values are for single studs; for repetitive use, multiply Fb by 1.15.
- F_t Tension parallel to grain, psi
- F_v Horizontal shear, psi
- $F_c\parallel$ Compression parallel to grain, psi
- $F_c\perp$ Compression perpendicular to grain, psi

Finger-jointed studs have the same design values as the species from which they are made; values for LSL and LVL are significantly higher. Bear in mind, however, that tall walls and heavy loads create specific structural requirements, so these applications should always be engineered.

Engineered Studs for Tall Walls

species, but because the defects have been cut out, the studs are more dimensionally stable and are less prone to split and warp. And because the grain is interrupted by glue joints, finger-jointed studs are less likely to twist and bow. In addition to the obvious advantage of straight, stable 2-bys in long lengths, finger-jointed studs have no wane and no large knots, a real benefit when toe nailing. And while the surface of finger-jointed studs may not be finished — for example, there may be glue smears and the joints may be offset by as much as 1/16 inch — these cosmetic blemishes don't affect structural integrity.

Prices vary, but in general, finger-jointed studs cost about the same as solid-sawn lumber of the same species.

LVL

By now, most builders are familiar with LVL beams and headers, but not many have used the material for studs. Cost is one reason — LVL studs are nearly twice as expensive as solid-sawn studs. But the extra expense is worth it for some applications. LVL studs, which are manufactured and sold by Louisiana-Pacific under the name Gang-Lam Studs, are particularly useful for bearing walls under large compression loads — the modulus of elasticity is nearly twice that of solid-sawn and S-P-F finger-jointed studs, and about 50% higher than that

of LSL and SYP finger-jointed studs.

Moisture can be a problem. LVL studs leave the factory at about 5% moisture content, so they are likely to wick up moisture on site. As with headers, studs may bow if one side gets significantly wetter than the other. To prevent this, you should take precautions when storing LVL, and dry-in the structure as soon as possible.

LSL

Laminated strand lumber studs are manufactured by Trus Joist Macmillan as part of the company's Timberstrand Frameworks Wall System. In a manufacturing process similar to the one used for Parallam beams, strands of aspen or yellow poplar are mixed with a polyurethane adhesive and formed into billets in a steam injection press. The 3 1/2- and 5 1/2-inch-thick billets are then sawn into 11 1/2-inch strips to make nominal 2x4 and 2x6 studs of varying lengths.

The finished product is uniform and very stable. However, TJM's product literature states that the studs are "not treated and are intended for use in a protected (enclosed) condition where the maximum moisture content does not exceed 19%." In other words, Timberstrand studs need to be kept dry. We tested a 12-inch-long sample by partially immersing it in water overnight. The material initially swelled almost 3/4 inch in thickness, but after about a week of drying, it returned to within 1/8 inch of its original dimension. While an LSL stud is unlikely to get this kind of prolonged soaking on site, it's prudent to store this type of framing material under cover and to get it dried in as soon as possible after installation.

As part of an engineered wall system, Timberstrand should never be substituted one-for-one for conventional framing materials. In addition, the studs may require special treatment, depending on the situation. For example, there are strict requirements for blocking in tall walls: Walls between 10 and 18 feet tall require one row of blocking at midspan; walls between 18 and 22 feet tall require two rows of blocking, one

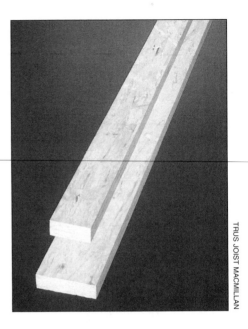

each at the third points. Let-in bracing is allowed as long as the maximum depth of cut in the stud is 7/8 inch; studs that are notched in the middle third must be doubled.

Timberstrand studs cost about 50% more than solid sawn material of the same size. As with finger-jointed studs, this is one reason the builders we spoke with confined use of LSL studs to tall walls and to interior partitions that needed to be as flat as possible to accept cabinetry—typically in kitchens and bathrooms.

Callback Savings

Most builders who use engineered studs, use them selectively because of the higher cost. Some find that engineered studs more than pay for themselves. One Minnesota contractor uses engineered studs for all exterior and interior walls behind kitchen and bathroom countertops, and for all door trimmers. With dimensional studs, walls have moved up to 1/4 inch, leaving an unsightly gap. Using engineered studs, he saves on callbacks to fix gaps behind backsplashes or sagging doors caused by shrinking trimmers. He says that saving just one callback easily justifies the cost of using engineered studs in selected parts of the house.

In fact, he would use engineered lumber throughout the house if the cost difference were lower. Right now the added cost for engineered lumber ranges from

20% to 50% more than dimensional studs. But, he says, engineered lumber pricing is much more stable. Like other builders, he started using engineered studs a few years ago when lumber prices were high. There was little difference in cost, but because supply was low he had to discontinue using it for four or five houses. That's when he especially noticed a big difference in quality between the engineered stock and the dimensional stock, and has been using them selectively ever since.

JLC Staff

Sources of Supply

Georgia Pacific Corp.
Atlanta, GA
404/652-4000
www.gp.com

Louisiana-Pacific
Portland, OR
800/648-6893
www.lpcorp.com

Trus Joist
Boise, ID
208/395-2400
www.trusjoist.com

Chapter Thirteen
STEEL FRAMING

Steel Beam Options 230

Strengthening Beams with Flitchplates . .232

Attaching Steel Beams and Columns 234

Installing a Steel Moment Frame 236

Hybrid Framing with Wood and Steel . . .238

Steel Beam Options

From time to time, we face a situation where we just don't have room for enough wood to hold things up — for example, where a second-floor girder meets a wall directly above a sliding glass door. This girder supports second-story floor joists as well as a bearing wall carrying the roof structure. This requires tons of support for the end of the girder. Because the girder rests above the slider, a beam is needed to carry the girder's load to solid wall framing at either side of the door. But there just isn't room for a wood header large enough to carry the point load from the girder.

So we use a piece of steel instead. A properly sized steel beam has sufficient strength to carry the load a few feet over from the head of the glass door to where the wall can provide sufficient vertical support. It may be heavy stuff, but a little steel can hold up a lot of building. When we don't have enough room to do it with wood, steel is an attractive option.

Wood vs. Steel

Let's compare wood and steel. One foot of 2x12 weighs about 4 pounds and costs approximately 70¢. A steel beam of the same depth (11 1/4 in.) and strength would be only about 1/10 of an inch thick — 1/15th of the volume of the 2x12. This steel beam would also weigh roughly 4 pounds per foot and would cost about $1.80. (It was a big surprise to me to find out that wood is just as strong as steel, pound for pound.) However, the steel beam is about 35% stiffer than the wood one; which means that the sag in a steel beam is only about three quarters of the sag in an equivalent timber.

Besides the numbers, there are several other clear differences between wood and steel. Unless you have a lot of spare time, you can't cut steel with hand tools, or even with light power tools. You can't pound nails into steel. However, welded joints in steel are quite stiff, whereas wood joints flex. Steel also doesn't warp and change dimensions with humidity, although it does expand with heat. And of course, steel doesn't rot, but then wood doesn't rust.

All of this suggests the obvious: Build houses of wood and use a little steel in the tricky spots.

When Wood Won't Work

Let's look more closely at the example in Figure 1. The girder is 20 feet long. It's supported by an interior bearing wall at one end, and by the door header at the other end.

The girder supports loads from the roof and from the second floor. In my area of western Oregon, the roof design load is 35 psf — 25 pounds for snow and 10 pounds of dead load for the roof structure and covering. Since the house is 30 feet wide, the total load from the roof onto the girder is:

P1 = 20 ft. x 15 ft. x 35 psf = 10,500 lb.

The load from the floor onto the girder is calculated in generally the same way. The floor load is 40 psf for the live load and 10 psf for the floor itself, for a total load of 50 psf. The girder carries half of the load on each side (the other half is carried by exterior walls), so the load from the floor onto the girder is:

P2 = 20 ft. x 15 ft. x 50 psf = 15,000 lb.

So the total load on the girder is

P1 + P2 = 25,500 lb.

Since each end of the girder carries half this load, the force on the header is 12,750 lb.

In addition to the girder point load, the header also carries some weight from the exterior wall located directly above it all the way to the roof. For an all-wood structure, this dead load — the combined weight of the wall materials — can often be safely ignored because the plywood-sheathed wall acts like a huge box beam and safely carries these vertical loads to the foundation. There are cases — with stucco or brick cladding, for example, or extra-long headers — where the weight of wall materials should be considered.

We need a header that can span 6 feet. If the header were select structural Doug fir-larch, five 2x12s would be required to support it. With a weaker lumber like No. 1 hem-fir, you would need seven 2x12s. But in our example,

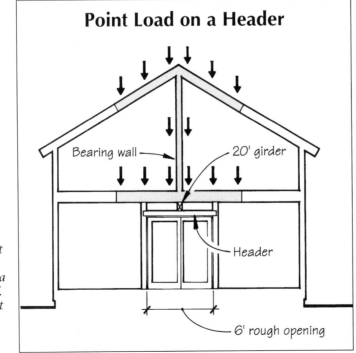

Figure 1. *In this example, a 20-foot first-floor support girder also carries a load from the roof. The resulting point load onto the window header is 12,750 pounds.*

230   *Chapter Thirteen: Steel Framing*

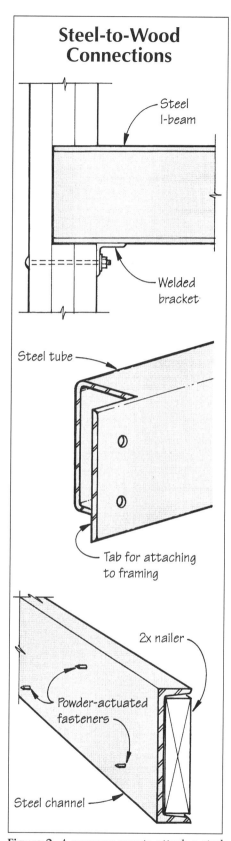

Steel-to-Wood Connections

Figure 2. *A common way to attach a steel header to wood framing is to have the steel yard weld on an attachment angle bracket (top). The steel cutter can also leave an attachment tab on the ends of a piece of steel, as with the tube shown here (middle). On site, the quickest method is to fasten a 2x nailer to the steel using powder-actuated nails (bottom).*

Table 1. Structural Steel Headers

Point load, lb.	Span 3'	4'	5'	6'
5,000	T3x3x5/16 (1.88)	T4x4x3/16 (2.50)	T4x3x1/4 (3.13)	C6x8.2 (3.75)
10,000	C6x8.2 (3.75)	C6x10.5 (5.00)	W6x12 (6.25)	T6x3x3/8 (7.50)
15,000	C6x13 (5.63)	T6x3x3/8 (7.50)	T7x3x3/8 (9.38)	W8x15 (11.25)
20,000	T6x3x3/8 (7.50)	W6x16 (10.00)	W10x15 (13.00)	W10x17 (15.00)

This table gives examples of standard steel (A36) shapes and sizes that will work as headers to carry heavy point loads where height is limited. Each of these pieces of steel can be worked into a 2x4 wall, although the W beams are a full 4 inches wide and require a little work to fit. The number in parentheses is the section modulus required for the given load and span; the steel section above it is the shallowest and lightest piece of steel that will work. Substitutions may have to be made because most steel yards do not carry every size and shape.

there's no room for 2x12s. Because of the door height and the depth of the girder, there's only 8 inches of room left for the header — making steel an obvious solution.

Steel to the Rescue

For headers, there are three steel shapes that work well in 2x4 framing: tubes, channels, and 4-inch-wide flange I-beams (Figure 2). My favorite is the tube, because it can have a two-by nailed to either the wide face or the top, depending on the situation. The channel can be used this way as well and is sometimes a little lighter. Most steel yards have a variety of angles and wide-flange beams available, though these are more difficult to fit into the wood framing. Flitch plates are another common way to strengthen built-up wood headers (see "Strengthening Beams with Flitchplates", next page).

In our design example, a W8x15 does the job. Table 1 gives a range of steel options for a variety of spans and loads. The table is not meant to be used as a substitute for proper structural design — steel always requires careful engineering. The point is to show that many options exist. Much depends on what your local steel yard stocks — by substituting, you can generally find a shape and size that will work.

Installing Steel Headers

Headers should be supported by at least double 2x4 jack studs for up to a 12,000-lb. load, and triple jacks for loads above 12,000 lb. up to 18,000 lb. The header should sit directly on the end grain of the jack studs. Don't use wood shims under the header — the load is perpendicular to the grain and tends to crush the wood.

There are several ways to attach steel to wood, depending on the circumstances. I mention nailing 2x4s to tubes and channels — this is done with powder-actuated nails. The nailed-on pieces provide a tie-in point for the surrounding wood framing.

The steel can also be predrilled at the yard for tabs or L-brackets, or tabs can be welded on. For slim-profile tube headers where you don't have room to nail on lumber, ask the steel cutter to leave a tab sticking out at each end. Then notch the framing to fit and nail or lag the steel to the framing.

—Harris Hyman

Strengthening Beams with Flitchplates

Flitchplates (sometimes mistakenly called fish plates, or flinch plates) are steel plates sandwiched between wooden members to provide increased strength or stiffness. Flitchplates are primarily used in header assemblies when common framing lumber is not strong enough or stiff enough to carry the loads. In such cases, steel beams, laminated veneer lumber (such as Microllams), or glue-laminated beams can also be considered. But in some situations, flitchplates can offer some advantages over these materials:

- Because the steel in a flitch beam is sandwiched between wood framing lumber, the beam can be readily connected to other wood framing with nails and lag screws.
- As headers over multiple doors, wide sliders, or window assemblies, flitchplates are pretty much free of "creep," or the slow sagging that often affects wooden headers and results in doors and windows binding or finish materials cracking. This is the most common reason I choose flitch beams.
- Flitchplates can be assembled in place, reducing the lifting weight of individual beams.
- Flitch beams can be easily retrofitted into existing wood-framed structures to solve problems, such as large notches left in the framing by a plumber (see Figure 3) or sagging members.
- The fire resistance of a flitchplate is generally high.

There are also some disadvantages:

- The cost of flitchplates is usually higher than the cost of most alternatives. Steel is sold by weight, which is often substantially greater in a flitchplate than in a wide-flange beam of equal capacity. Typically the weight of a flitchplate will be 50% to 100% greater than a wide-flange beam.
- Flitchplates are typically thicker than the web of a wide-flange beam, so it is harder to cut holes for pipes and wires.
- When floor framing connects to the side of a flush-framed flitch beam, you may have to use joist hangers, whereas if you are connecting to a flush-set wide-flange steel beam, the loads can be carried directly by the bottom flange.
- End connections of flitch beams require special attention (and lots of bolts!) to properly distribute the load into the fibers of the wood members.

Flitch Beam Makeup

A flitch beam usually consists of two wooden members, one on each side of a single steel plate, or three wooden members with two steel plates, all bolted

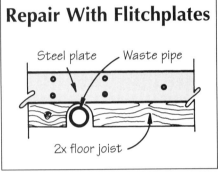

Figure 3. *Flitchplates are useful for repairing existing wood beams, like this floor joist that has been poorly notched for a plumbing pipe.*

Table 2. Equivalent Strengths: Flitch Beams vs. Built-Up Wood Beams & Steel I-Beams

Single Steel Flitch Beam		Built-Up Lumber Equivalents*	Steel I-Beam Equivalents	Double Steel Flitch Beam		Built-Up Lumber Equivalents*	Steel I-Beam Equivalents
Wood	Steel			Wood	Steel		
2x6	5" x 1/4" 5 x 3/8 5 x 1/2	2.3 - 2x6s 3.5 - 2x6s 4.7 - 2x6s	– – –	2 x 6	5" x 1/4" 5 x 3/8 5 x 1/2	4.7 - 2x6s 7 - 2x6s 9.4 - 2x6s	– – –
2x8	7 x 1/4 7 x 3/8 7 x 1/2	2.8 - 2x8s 4.2 - 2x8s 5.6 - 2x8s	– – W6x9	2 x 8	7 x 1/4 7 x 3/8 7 x 1/2	5.6 - 2x8s 8.4 - 2x8s 11.3 - 2x8s	W6x9 W8x10 or W6x12 W8x13
2x10	9 x 1/4 9 x 3/8 9 x 1/2	2.9 - 2x10s 4.3 - 2x10s 5.8 - 2x10s	– – W8x10	2x10	9 x 1/4 9 x 3/8 9 x 1/2	5.6 - 2x10s 8.4 - 2x10s 11.5 - 2x10s	W8x10 W8x15 W8x18
2x12	11 x 1/4 11 x 3/8 11 x 1/2	2.9 - 2x12s 4.4 - 2x12s 5.8 - 2x12s	– – W10x12	2x12	11 x 1/4 11 x 3/8 11 x 1/2	5.8 - 2x12s 8.8 - 2x12s 11.7 - 2x12s	W10x12 W10x17 W10x22

* The strength equivalent shown for built-up lumber beams assumes Doug-fir-larch (E=1.6, Fb=1,200).

tightly together. It is seldom necessary to use a torque wrench to bolt the beams together, as long as the bolts are all cinched down snugly.

The steel is put in the middle for two reasons. First, this increases fire resistance. Wood can resist fire longer than steel, which softens very quickly in a fire. Second, the steel is not likely to buckle if placed between two wood members. A thin plate of steel will wrinkle and buckle under relatively small loads unless stabilized.

I usually specify a flitchplate that is slightly narrower than the lumber in a beam, so if the lumber has a large crown, the steel doesn't overhang the wood and interfere with the drywall. I use an 11-inch flitchplate with 2x12s, a 9-inch plate with 2x10s, etc. However, because the steel is slightly smaller than the wood, the end connections must be engineered to carry specific loads to prevent the wood at the bearing points from crushing.

As the case studies show, three to six bolts are usually required. This calculated bolting schedule is required for the bolts to transmit the load from the plate onto the wood members, which in turn bear on the posts.

As a rule of thumb, the bolt holes should be placed 2 inches from all edges of a beam. Use the flitchplate as a template to precisely align the holes in the wood. Insufficient bolting and poorly sized bolt holes will cause the wood around the bolts to crush and the plates to settle. In extreme cases, the steel can knife into the top of the post.

An alternative to an extreme end-bolting schedule is to hold the flitchplates flush to the bottom of the beam and insert a steel bearing plate on top of the post. In this case, however, you can't toenail the beam into the post. But don't ignore this connection for this reason. Use a steel tie plate or angle bracket to secure the beam to the post.

The steel plate and lumber dimensions for flitch beams are based upon loading. When sizing the flitch beam, I ignore the strength of the wood since it is much less than the strength of the steel. Also, wood is subject to creep,

Case Study 1

Note: *To carry the heavy loads illustrated on the left, the flitchplate must be sized and bolted as shown at right.*

Case Study 2

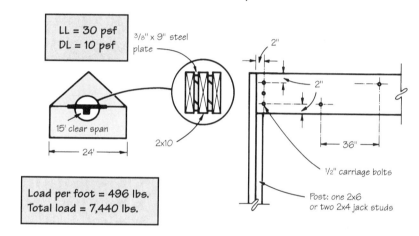

Note: *Although a relatively heavy flitch beam is required due to the large span, the load it carries is relatively small, so the bolting schedule at the ends is not too extreme.*

Case Study 3

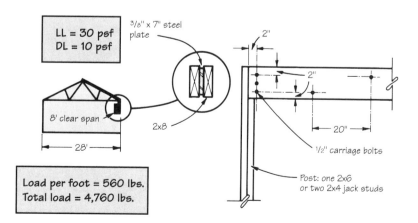

Note: *In this case, flitchplates would be chosen over other beam materials because of the vertical height restrictions of a large window or a low ceiling. To get the same load-carrying capacity from a built-up lumber beam would require three 2x12s.*

Strengthening Beams with Flitchplates

shrinkage, and crushing — problems that steel is not subject to. However, I normally do not include the dead weight of the wall construction in sizing the steel, but assume it to be carried by the wood members while the steel carries the floor and roof loads.

Table 2 (page 232) shows the equivalent strengths of various flitch beams with conventional built-up wood beams or wide-flange steel beams. I have not attempted to describe how to calculate the sizing and bolting requirements of flitchplates. This involves relatively complex calculations that should not be attempted by a beginner. My advice is to talk to a professional engineer.

—*Robert Randall*

Attaching Steel Beams and Columns

Even with the advent of engineered lumber beams, steel is still a good choice for carrying heavy loads: It's stronger, lighter, and typically less expensive than engineered lumber beams. The main problem for home builders is how to integrate the steel into the wood frame. In many cases, builders will go to great lengths to attach a steel I-beam to a wood post using scabbed-on wood blocking when it would be simpler — and stronger — to use a steel post and the kind of steel-to-steel connections common in commercial construction.

In this section I'll explain the right and wrong ways to attach steel beams, and give some options for making strong connections.

Column-to-Beam Connection

For starters, I always advise against supporting any steel member on wood columns of any type. In the first place, the "rolling tolerances" allowed in the manufacture of the steel beam (Figure 4) mean that the two flanges may not be parallel to each other, or they may be parallel to each other but not at 90 degrees to the web. The result is uneven bearing on top of the wood post, which can cause localized crushing of the wood fiber. (For more on rolling tolerances, consult the *Steel Construction Manual*, published by the American Iron and Steel Institute, 800/797-8335.)

Second, the wood post may not be cut squarely, so that even if the flanges of the steel beam were perpendicular to the beam web and parallel with the ground, they might not have full contact bearing on the top of the wood column.

The best connection is steel to steel. The simplest way to do this is to have a steel cap plate prewelded on top of a steel column, with bolt holes prepunched in both the plate and the beam's bottom flange. In spite of the rolling tolerances in the I-beam, the bolts will draw the cap plate into full contact with the flange as they are tightened. Another option, more common in commercial construction, is to have a plate welded onto the column that provides for a bolted connection through the I-beam's web.

Steel columns are relatively easy to integrate into wood framing. A 3-inch-diameter standard column will fit in a stud wall, as will a TS 3x3 or TS 3 1/2 x 3 1/2 tube.

Column-to-Footing Connection

Steel beams are frequently used as central girders in the basement or garage. In either case, the post usually bears on a footing or thickened slab. For attaching to concrete, use a column with a prewelded base plate, punched for at least two (preferably four) 1/2-inch bolts. The bolts are set in the fresh concrete, using a plywood template for position.

Some builders prefer to use expansion anchors because they allow the slab to be finished without interference from embedded bolts. When using expansion

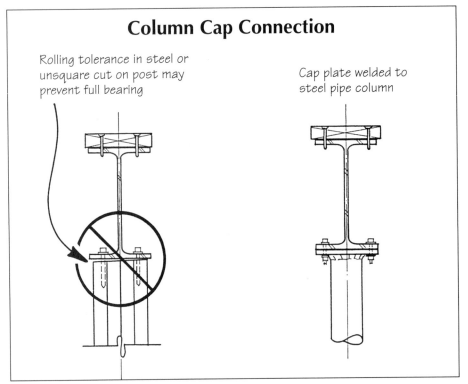

Figure 4. *The author recommends against supporting steel beams on wood posts (left). Rolling tolerances in the steel or an unsquare cut on the post could result in uneven bearing. Also, the lag screw into the post's end grain is a weak connection, while the blocking required to make a strong connection is much more time-consuming than simply bolting a steel column's cap plate to the bottom flange (right).*

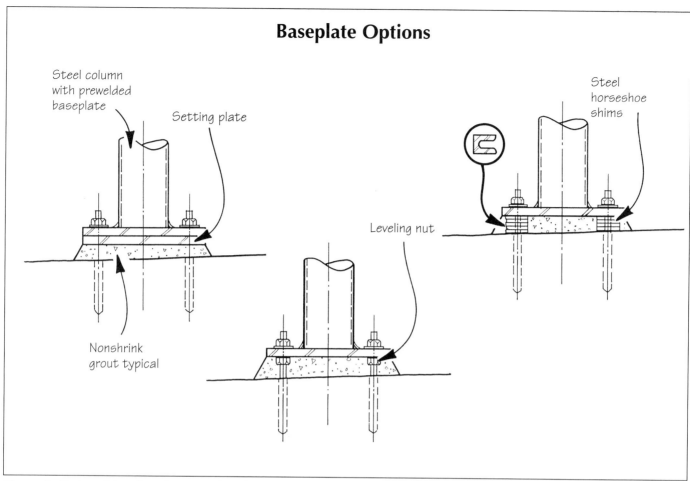

Figure 5. *On large jobs where there are a lot of columns, it may be fastest to level a separate loose setting plate in a bed of nonshrink structural grout (left). Then, after the grout has set, the column can be quickly set in place. On smaller jobs, you can plumb the column as it is installed, using either leveling nuts (middle) or shims (right). Then use nonshrink grout to fill the void under the baseplate.*

anchors, make sure to maintain the manufacturer's required "minimum edge distance" — the space from the bolt centerline to the edge of the concrete. Otherwise, the expansion anchor may fracture the concrete. When this edge distance cannot be maintained, you may be able to use an epoxy system, which doesn't create horizontal pressure. However, epoxy systems also have minimum edge distance requirements that must be followed. Also, cold temperatures may prevent the use of epoxy anchors.

Setting Columns

There are three methods I commonly recommend for installing the column (Figure 5). With the first method, a loose 1/4-inch-thick setting plate the same size as the baseplate is first set over the bolts into a 1/2- to 3/4-inch-thick bed of nonshrink grout. (Nonshrink grout is a special structural grout with high compressive strength, available from concrete suppliers.) The grout bed is slightly higher than needed so that the loose baseplate can be tapped into place and leveled with a carpenter's level. The nuts are then used to snug the baseplate in place while the grout sets. This method is popular with contractors on big commercial jobs because it enables them to quickly and accurately set large numbers of level baseplates. It's then easy to install the columns: You simply remove the nuts, place the prewelded baseplate of the column over the bolts, and retighten the nuts, fastening the two plates together.

On residential jobs where there are only a few columns to set, it may be easier to plumb the column as it is installed. The column with its prewelded baseplate is set onto the embedded bolts, then plumbed using either leveling nuts or "horseshoe" shims. (Horseshoe shims are usually available from your steel supplier.) Once the column is in final position, it is bolted into place and nonshrink grout is used to fill the void under the plate.

When using either shims or leveling nuts, I recommend allowing space for at least 1 inch of grout beneath the baseplate, and preferably 1 1/2 inches. This makes it easier to work the grout under the baseplate and around the shims or leveling nuts.

By Stuart Jacobson, P.E., a consulting structural engineer in Northbrook, Ill.

Installing a Steel Moment Frame

New custom-designed oceanfront homes all seem to share one feature: lots of glass looking out on the water. In the case of one house we recently built, the architect filled most of the east side of the house with an expansive two-story window-wall (Figure 6), so no effective plywood shear walls could be incorporated. To achieve lateral stability on that side of the house, we integrated a structural steel "moment frame" into the wood framing.

A Moment for Definitions

The word "moment" is a physics term adopted by engineers to refer to "the measure of a force with reference to its effect in producing rotation." (This is the same thing your auto mechanic calls torque.) A moment is a product of a force and a distance, and is typically measured in foot-pounds. As the distance gets greater, the moment gets larger. That's why you use a long ratchet handle to unscrew a tight bolt: The force of your arm is multiplied by a greater distance, increasing the moment you're applying to the bolt. But what's this got to do with a wall?

Wood-framed residential structures typically rely on plywood sheathing to resist lateral wind loads. As every builder knows, when you stand up a framed wall with no sheathing on it, it takes practically no force at all to rack that wall. But once you nail on the sheathing, the wall is rock-solid against racking.

When a lateral force is applied at the top plate of an unsheathed wall and the wall racks, the studs rotate with respect to the top plate (Figure 7). A plywood-sheathed wall, acting as a shear wall, removes the rotational force. Instead, the plywood, stiffened by the wall studs, safely transfers the lateral force to the foundation.

Plywood shear wall sections generally need to be at least 4 feet long to be effective. But what do you do with a wall that's so broken up with doors and windows that there's no room for plywood? It's difficult to achieve a rigid structure with unsheathed wood framing alone. Even if you use steel connectors and heavy timbers or builtup posts and beams, the wood itself still has a tendency to crush and split, allowing rotation at the joints. (One exception is a timber frame with knee braces.)

Steel to the Rescue

Steel moment frames are typically used in low- to mid-rise steel-framed commercial buildings, and can be adapted for residential construction. A moment frame consists of at least two columns and one beam. The welded connections where the beams join the columns have the full strength of continuous steel, so they can resist rotational forces. These moment connections work together with the stiffness of the steel members to prevent the frame from racking.

The moment frame for this house had six columns and ten beams, and supported the top two floors on the east face of the 55-foot-wide house. The complexity of the steel frame was typical of the oceanfront homes we build, but the techniques we used and problems we encountered would apply to simple projects as well.

A steel moment frame, as you might imagine, is considerably more complicated than using a single steel beam in a garage. The posts and beams all have to be carefully sized, connection and reinforcing plates welded on, and the steel punched in various places for bolting on

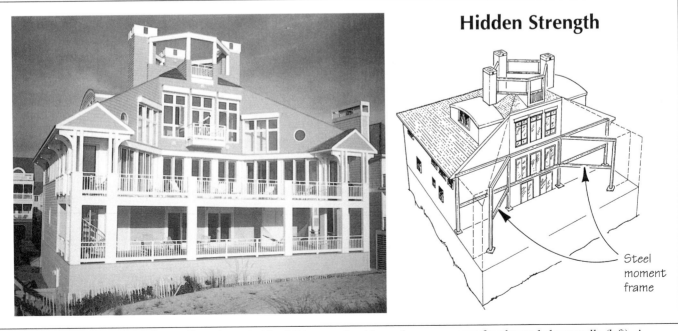

Hidden Strength

Figure 6. *The expanse of window wall on the east side of this oceanfront home leaves no room for plywood shear walls (left). A structural steel moment frame embedded in the framing (right) handles large lateral wind loads.*

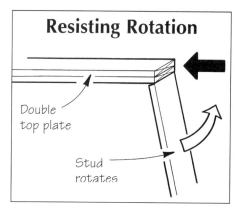

Figure 7. *When a stud wall racks, the studs rotate with respect to the plates. A moment connection resists this type of rotation.*

Figure 8. *The author erected the moment frame in two main sections, then dropped connecting beams into place.*

wood members.

Although we erected the steel ourselves, we had an experienced local shop fabricate the necessary components. The fabricator worked with the architect's and the engineer's drawings to create a set of shop drawings. These showed the size and layout of each of the parts, as well as all the necessary bolt holes. We carefully reviewed the drawings, comparing all three sets to make sure that nothing was overlooked. After our final okay of the drawings, fabrication began.

Meanwhile, we framed the first-floor deck and lined up a crane for the big day. When the steel arrived, we first unloaded the individual pieces and sorted them on the plywood deck. The fabricator sent along a set of erection drawings. Each piece of steel was numbered and each end labeled so that matching the parts to each other was simple.

The primary section of frame was a line of four columns supporting two tiers of three beams each, one tier at the second-floor level and one at the third-floor level. We assembled these as two separate sections, with connecting beams to go between them once they were stood up (Figure 8). We bolted these sections together, hand tight, while they lay on the deck. Then, using the crane, we stood up each section and lag-screwed the column baseplates to solid blocking in the floor (or in some cases, to the tops of pilings), using four $1/2$-inch lag screws per base, per the engineer's specs. Next, we braced the columns using long 2x4s fastened to the steel with big C-clamps. Once these two sections were up, we set the connecting beams between them. Finally, we set the two outlying columns and beams that came off the main frame at an angle.

Because the frame stood two stories tall, it was a little tricky to plumb. We used our 8-foot level to plumb and brace the bottom 10 feet of the frame. We then welded the moment connections at the first-floor level, and built the first-floor walls around the frame. We next built the second-floor platform, checked the upper sections of the moment frame for plumb, and welded the upper connections.

Attaching Wood

The final headache was to actually integrate the steel into the wood-framed walls and floors of the house. As much as possible, we had the steel prepunched so we could simply bolt 2-bys all around. Columns in walls received nailers on both sides. The exterior sheathing and the interior drywall then bridged the column from nailer to nailer. We attached nailers to the top and bottom flanges of I-beams and filled the webs with blocking. Some of the columns were square tubes, which can't be prepunched. Instead of drilling the tubes, we attached the wood framing with powder-actuated fasteners. This was expensive and noisy, but quite effective.

The importance of these wood-to-steel attachments should not be underestimated. The proper transfer of lateral forces from the floor and roof diaphragms to the steel moment frame depends on their connection to the steel members. If these connections are weak, the moment frame can't effectively brace the entire structure.

—*Patricia Hamilton*

Hybrid Framing with Wood and Steel

My company has almost 20 years of design/build general contracting experience in a variety of commercial and residential work. We've always used steel framing in commercial applications and wood framing in residential projects. However, given the uncertainty in lumber prices in the early 90's and the comparatively low price of steel framing, we decided to try a hybrid framing system in our multifamily residential work.

Steel Framing

Lightweight steel framing is cold-formed — the studs, joists, track, and accessories are manufactured by brake-forming and punching galvanized coil and sheet stock. There are a variety of manufacturers, with plants located throughout the U.S. (see box at end of section). You can typically find steel framing through your local commercial drywall and finish supplier. Find out what brands are available in your area — structural capacities and available sizes and gauges vary among the manufacturers — and make sure that you design according to the product literature of the manufacturer you buy from.

Sizes. Steel framing members are available in a wide variety of sizes and gauges (see Table 3). Most manufacturers use color-coding to prevent the different gauges from being mixed at the job site. Pricing of steel framing is always quoted per thousand lineal feet, though you can buy less.

Advantages of Steel Framing

The advantages of steel are obvious. Depending on lumber prices at the time, 25-gauge steel studs may be as much as 50% less expensive than 2x4s. In a recent project, we saved over 35¢ per lineal foot on steel joists as compared with wood 2x12s.

In strength, steel framing holds no comparison — its allowable bending stress values typically range from 21,000 to 33,000 psi (Table 4). By comparison, the bending fiber stress (Fb) for select structural Douglas fir is 1,800 psi (ordinary No. 2 SPF is a lowly 875 psi).

Steel is a processed, not an organic material. As such, it arrives at the job site straight and consistent — no checks or knots to work around. While it expands and contracts like anything else, steel won't swell or move from changes in moisture content.

Design flexibility. With steel framing, the variety of sizes and gauges gives the builder great flexibility in determining cost-effective framing solutions. Take, for example, a situation where you want to maximize ceiling headroom. With wood framing you can use a narrower joist but you'll probably have to decrease the on-center spacing, increasing both labor and material costs. With steel, you can use a narrower joist but in a heavier gauge, and keep the joist spacing the same. And since the cost of steel framing material is primarily based on weight, this solution will cost you no more, either in labor or material.

There is little limitation in the length of steel framing. If desired, joists and studs can be fabricated up to 40 feet in length. Given sufficient quantities, wholesale distributors can also cut joists and studs to exactly the length you need.

Disadvantages. Steel does have its disadvantages. The standard 25-gauge non-loadbearing studs are flimsy to work with. And for those not accustomed to working with the material, steel is sharp and will slice your skin if you don't handle it properly. We always ask for "hemmed" track — it's much safer than unhemmed (Figure 10).

Steel is also hard to cut, and many of the tools in your toolbox just won't work on the stuff. Eye protection — a good idea when cutting lumber — is imperative when cutting steel (Figure 11).

Probably the biggest drawback with steel is its lack of insulating qualities. Steel studs are about 400 times more heat-conducting than wood studs. Used in exterior applications, steel studs can reduce overall R-values by up to 50%. A wholesale changeover to steel framing would result in a drastically inferior thermal envelope. Also, with exterior steel studs, fastening the typical residential exterior finishes and blocking for trim would take longer.

Hybrid Solution

Our solution has been to integrate wood and steel: We use wood framing

Table 3: Sizes of Steel Framing

Studs
Widths: 1⅝"*, 2½", 3⅝", 4", 6"
Gauges†: 25*, 20, 18, 16, 14, 12

Joists
Widths: 4", 6", 8", 9¼", 10", 12", 14"
Gauges†: 18, 16, 14, 12

* Non-loadbearing uses only
† Gauge equivalents: 25=.019", 20=.0346", 18=.0451", 16=.0566", 14=.0713", 12=.1017"

Table 4: Strength of Steel vs. Wood Joists

Required joist size to frame 16-foot span, 16 inches on-center, for 40-lb live load + 10-lb dead load @ L/360 deflection:

Joist Type	Size	Cost	E*(psi)	Fb†(psi)
Steel (Marino brand)	8" 16 ga.	$1.03/lf	29,000,000	33,000
No. 2 SPF	2x12	$1.32/lf	1,300,000	875
Select Struct. Doug Fir-larch	2x10	$1.35/lf	1,800,000	1,800
Wood I-Joist (TJI 15SP)	9½"	$1.43/lf	N/A	N/A

Note: The prices in this chart are contractor's prices for northern New England as of November 1993.
* Modulus of elasticity.
† Max. fiber stress in bending.

for exterior walls and the roof and steel framing for floor systems and interior partitions.

Wood studs provide better insulating values, as well as a nail base for exterior finishes. Studs are primarily compression members and don't need particularly high fiber-stress values. Quicker-growth and structurally inferior softwoods work just fine here.

For floor systems, however, you need a framing material with a high strength-to-weight ratio, one that's rigid, and is available in long lengths. Steel joists fit the bill perfectly. For girders, beams built up of multiple members are again much lighter and stronger in steel than in wood, and they won't cup or warp.

In the interior, we use steel studs for both bearing and nonbearing walls. Once you're used to it, you can frame faster with steel than with wood because the material is lighter and easier to handle. For non-loadbearing partitions, we use standard 8-foot steel studs and hold them up slightly from the bottom track to match the 8-foot 1-inch exterior walls that we get using precut wood studs. For interior bearing walls, we either buy studs precut to exact length or cut them on site.

We've also found that you can attach baseboard to steel studs by nailing with galvanized finished nails instead of installing a lot of blocking. We block only at those points where we need nailing for a short length of baseboard.

Steel floor joists. In framing floor systems, there are several points to keep in mind. If you plan carefully, you can order precise lengths of the various members your specific project requires. There is no need to pay for excessive waste. You can minimize thermal bridging by holding the track (band joist) away from the outside edge of the foundation wall, then filling the gap with fiberglass insulation or rigid foam board (Figure 12). Just make sure you maintain the typical minimum end bearing of 1½ inches for the joists.

Also, at end bearing points you have to install web stiffeners (Figure 13). Because steel has such high tensile strength, individual members do not normally approach a bending failure. Instead, steel joists will fail by folding at bearing points, where shear forces are greatest. Web stiffeners help resist shear and must be used at ends and anywhere along the joist span where there are point loads or loadbearing walls above. Since we use wood studs in the exterior walls, we cut our web stiffeners from stud scraps. They must be cut to fit tight, then slid into the joist end before it is

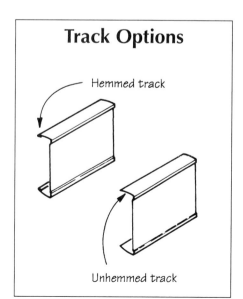

Figure 10. *Hemmed track is much safer to work with than unhemmed.*

Figure 11. *Framing with steel requires a special set of tools: (1) variable speed drill and screw gun (2) hearing protectors (3) vise clamps (4) metal snips (5) light-gauge metal punch (6) metal cut-off saw (7) magnetic level (8) circular saw with abrasive metal blade (9) safety glasses (10) right-angle drill.*

Hybrid Framing with Wood and Steel

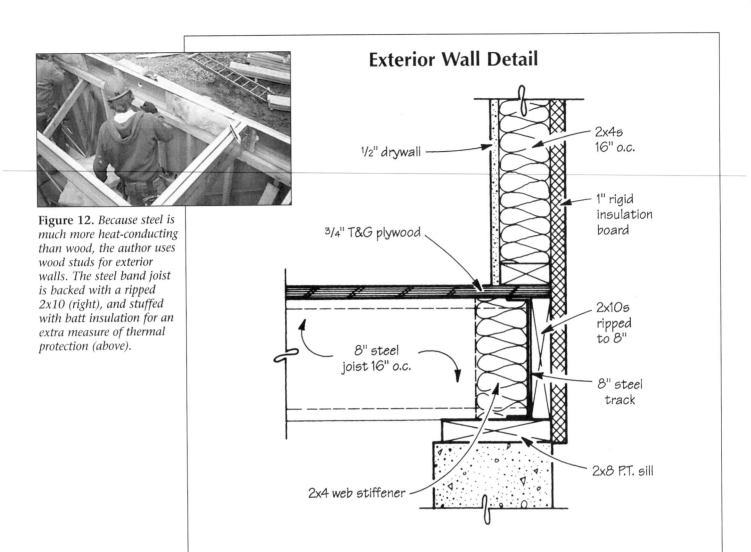

Figure 12. *Because steel is much more heat-conducting than wood, the author uses wood studs for exterior walls. The steel band joist is backed with a ripped 2x10 (right), and stuffed with batt insulation for an extra measure of thermal protection (above).*

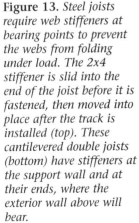

Figure 13. *Steel joists require web stiffeners at bearing points to prevent the webs from folding under load. The 2x4 stiffener is slid into the end of the joist before it is fastened, then moved into place after the track is installed (top). These cantilevered double joists (bottom) have stiffeners at the support wall and at their ends, where the exterior wall above will bear.*

put in place. Otherwise, once the track is on, there's no way to get the web stiffeners in. Also, avoid holes 12 inches from the end bearing point.

Subflooring. You attach plywood subflooring to steel joists as you do to wood joists. Use a heavy-duty construction adhesive, like Hilti's CA3400, that's approved for use with steel. Be sure to wipe down the joist flange with paint thinner to remove the oils used in the manufacturing process or the adhesive may not stick. For attaching subflooring, we use Marino's Teks screws with wings. The wings on the screw tip clear out the plywood shavings so the threads don't clog. When the screw hits the steel, the wings shear off. These screws cost more but they save time.

Bridging is more straightforward with steel joists than with wood. The plywood subflooring braces the top flange, and metal strapping braces the bottom flange. There is no need for full-depth

For More Information

Industry Trade Groups:

American Iron and Steel Institute (AISI)
Washington, DC
202/452-7100
www.steel.org

National Association of Architectural Metal Manufacturers (NAAMM)
Chicago, IL
312/332-0405
www.naamm.org

Manufacturers of Steel Framing:

CEMCO
City of Industry, CA
800/775-2362
www.cemcosteel.com

Dale/Incor Industries
Ft. Lauderdale, FL
954/772-6300
www.daleincor.com

Dietrich Metal Framing Inc.
Pittsburgh, PA
412/281-2805
www.dietrichmetalframing.com

Marino/Ware
South Plainfield, NJ
908/757-9000
www.marinoware.com

Unimast Incorporated
Schiller Park, IL
800/654-7883

---

blocking at midspan. We use standard 20-gauge drywall furring channel — "hat" channel — for strapping. The channel goes up quickly and provides a good backing for the ceiling drywall. If you try to screw the drywall directly to the thicker steel of the joist, the self-tapping drywall screws will be pulled right through the paper face.

Joist-to-joist connections are easy with standard available hangers, or you can have simple 12-gauge plates fabricated.

What'll the Subs Think?

Subcontractors have various reactions when first confronted with steel framing. The prepunched cutouts in steel joists make wiring go quicker but they don't usually line up perfectly for straight pipe runs. Plumbers and electricians with commercial experience use hand punches to quickly make holes. Electricians have to exercise caution when running conventional Romex wire. Snap-in plastic grommets, available for standard-size prepunched holes, must be used to protect the wire sheathing.

But ultimately, the cost-effectiveness of hybrid framing with steel depends on acceptance of the material by the people who will install it. There is no point in trying this technique unless you embrace it wholeheartedly. Even with good planning and the inherent advantages of steel, our initial labor costs increased by a factor of 25%, because screws simply take longer to install than nails. With experience, we've been able to trim the increased labor costs to 10% to 20% above all-wood construction. But since steel framing materials are less expensive than wood framing, for now we are still able to match or beat our costs for all-wood frames. Hopefully, the steel framing industry will in time develop pneumatic fasteners and other products to help us bring our labor costs down as well.

By Tim Duff, a partner in Kessel/Duff Construction, a design/build firm in Williston, Vt.

Authors' Contributions

Robert Randall, P.E., is a professional engineer licensed to practice in New York, New Jersey, Connecticut, and Wisconsin.
- Simple Approach to Sizing Built-Up Headers 14
- Misplaced Load Paths 22
- Using Metal Connectors 29
- Frequently Asked Framing Questions 32
- Holding the Roof Up 42
- Framing With a Raised Rafter Plate 43
- Resisting Wind Uplift 45
- Straight Talk About Hips and Valleys 47
- A Fix for Bouncy Floors 93
- Framing for Corner Windows 116
- Site-Built King Truss for Ridge Beam Support 205
- Strengthening Beams With Flitchplates 232

Paul Fisette is a wood technologist and director of the Building Materials Technology and Management program at the University of Massachusetts in Amherst.
- How To Read Span Tables 2
- Calculating Loads on Beams and Headers 8
- Structural Sheathing: Plywood vs. OSB 24
- Bracing Foam-Sheathed Walls 113
- Sizing Engineered Beams 211

Frank Woeste, P.E., is a professor at Virginia Tech University in Blacksburg, Va.
- Beyond Code: Preventing Floor Vibration 6
- Sizing Stiff Floor Girders 17
- Frequently Asked Framing Questions 32
- Critical Bracing for Piggyback Trusses 199

Carl Hagstrom is a builder from Montrose, Pa., and a contributing editor with The Journal of Light Construction.
- Layout Tricks for Rough Openings 98
- Laying Out Unequally-Sloped Gables 127
- Joining Unequally Pitched Hips and Valleys 131
- Building Doghouse Dormers 142

Harris Hyman, P.E., is a civil engineer in Portland, Ore.
- Sizing Simple Beams 10
- Strength of Nails 26
- Stiffening Garage Door Openings 56
- Overhanging Decks: How Far Can You Go? 79
- Steel Beam Options 230

Christopher DeBlois, P.E., is a structural engineer with Palmer Engineering in Chamblee, Ga.
- Frequently Asked Framing Questions 32
- Deck Support: Making the Crucial Connections 78
- Building Stiff Two-Story Window Walls 109

Jim Hart is a foreman in Mountain View, Calif. and a former editor with The Journal of Light Construction.
- Shear Wall Construction Basics 60
- Installing Seismic Framing Connectors 64
- Roughing In for Kitchens and Baths 158

Scott McVicker, P.E., is a consulting structural engineer in Palo Alto, Calif.
- Frequently Asked Framing Questions 32
- Hold Down Problems and Solutions 70
- Fixing Shear Wall Nailing Errors 73

Dan Dolan, P.E., is a professor at Virginia Tech University in Blacksburg, Va.
- Beyond Code: Preventing Floor Vibration 6
- Sizing Stiff Floor Girders 17

Don Dunkley is a framing contractor in Cool, Calif.
- Tips for Installing Hold-Downs 66
- Framing Recessed Ceilings 181

Patricia Hamilton is owner of Boardwalk Builders in Rehoboth Beach, Del.
- On Site With Parallam 217
- Installing a Steel Moment Frame 236

Ned Murphy is a contractor and owner of E.J. Murphy Builders in Framingham, Mass.
- Floor Framing with Wood I-Joists 85
- Working With Laminated Veneer Lumber 214

John Siegenthaler, P.E., is a building systems designer in Holland Patent, N.Y.
- When Columns Buckle 19
- Wood I-Joist Fundamentals 210

Rick Arnold is a builder in North Kingstown, R.I.
- Installing Gable Roof Trusses 188

Paul Bartholomew is owner of PFB Company, Inc., a truss installation company in Raleigh, N.C.
  Installing Hip Roof Trusses..................................193

Paul DeBaggis, a former builder, is an instructor at Minuteman Technical School in Lexington, Mass.
  Truss Bracing Tips 197

Richard Dempster is a builder and remodeler in Asheville, N.C.
  Setting Floor Trusses: Three Men and a Crane 92

Eric Dickerson is a long-time framing sub, and owner of a general contracting company in Ridgway, Colo.
  Framing Rake Walls 102

Tim Duff is a partner in Kessel/Duff Construction, a design/build firm in Williston, Vt.
  Hybrid Framing with Wood and Steel 238

Curtis Eck, P.E., is a technical representative for Trus Joist MacMillan in Seattle, Wash.
  Roof Framing with Wood I-Joists 145

Rob Dale Gilbert is a custom builder in Lyme, N.H.
  Layout Basics for Common, Hip,
    and Jack Rafters 122

Eliot Goldstein is an architect in West Orange, N.J., and serves on the American Forest and Paper Association committee that writes guidelines for engineered wood structures.
  Building With Glulams..220

Don Gordon, of Gordon Fiano, is a builder of custom homes in Santa Barbara, Calif.
  Flat Roof Framing Options 151

Chuck Green is owner of Four Corners Construction in Ashland, Mass.
  A Second Story in Five Days..................................166

Mike Guertin is a builder in East Greenwich, R.I.
  Installing Gable Roof Trusses 188

Will Holladay is author of A Wood Cutter's Secrets and a freelance roof framing contractor.
  Stacking Supported Valleys 136

Stuart Jacobson, P.E., is a consulting structural engineer in Northbrook, Ill.
  Attaching Steel Beams and Columns..........................234

Richard Mayo, P.E., is a certified professional engineer and director of Graduate Studies at the Del E. Webb School of Construction at Arizona State University.
  Earthquake Design 58

Bill Nebeker is a framing subcontractor in Crestline, Calif.
  Site-Built Panelized Walls 105

Gary Rowland is a general contractor in the Atlanta area.
  Making Room at the Top 202

Rick Schneider is a builder and remodeler in Addison County, Vt.
  Case Study: A Retrofit Ridge Beam 169

David Schwartz is a builder and remodeler in Overland Park, Kansas.
  Tying Into Existing Framing 162

Mike Shannahan is a carpenter in La Porte, Texas.
  Jacking Old Houses 172

Mike Stary, of Stary Construction, is a builder in Leucadia, Calif.
  Fast Fascia Techniques 178

Robert Thompson is a construction supervisor for Framing Square Construction Co. Inc., in Albuquerque, N.M.
  Framing a Simple Radius Stair 183

Silas Towler is a builder and remodeler in Addison County, Vt.
  Case Study: A Retrofit Ridge Beam 169

Charles Wardell is a former editor at The Journal of Light Construction.
  Floor Framing With Open-Web Trusses 87

Philip Westover, P.E., is a structural engineer in Winchester, Mass.
  Plywood vs. Let-In Bracing....................................54

Roger Whitaker is a builder in Hartford, Pa.
  Leveling the Deck 82

Ron Whitaker is a builder in Hartford, Pa.
  Leveling the Deck 82

Index

A

Actual compressive stress, figuring, 20
Attic trusses, 202-204
Additions
    dropped footing, to match existing, 164
    foundation height for, 164
    framing, tying into existing, 162-166
    layout heights for, 84
    level and plumb framing, 163
    second-story to ranch house, 166
    See also Remodeling.
Air nailers. *See* Pneumatic nailers.
Anchor bolts
    installing out-of-plumb, 68
    layout for hold-downs, 68
    types used with hold-downs, 67

B

Backing the hip, 135
Band joist
    shimming to level, 83
    tying deck to, 78-79
Bathrooms, rough-in, 159-160
Beams
    built-up, sideloading, 36
    built-up, sizing, 17-18
    calculating loads on, 8-10
    engineered lumber options, 211-216
    flitchbeams, 232-234
    sideloaded, nailing, 36. *See also* Nailing.
    simple, with uniform load, 13
    sizing, simple, 10-14
    splices over posts, 38
    steel, 230-231, 234-235
    See also Engineered lumber.
Bearing walls
    misaligned with foundation, 22
    two-story framing, 109-113
Bending stress, in simple-span beam, 80
Birdsmouth cut, hip rafter, 134
Bouncy floors. *See* Floor vibration.
Bracing
    floor trusses, 92-93
    foam-sheathed walls, 113-116
    let-in, 54, 114
    plywood, 54-55, 57, 58, 115
    posts, 19
    roof trusses, 197-201, 203-204
    seismic, 54-55, 58-59
    tall walls, 110-112
    See also Temporary bracing.
Buckling
    of columns, 19-21
    of tall walls, 110
    of temporary posts, 21
Built-up beams. *See* Beams, built-up.

C

Cantilevers, in exterior decks, 79-80
Cathedral ceilings
    roof loads on interior walls, 22
    structural ridge at dormer, 205-206
    structural ridge with I-joist rafters, 146
    with hip roofs, 50
Cathedral shed roof, and rafter thrust, 36
Ceiling joists
    resisting thrust, 42-43
    sizing, 6
    thrust in shed, 36
Center girder
    calculating uniform load, 10
    sizing, 17-18
    tributary load on, 11
Cheek cut, 135-139
Clear span
    defined, 3, 5
    maximum for joists, 7
Coffered ceilings
    below floor joists, 183
    ceiling box, 182
    framing, 181-183
Coil strap. *See* Metal connectors.
Collar ties
    not needed in gable roofs, 37
    vs. eaves ties, 42-43
Columns
    avoid wood with steel beam, 234
    buckling, 19-21. *See also* Posts.
    setting on slab, 235
    stability factor, 20
    steel, 234-235
Compression, perpendicular to the grain, 6
Corner windows
    cantilevered headers for, 117-119
    framing for, 116-119
Cranes
    lifting floor trusses, 92
    lifting gable truss, 191
    lifting panelized walls, 108

lifting roof for second-story addition, 166-169
setting storage trusses, 202-203

D

Decimal conversion, 127
Deck bands, bolting schedule for, 79
Decks
    collapse of, 78
    fastening to house, 78-79
    overhanging, 79-80
Deflection
    and vibration in floors, 6-7
    formula, 13
    in girders, 17-18
    in LVL headers for corner windows, 118
    in simple-span beam, 79
    limits, 2-3
    of plywood vs. OSB, 35
    overhanging decks and, 79-80
Diagonal bracing
    for foam-sheathed walls, 115
    in seismic construction, 58
    See also Bracing; Seismic bracing.
Doors
    garage, framing rough openings for, 99-100
    hinged, rough openings for, 99
    pocket, rough openings for, 99
Dormers
    doghouse, flashing, 145
    doghouse, layout and framing, 142-145
    king truss support, 205-207
Drywall
    shear strength of, 37, 116

E

Earthquakes
    design for, 58-59
    forces in, 60
    shear wall basics, 60-64
See also Seismic bracing; Shear walls.
Eaves ties
    and lateral thrust, 42
    and roof collapse, 42
    nailing and bolting schedule for, 43, 207
    tying raised rafter to, 44
    See also Roof loads.
EIFS
    moisture problems over OSB, 25
Engineered lumber
    beam choices, cost comparison, 213
    beam loading examples, 212
    glulams, 220-223
    laminated-veneer lumber, 117-119, 170, 215-216, 226
    Parallam, 217-219
    sizing beams, 211-214, 215
    wood I-joists, 85-87, 145-151, 210-214
    See also Glulams; Laminated-veneer lumber; Parallam; Wood I-joists.
Engineered studs
    design values for, 225
    finger-jointed, 224-226
    for tall walls, 224-227
    LSL, 226
    LVL, 226

F

Fascia
    cutting to length, 178
    hanging barge boards, 179-180
    installing, 178-180
    lining up with unequal pitches, 132
F_b, 2-6, 11, 12
Feet-inch conversions, on ordinary calculator, 104
Fink trusses, 203
Fireplace opening, 101-102
Flat roofs
    double pitched roof, 153, 155
    rip strip technique, 151-154
    single-slope roofs, 152
Flitchplates
    beam makeup, 232-234
    case studies, 233
    repair with, 232
    strength, vs. wood, steel, 232
Floor design software, 8
Floor framing
    bouncy, fixing, 93-95
    leveling, 82-84
    overloading, 22
    sizing joists, 5
    with open-web trusses, 87-93
    with wood I-joists, 85-87
    See also Floor joists; Joists; Open-web trusses; Point loads; Wood I-joists.
Floor girders, sizing stiff, 17-18
Floor trusses
    design and planning, 88-89
    firestopping, 91-92
    framing stairwells, 90-91
    metal-plate-connected, 7
    ordering, 89
    setting with crane, 92
    subflooring, thickness over, 93
    truss types, 88

vibrational performance, 7
Floor vibration
    analyzing, 14, 18
    calculating vibration, 18
    fixing with stressed-skin, 93-95
    girder stiffness, 17
    in floor trusses, 7
    preventing 6-8
Foam sheathed walls, bracing, 113-116
Foundations
    footing forms, 84
    for additions, 164
    layout for addition, 84
Foundation connectors, 65-69
Foundation straps, 68-69
Framing. *See* Engineered lumber; Floor framing; Roof framing; Rough openings; Trusses.
Framing square, to lay out hips and jacks, 122-126

G

Gable end walls. *See* Rake walls.
Gable roofs, layout for unequally-sloped, 127-130
Gable roof trusses
    bracing, 191
    installing, 188-193
    layout on the ground, 189
    lifting with crane, 191
    planning, 188-189
    sheathing, fascia, and bracing, 192-193
Garage doors, sizing headers over, 56
Garages
    resisting racking on, 57
    wind loads on, 56
Girders
    calculating uniform load, 10
    sizing, 17-18
    tributary load on, 11
Glulams
    building with, 220-223
    connectors for, 222
    custom, 221
    moisture problems, detailing for, 223
    using a crane to set, 221
    See also Engineered lumber.
Gypboard. *See* Drywall.

H

Headers
    calculating loads on, 8-10, 14-15
    choosing stock, 16
    framing, 16
    sizing built-up, 14-17
    See also Load factors.
Hip rafters
    apex load, support of, 48, 50
    birdsmouth cut, 134
    calculating tributary load on, 47
    cheek cuts, 136-139
    end loads, calculating, 49
    end loads, support post, 48
    layout of, 123-125
    rafter truss, 48, 51
    sizing, 49, 51
    tension tie to prevent spreading, 51
    tributary area, 47
    with cathedral ceilings, 50
    See also Valley rafters.
Hip roof trusses
    anatomy of, 194
    installing, 193-197
    layout, 194
    ordering, 193-194
    stacking, 195
    step-down blocking, 196-197
Hold-downs
    installation tips for, 66
    problems and solutions, 70-72
    types of, 67
Hurricane anchors
    to reinforce rafter-to-plate, 70
    wind ties at corner window, 119
    with raised-heel rafters, 44
    See also Metal connectors; Strap ties.
Hydraulic jacks
    for jacking houses, 172
    for lifting LVL beam, 170

I

I-joists. *See* Wood I-joists.
Inflection point, 38
Insulation, space with raised rafter plate, 44

J

Jacking
    needle beams, 173, 176
    old houses, 172-176
    panelized walls, 107-108
    using pipe staging, 174-176
Jack rafters
    cutting, 125-126
    layout, for unequally pitched roofs, 136
    loads, 47

Index

stepping off, 126
Joist hangers
  at flush-framed beam, 42
  at ridge for I-joists, 146-147, 150
  double-shear type, 29
Joists
  loading, 3
  sizing and code requirements, 4-5
  splicing at center beam, 42
  stiffness requirements, 2
  strength of, Fb, 2
  See also Floor joists.

K

King truss
  bolting schedule for, 207
  calculating loads for, 205-206
  for ridge beam support, 205-207
  installation details, 206
  point loads on, 207
Kitchen
  range hood ductwork, 161-162
  rough-in, 161-162
  sink dimensions, 160
Kitchens and baths
  roughing in for, 158-162
Kneewalls
  angled, to support LVL beam, 171

L

Laminated-veneer lumber
  as engineered studs, 226
  as ridge beam, 170
  beam sizing chart, 215
  connectors for, 216
  ganging, 216
  headers for corner windows, 117-119
  use in I-joists, 211
  vs. other options, 213
  weather protection, 216
  See also Engineered lumber.
Lateral loads
  from wind, 54, 56-57
  seismic, 60
  See also Seismic bracing.
Layout
  common rafters, 122-124
  doghouse dormers, 143
  jack rafters, 125-126
  of trusses, 189-190, 194-196
  rough openings, 98-102

  radius stair, 184-185
  rake walls, 103-104
  supported valleys, 136-141
  unequally sloped gables, 127-130
  unequally pitched hips and valleys, 131-135
Lead carpenter, using for k&b rough-in, 158
Let-in bracing
  for foam-sheathed walls, 113-114
  strength of, 54
  vs. plywood sheathing, 54-55
  See also Seismic bracing.
Leveling foundations, 82-84
Liquefaction, 59
LL ratios, figuring for roof slopes, 140
Load factors, table for headers, 15
Load path
  example, 9-10
  misplaced, 22
Loads
  calculating for beams and headers, 8-11
  excessive on floor, 22
  on hips and valleys, 47-49
  on overhanging deck, 80
  point loads, 22, 83, 207, 230
  roof, 42, 44
  snow, 35
  uniform vs. point, 9
  See also Point loads; Roof loads; Snow loads.
Lumber shrinkage. *See* Wood shrinkage.
Lumber design values, 18
LVL. *See* Laminated-veneer lumber.

M

Masonry piers, new under jacked house, 174
Maximum allowable bending stress, "Fb," 11
Maximum bending moment, "M," 11
Metal bracing, for foam-sheathed walls, 114-115
Metal connectors
  coil strap, 30, 44
  double-shear nailing, 29
  for reinforced ridge beam, 30
  for seismic framing, 61, 64-72
  for sill repair,
  hangers to avoid, 31
  heavy-duty angles, 31
  hurricane ties, 44
  strap ties, 29-30, 44, 45, 106
  twist straps, 30, 44
  See also Seismic framing connectors.
Modulus of elasticity, "E," 2, 13, 35
Moment frames, 58-59, 112-113

N

Nail guns. *See* Pneumatic nailers.
Nailing
    of built-up beams, 36
    of shear walls, 61-62, 64
    shear walls, errors, 73-75
    wood I-joists, 86-87
Nails
    box vs. common, 28-29
    design values, 45
    for deck ledgers, 26, 78-79
    for shear walls, 61-62, 74-75
    holding power and metal side plates, 27
    lateral loading, 26, 34, 46
    pneumatic for shear walls, 61-62
    pneumatically driven, guidelines for, 28-29
    proper setting in shear walls, 74
    sideloaded beams and, 36
    spiral shank, 27
    toe-nails vs. end nails in studs and joists, 34
Nail size
    safe loads and, 27
    shear values and, 55
Nail strength, 26-29
National Design Specification for Wood Construction, 34, 39, 46
Needle beams, for house jacking, 173, 176

O

On-center spacing, reducing to prevent bounce, 6
Open-web trusses
    bracing bottom-chord, 91
    design of, 88
    firestopping and, 91
    floor framing with, 87-93
    lally column support, 90
    ordering, 89
    stairwell details and, 90
    strongbacks for, 7-8
    truss types, 88
    use of cranes with, 89, 91, 92
Oriented strand board. *See* OSB.
OSB
    as roof sheathing, 25-26
    as subflooring, 25
    as wall sheathing, 25
    composition of, 24
    over roof rafters, 35
    problems with EIFS, 25
    vs. plywood, as structural sheathing, 24-26
    wet conditions and expansion, 24
    See also Structural sheathing.

P

Panelized walls, site built, 105-108
Parallam
    connectors for, 219
    installing, 217-219
    vs. Glulam and steel, comparison, 219
Parallel-chord trusses. *See* Open-web trusses.
Piggyback trusses
    bracing for, 199-201
    retrofit bracing for, 201
    roof collapse and, 199-201
Pipe staging, using for jacking, 174-176
Plywood
    as corner bracing, 115-116
    as shear wall, nailing table, 55
    as structural sheathing, 54
    horizontal vs. vertical installation, 33
    in built-up headers, 33
    over roof rafters, 35
    to hold down band joist, 46
    to strengthen bouncy floors, 93-95
    vs. let-in bracing, 54-55
    See also Shear wall; Structural sheathing.
Plywood corner bracing, for foam-sheathed walls, 115-116
Pneumatic nailers, 27
Pneumatic nails, for shear walls, 61-62
Point loads
    on first-floor framing, 22
    on headers, 230
    on king trusses, 207
    on wood I-joists, 86
    shimming under, 83
    vs. uniform loads, 9
Porch deck construction, 26-27
Posts
    beam splices over, 38
    bracing to prevent buckling, 19
    buckling of temporary, 21
    See also Columns.
Pressure-treated lumber, strength of, 39
Purlins, used with roof struts, 38
Pythagorean Theorem, 123

R

Racking
    bracing to prevent, 54-55
    in foam-sheathed walls, 113-116
    of garages, 56
    resisting in tall walls, 110-112
    See also Plywood; Seismic bracing; Shear walls; Wind loads.

Radius stairs
    code requirements, 183
    finish work, 186
    framing, 183-186
    layout, 184-185
    pony walls, 185

Rafter layout
    backing the hip, 135
    commons, 122-124
    hips, 124, 132-136
    jacks, 125-126, 135-136
    unequally pitched hips and valleys, 131-135
    unequally sloped gables, 127-130
    valleys, 136-141
    See also Roof framing; Roof layout.

Rafters
    bracing with struts, 38
    doubling at sides of skylight opening, 32
    for coffered ceilings, 182
    layout of commons, 122-124
    resisting thrust in, 42-43
    span, calculating, 5
    stiffness and strength requirements, 2
    strengthening with purlins and struts, 38
    tie-in at raised plate, 43-46
    See also Rafter layout.

Rake walls
    feet-inch conversion on ordinary calculator, 104
    framing, 105-105
    layout, 102-104
    math for, 103-104

Raised rafter plate
    resisting wind uplift, 45-47
    tying rafters into, 43-44

Recessed ceilings, framing, 181-183

Remodeling
    jacking old houses, 172-176
    kitchen and bath rough in, 158-162
    second-story addition, 166-169
    tying into existing framing, 162-166
    See also Additions.

Ridge beam
    king truss support of, 205-207
    load per foot of beam, 10
    reinforcing existing, 30, 206
    retrofit of, 169-171

Rigid foam, bracing walls sheathed with, 113-116

Rim joist. *See* Band joist.

Rip-strip jig, 154

Roof collapse, and improper truss bracing, 199-201

Roof connectors, 43-45, 45-46, 70

Roof framing
    common rafters, 122-124
    flat roofs, 151-155
    hip layout, 47-51, 122-126
    jack rafters, cutting, 125-126
    joining unequally pitched hips and valleys, 131-135
    line length ratios for roof slopes, 140
    raised rafter plate detail, 43-45
    reinforcing before lifting, 168
    supported valleys, stacking, 136-141
    with wood I-beams, 145-151
    See also Hip rafters; Valley rafters.

Roof layout
    common rafters, 122-124
    doghouse dormers, 143
    jack rafters, 125-126
    saltbox-style, 130
    stacking supported valleys, 136-141
    supported valleys, 136-141
    terms, 127
    unequally pitched hips and valleys, 131-135
    unequally sloped gables, 127-130
    See also Hip rafters; Rafters; Roof framing; Valley rafters.

Roof loads
    and eaves tie failure, 42
    and thrust, 42, 44
    hips and valleys, 47-49
    snow loads for, 35
    wind uplift, 42-46
    See also Eaves ties.

Roof raising, lifting with crane, 166-169

Roof trusses
    bracing, permanent, 192-193, 200-201
    bracing, retrofit, 201
    bracing, temporary, 192, 197-199, 200
    collapse of piggyback, 199-201
    crane to lift, 191
    fascia installation, 192-193
    Fink truss, 203
    hip trusses, 193-197
    installing, 188-193
    layout on ground, 189-190, 194-196
    ordering, 188-189, 193
    piggyback, bracing required, 199-201
    storage trusses, 202-204
    Truslock spacing tools, 192, 204
    See also Gable roof trusses; Hip roof trusses; Truss bracing.

Rough openings
    for doors, 99
    for extension jambs, 100
    for fireplaces, 101-102
    for stairs, 98
    for toilet drops, 100-101
    in window walls, 100
    layout tricks for, 98-102

S

Safety factor, in wood construction, 34
Scaffolding
    to erect trusses, 191
    using pipe staging for jacking, 174-176
Screw jacks, for jacking buildings, 172
Scribing, band joists, 83
Section modulus, 11-12
Seismic action, 58, 60, 70
Seismic bracing
    earthquake design and, 58-59
    plywood vs. let-in bracing, 54-55
    See also Earthquake design; Shear walls.
Seismic framing connectors
    anchor bolts, 67-68
    foundation straps, 68-69
    hold-down errors, 70-72
    hold-downs, 61-62, 65-68
    installing, 64-70
    locating on blueprints, 65
    roof connectors, 70
    typical details, 63
    wall connectors, 69-70
    See also Metal connectors.
Shear
    checking for, 12
    definition, 12
    resisting in garages, 56-57
    resisting in two-story walls, 109-113
Shear panels, for bracing window walls, 112
Shear walls
    construction of, 60-64, 70
    definition, 58
    estimating, 61-62
    hold down problems, 70-72
    installing connectors, 64-70
    materials, 60
    nailing errors, 73-75
    nailing requirements, 64
    nails and connectors, 61-62
    pneumatic nailing of, 61
    typical details, 63
Sheathing
    as diaphragm, 57, 60
    as shear panel, 110, 112
    as shear wall, 60
    plywood vs. OSB, 24-26
    to prevent buckling, 19
    See also OSB; Plywood.
Shed roof, and rafter thrust, 36
Sheetrock. *See* Drywall.
Shimming
    avoid under steel header, 231
    to level foundations, 82-83
    under point loads, 83
Shrinkage, of wood I-joists, 87
Siding, installing on panelized walls, 107
Sills
    jacking and repairing, 172-176
    shimming to level, 82-84
Site-built panelized walls, 105-108
Slenderness ratio, 19
Snow loads, calculating for sloped roofs, 35
Soft story, 59
Solid-sawn joists, design and stiffening, 6
Span tables, how to read, 2-6
Squash blocks, on wood I-joists, 86
Stairwells
    framed with floor trusses, 90-91
    rough openings for, 98-99
Steel beam
    attaching wood nailers, 231, 237
    attachment to column, 234
    avoid wood columns with, 234
    flitchplates, 232-233
    header span table, 231
    headers, installing, 231
    options, 230-231
    vs. wood, 39, 230
    See also Structural steel.
Steel channels
    to reinforce tall studs, 110
    to reinforce wood header, 231
    vs. wood studs, 110
Steel columns
    connection to beam, 234
    connection to footing, 234-235
    setting on slab, 235
Steel connectors. *See* Metal connectors.
Steel framing
    advantages of, 238
    hybrid with wood, 238-241
    joists, need web stiffeners, 240
    sizing table, 239
    strength vs. wood, 239
    track options, 239
Steel moment frame
    definition of, 236
    for tall walls, 112-113
    installing, 236-238
    integrating into wood system, 237
Steel plates. *See* Flitchplates.
Steel studs
    use in hybrid system, 238-239
    See also Steel framing.
Stepped footings, for panelized walls, 105
Stepping off rafter lengths, 123-124
Stiffness
    floor girders, 17-18

Index

floor joist requirements, 2
of solid-sawn joints, 6
of steel channels, 110
Strap ties
for corner windows, 119
for panelized walls, 106
in raised rafter plate detail, 44-45
uses, 29-30
See also Hurricane anchors; Metal connectors.
Stressed-skin floor system, to stiffen bouncy floors, 93-95
Strongbacks, with open-web trusses, 7-8, 92
Structural ridge
load per foot of beam, 10
retrofitting, 169-171
support with king trusses, 205-207
Struts, to reduce rafter span, 38
Studs
buckling, 21
engineered, 224-227
steel, 238-239
tall, 102-105
Subflooring, at 24 inches on-center, 93-94
Supported valleys, stacking, 136-141

T

Tall walls
engineered studs for, 224-227
framing for, 102-105
Temporary bracing
buckling and collapse, 21
of trusses, 192, 197-199, 200
See also Roof trusses.
Toilet clearances, 160
Tray ceilings, framing, 181-183
Tributary load
for hips and valleys, 47-49
on girder, 10-11
Truslock spacing tools, 204
Truss bracing
improper and roof collapse, 199-201
of floor trusses, 7-8, 92-93
of piggyback trusses, 199-201
permanent, 192-193, 200-201
retrofit, 201
temporary, 192, 197-199, 200
See also Roof trusses.
Truss roofs
bracing, 191, 197-199
hip, 193-199
installing, 188-193
layout on the ground, 189
lifting with crane, 191

loads, 9
planning, 188-189
sheathing, fascia, and bracing, 192-193
storage trusses, 202-204
See also Hip roof trusses; Roof trusses; Truss bracing.
Tub/shower rough-in, 158
Two-story walls. *See* Tall walls.

U

Uniform loads
on roofs, 9
on simple beam, 10
vs. point loads, 9

V

Valley rafters
apex load, support of, 48, 50
calculating end loads for, 49
calculating tributary load on, 47
layout for unequally pitched, 136
rafter truss, 48, 51
sizing, 49, 51
support post, 48
tributary area, 47
Valleys
connection to second valley, 139-140
disappearing type, 137, 141
stacking supported, 136-141
Vanity rough-in, 159
Vibration, floors. *See* Floor vibration.

W

Wall connectors, 69
Wall jacks, lifting panelized walls with, 107-108
Walls
bracing foam-sheathed, 113-116
framing rake walls, 102-105
framing two-story, 109-113
panelized on site, 105-108
rough openings in, 99-100
See also Seismic bracing; Shear walls.
Wind bracing
at corner window, 119
in foam-sheathed walls, 113-116
of garages, 56-57
of roofs to walls, 43-45, 45-46, 70
of tall walls, 110-112
plywood vs. let-in, 54-55
to resist uplift, 45-47

See also Plywood; Seismic bracing;
   Shear walls; Wind loads.
Wind lift calculation, 46
Wind loads
   and racking in frame walls, 113
   calculating lateral, 54-55
   on garage wall, 56-57
   on tall window walls, 110
   uplift forces, 42-46
   See also Seismic action; Shear walls; Wind bracing.
Wind uplift, resisting, 45-47
Windows
   corner type, 116-119
   extension jamb framing, 100-101
   installing in panelized walls, 106-107
   See also Headers.
Window walls
   design table for tall, 111
   framing, 110
   layout of, 110
   rough openings in, 100
   stiffening two-story, 109-113
Wood design values, 5

Wood I-joists
   bending stress, 210
   blocking, 85
   cantilevers, 86
   crosscutting template for, 85
   cutting, 85, 211
   floor framing with, 85-87
   fundamentals, 210-214
   horizontal shear in, 210
   maximum bending stress, 211
   nailing, 85
   preventing vibration in, 7
   roof framing, header details, 151
   roof framing, mistakes, 150
   roof framing, plate details, 147
   roof framing, ridge details, 146-147
   roof framing, soffits and overhangs, 149
   roof framing with, 145-151
   shrinkage, 87
   squash blocks, 86
   storage and handling, 85
   See also Engineered lumber.
Wood structural panel. *See* OSB; Plywood.

NOTES

NOTES

Notes

Notes